be living in California numbered 7. At the end of 1849, San Francisco had exploded to nearly 25,000; Sacramento was a city of 12,000; and by the early 1850s, 20,000 more Chinese had arrived. In 1849 alone 85,000 men and women flocked to northern California, 23,000 from countries other than the United States.

Within two years, San Francisco had become a major seaport, the city filled with three-story brick buildings and thousands of masts lining the waterfront. A year later, dirt roads crisscrossed Telegraph Hill, and a solid line of houses ran halfway up. A lot facing Portsmouth Square sold for $16.50 in 1847, went for $6,000 the following year, and brought $45,000 six months later. The cost of lumber shot up twenty-five times, and still there were shortages. Labor wages escalated from a dollar a day, to ten dollars, to twenty dollars, and then to thirty dollars.

By the mid-1850s, San Francisco was a city of seventy-five thousand, supporting five hundred saloons and twice as many gambling dens. Each day thirty houses went up, two people died by knife or gun, and one fire broke out. Her more prominent citizenry sported the latest fashions from Paris and filled two-thousand-seat theaters nightly.

Richard Henry Dana had sailed into a pristine San Francisco Bay aboard a hide ship in 1835 and later described it in his classic, *Two Years Before the Mast*. "If California ever becomes a prosperous country," wrote Dana, "this bay will be the centre of its prosperity." Yet at that time, besides the ruins of the presidio and an almost deserted mission, in the whole area, smoke rose from the chimney of a single fur trader's shanty on the far eastern shore.

Twenty-four years later, in 1859, Dana returned. He arrived at midnight aboard a steamer and took a room at a hotel, which as close as he could ascertain, stood near the spot where he and the crew had beached the boats from the hide ship. The site had changed. "I awoke in the morning," wrote Dana, "and looked from my windows over the city of San Francisco, with its storehouses, towers, and steeples; its court-houses, theatres, and hospitals; its daily journals; its well-filled learned professions; its fortresses and light-houses; its wharves and harbor, with their thousand-ton clipper ships, more in number than London or Liverpool sheltered that day, itself one of the capitals of the American Republic, and the sole emporium of a new world, the awakened Pacific."

Occupying an outpost of prosperity still in the far reaches of the continent, San Franciscans communicated with the rest of the world by steamship. Along their broad promenades the steamers carried mail and merchandise and new settlers, and brought news and ideas and fashion from the outside world. California became a state on September 9, 1850, but no one in California knew until six weeks later, when the *Oregon* steamed into San Francisco Bay draped in banners and national flags and firing its big guns.

From 1849 to 1869, 410,000 passengers traveled west over Panama, and another 232,000 journeyed back east. Of those who crossed the Great Plains on foot, most returned by sea. The Panama route was the quickest and the safest, and the ships carried passengers of means, persons helping to shape the American West, and for the first twenty years they carried nearly every ounce of California's precious export, the only thing besides land that California had and everyone else wanted: gold. Officially cataloged, duly recorded and delivered, $711 million in gold passed over the Panama route, and $46 million went via another route later established across Nicaragua. On steamer day, crowds of merchants, shippers, passengers, and well-wishers packed the wharf and scurried among hand carts and wagons, coaches and cabs, past agents of the press gathering information for the shipping register. It was a time of settling accounts, remitting to eastern creditors, and taking stock of mercantile affairs, "a time of feverish activity," noted one merchant. Goods and gold had to be properly consigned with official receipts, first- and second-class staterooms had to be readied, barrels of beef and flour had to be loaded, and the baggage for five hundred passengers leaving San Francisco for several months had to be stowed. Below, coal tenders ran coal from the bunkers to the furnace amidships, stoking the fires for departure. Every two weeks, a steamer departed San Francisco, with cargo and passengers bound for the East, and carrying a commercial shipment of gold weighing close to three tons.

THE MORNING OF August 20, 1857, the sidewheel steamer SS *Sonora* lay tight to the wharf on Vallejo Street, her gangways aflow with human bodies dodging big trunks, little trunks, valises, carpetbags, bedding, and bundles. Men in long jackets and stovepipe hats conversed in clusters. Light breezes off the bay caused the hooped dresses of the women to dance about their waists. From the heart of the city, a wedding procession wound toward the wharf, a horse-drawn carriage surrounded by the wedding party. When the procession arrived, the bride and groom alighted from the carriage and mounted the gangway, the bride still in her wedding gown. She was Adeline Mills Easton, the petite, vivacious sister of Darius Ogden Mills, who later founded the Bank of California and was one of the richest men in the state. Addie's husband, Ansel Easton, had immigrated to California in early 1850 and built a fortune selling furnishings to the new steamship lines. He now raised thoroughbred horses on his fifteen-hundred-acre estate south of San Francisco. As they hurried up the gangway swinging baskets of wedding gifts and carrying hampers of wine and sweet cakes, the wedding party swept Ansel and Addie across the promenade deck with wishes for a bon voyage and a happy life together.

Approaching the quay was another young couple easily recognized in San Francisco, the famous minstrel and actor Billy Birch and his bride of one day, Virginia. Recently, the newspaper *Alta California* had applauded Birch as "the bright, particular star of the San Francisco Minstrels." He sang "The Grape Vine Twist" and "I'm Fatter Than I Wish to Be" and starred in farces like *The Rival Tragedians*. A year earlier, the theater critic for the *San Francisco Alta* had written that "the very sight of Billy Birch is enough to make a cynic laugh." Birch had just concluded a successful engagement at Maguire's Opera House and was on his way to join Bryant's Minstrels in New York City. His new bride, as one journalist described her, was "young, petite in form, and in personal appearance very attractive; added to this, she is possessed of a lively vivacity which renders her very interesting in conversation." As she walked the gangway to the *Sonora*'s deck, Virginia carried a small cage housing a yellow canary.

Another Birch in the crowd, no relation to Billy, was James Birch, a thick-chested man who had been a stagecoach driver in Providence, Rhode Island, and had trekked to California overland in 1849. Within five years, he had become president of the California Stage Company, then resigned to establish a stage line between Texas and California, which would complete the first transcontinental stagecoach route. The previous year, Birch's wife had given birth to a son, and to honor the birth, a friend in San Francisco had given Birch a sterling silver cup. With his family residing in Massachusetts, Birch carried the cup with him now to present to his infant son.

Among the clusters of men along the quay stood one man with thin hair carefully parted and slicked down, a large nose, and wide mutton-chops in a cotton candy cloud out two inches from his jowls: Judge Alonzo Castle Monson. A native of New York, Monson had graduated from Yale in 1840 and from Columbia Law School in 1844. Five years later he had migrated to California, one of the original forty-niners, and within three years took the bench in the geographic heart of the gold rush, Sacramento County. The *San Francisco Alta* claimed, "No more capable or efficient judge ever sat upon the bench in California." However, Judge Monson soared to legendary status in the gold country not for his intellect, but for losing his house in a poker game. As one newspaper discreetly put it, the judge "sported" to the limit.

The first-class passengers boarded at their leisure. Their three-hundred-dollar fare entitled them to a private cabin aft, where the ship rode smoother, a porthole looked out across the ocean, and the inner door opened onto the main deck dining saloon. Each private cabin contained three cushioned berths, one above the other, a locker, a mirror, toilet, washbowl, and water bottles and glasses. Carpet covered the floors, and layered damask and cambric curtains screened the berths.

Around the ticket office, nearly four hundred steerage passengers now clamored for the best berths forward in the hold. The hold was cramped and hot, the air damp, and the berths stacked three high, often no more than two feet side to side separating the tiers. A higher berth near a porthole to let in sunlight and fresh air made the trip in steerage tolerable.

One steerage passenger among the throng was Oliver Perry Manlove, a spirited young man who with three other men, a wagon, and four yoke of cattle, had set out from Wisconsin and crossed the prairie on foot in 1854. Manlove recorded every mile of the journey, a five-month search for grass and water to keep the animals alive, and for wood and game to keep the men warm and fed. In a train with three other wagons, they often walked twenty-five miles a day. Sometimes they passed wagon trains as long as six miles, three hundred wagons, their white canvas stained yellow with beeswax, a thousand head of cattle trailing behind. "This was life in earnest," wrote Manlove. "All rushing on to the Eldorado."

Manlove counted the miles, the Indians, the crosses marking the graves of those who had died along the trail: This one hit by lightning, that one drowned, the other one stricken by disease, another shot. During the five months, he counted 205 crosses.

In September, two days short of his twenty-third birthday, Manlove had arrived at Nelson Creek, which emptied into the middle fork of the Feather River, which joined the main Feather north of Marysville at one of the richest strikes of the gold rush, Bidwell's Bar. In a ravine only a few ridges south of Nelson Creek, three Germans had used penknives to pick $36,000 worth of gold out of cracks in the rock. News of the find had drawn thousands of other miners, several of whom washed $2,000 in gold in a single pan. A small party of men from Georgia pulled in $50,000 in one day.

Upon his arrival at Nelson Creek, Manlove wrote, "I had traded my rifle at a trading post for some clothing. This left me with only my satchel to carry, which contained my clothes with a Testament and a revolver—a six-shooter—strange company to be together. In my pocket book there was a half dollar, all the money I had in the world."

Like Manlove, most of the miners exhausted their money and supplies just getting to California, and they were dumbfounded at the cost of living and the difficulty of the work when they did. As promised, the gold was there, but the luck to find it and the labor to remove it had been greatly underestimated. To stay alive in the diggings, a miner had to find between a half ounce and an ounce of gold, between eight dollars and sixteen dollars, each day to keep abreast of the cost of living in the camps and lay a little aside for the trip home. But most miners averaged no more than a few cents to a few dollars, and that was after squatting on their haunches at the edge of a stream for ten hours and washing fifty pans of sediment.

For three years, Manlove watched men blow off fingers and hands with blasting powder, drink, scar each other in fistfights, read the Bible, scrape fingernails to the quick handling river rock, and wander from one claim to another, looking for profitable diggings, hoping and praying that the next shovelful would bring salvation. The miners wrote thousands of letters home, many telling of weariness, discouragement, and homesickness. The tales of rich finds were spectacular but few and always just over the next ridge.

When he left the diggings in July 1857, Manlove had been away from his Wisconsin farm nearly three and a half years. In that time he had sent a little money home, and he had a few hundred dollars left, just enough to pay for a berth in steerage on the fortnight steamer rather than walk back across the prairie.

LATE THAT MORNING at the quay, the decks of the *Sonora* erupted in a frenzy when the captain sounded the final departure bell and those not sailing tried to depart against the tide of those still struggling to board. When the uproar subsided, the captain ordered the crew to cast off, and a pilot boat led the *Sonora* into the bay past Alcatraz Island and the lighthouse. Then free of her pilot, the *Sonora* passed through the Golden Gate and steamed out upon the broad Pacific, heading south, carrying five hundred passengers, thirty-eight thousand letters, and a consigned shipment of gold totaling $1,595,497.13.

For fourteen days, the *Sonora* would steam south and east toward Panama, where her passengers would transfer to the new open-air railcars and shimmy for forty-eight miles to the Caribbean port city of Aspinwall. There the passengers would embark on the Atlantic steamer SS *Central America* bound for New York, a final trip of nine days across the Caribbean and up the East Coast, with an overnight call in Havana.

SHIP OF GOLD

Havana

⚓

Tuesday, September 8, 1857

T HE GAS LAMPS of Havana cast erratic ribbons of light out across the harbor, zigzagging among the dark silhouettes of more than a hundred ships at anchor. In the darkness, the SS *Central America* lay wrapped in the moist tropical air, her engines silent, her decks dimly lit and trod only by the night watch. In these predawn hours, her five hundred passengers slept with the ship motionless for the first time since departing Panama four days earlier.

High above the ships, at the mouth of the harbor, a massive brown escarpment called El Morro swept upward out of the sea. On top, the flag of Spain awaited the first light of day as it had ever since Columbus celebrated mass on the island three and a half centuries earlier. Then the first glimmer outlined El Morro, and slowly dawn touched the green hills of Cuba, following them down to the sea, as the flag of Spain brightened to crimson and gold, and the *Central America* emerged from the darkness as the biggest ship in the harbor.

She was sleek and black, her decks scrubbed smooth with holystones, her deckhouses glistening with the yellowed patina of old varnish. Along her lower wale, a red stripe ran nearly three hundred feet stem to stern, and three masts the height and thickness of majestic trees rose from her decks. Spiderwebs of shrouds and stays held her masts taut, and in moments she could sprout full sail, but she rippled with real muscle amidships: two enormous steam engines with pistons that traveled ten feet on each downstroke and turned paddle wheels three stories high. Between the paddle wheels, the funnel rose thick and black above all save the masts.

One of a new generation of sidewheel steamers, the *Central America* departed New York Harbor on the twentieth of each month, bound for Aspinwall, Panama, where she traded five hundred New York passengers bound for San Francisco for five hundred California passengers returning east. Since her christening in 1853 as the *George Law,* she had carried one-third of all consigned gold to pass over the Panama route. And in quantities rivaling her official gold shipments, unregistered shipments of gold dust and gold nuggets from the Sierra Nevada, and gold coins struck at the new San Francisco Mint, and gold bars, some the size of building bricks, had traveled aboard her in the trunks and pockets, the carpetbags and money belts of her passengers.

At sunrise the morning gun sounded from El Morro; trumpets blared and drums rolled from high on the fortifications, announcing to the international flotilla of ships that the harbor was now open for the business of the day.

Lighters immediately surrounded the *Central America,* the small boats filled with oranges and bananas and thin men wearing blue and white checkered shirts and hats made of straw. The boatmen spoke only Spanish, but they chattered and gesticulated, peddling their fruit for dimes thrown by the passengers, who in turn received oranges twice as large as any they had ever seen.

In another hour, the ship's bell resounded across the brightening harbor, and the captain ordered his crew to weigh anchor. Coal smoke and ashes rose from the funnel and roiled into the air over the afterdeck, the paddle wheels of the *Central America* churning the water white. With her bowsprit pointed onward as gracefully as the arched neck of a stallion, she glided through the mouth of the harbor beneath El Morro and out onto the sea, climbing to her cruising speed of eleven knots, the American flag rippling off the yardarm.

For many of her passengers, the final five days to New York would be the last leg of a long journey that began when news of the rich gold strike in California had first trickled east. "Many of us had been away for years," recalled Oliver Manlove. "We awaited the time of meeting our loved ones again. We were jubilant and made the old ship ring with our voices."

The *Central America* crossed the Tropic of Cancer, and with the green hills of Cuba shrinking above the whitened wake, the captain took her into the Gulf Stream, which he would follow most of the way to New York. The extra two-and-a-half-knot push lightened the work of his engines.

"As near as I can recollect," the second officer reported later, "we left Havana, Tuesday, September 8, 1857, at 9:25 A.M., and proceeded to sea, steering for Cape Florida, with fine weather, moderate breezes and head sea."

For half a day the seas remained clear and sapphire blue, the breeze in from the trade winds quarter and the surface smooth.

ANGLING NORTHEAST ACROSS the Straits of Florida, Captain Herndon followed the inner edge of the Gulf Stream, which flowed within a few miles of the Florida Keys, his course set for the point where the Keys broke loose of the mainland and arced westward. 'As the day wore on, the sun rose higher, blistering the sides of the ship. Tropical heat filled the hold, and the iron furnaces and boilers burned and bubbled at near capacity, running the temperature even higher.

Most of the passengers littered the weather deck, many of them still nursing mouth boils raised by the tropical sun and layers of raw skin peeling from their hands and faces. Some sat on wooden benches bordering the deck, some leaned over the rail, some coiled on top of the paddle guards, others sat in chairs or on seats under a large awning, and a few watched it all from the rigging. The air was so warm that even with the breeze many could remain in one spot for no more than ten minutes.

"The sky was bright overhead," noted Oliver Manlove, "while there was a slight ripple of the waves. But the hours were passing and by the middle of the afternoon quite a breeze was blowing. The waves were rising, dark and tossing, but were chopping up into little white hills that rose and fell."

That evening at sunset, the first- and second-class passengers took supper at the long tables and railroad benches in the dining saloon. Afterward, they retired topside again to stroll in the cooler evening breezes and amuse themselves with impromptu skits, or readings, or poems put to music and accompanied by a banjo, a guitar, or an old fiddle. Mostly they talked about loved ones and wondered silently how things had changed since they left their homes in the East.

While Captain Herndon entertained guests at his table, Manlove stood on deck, looking across the water, and recorded in his journal the end of their first day out of Havana. "The sun was shining brightly," he remembered, "and dropping down in the west with magnificent splendor, and when it reached the waves it was like a red fire upon them for a moment before it sank away, leaving a crimson flame above it in the sky."

CAPTAIN WILLIAM LEWIS HERNDON sat at the head of the captain's table, wearing thin gold spectacles. Gold epaulets hung from his shoulders. Married and the father of one daughter, Herndon was slight, and at forty-three balding; a red beard ran the fringe of his jaw from temple to temple. Though he looked like a professor or a banker more than a sea captain, he had been twenty-nine years at sea, in the Mexican War and the Second Seminole War, in the Atlantic and the Pacific, the Mediterranean and the Caribbean Sea. He knew sailing ships and steamers and had handled both in all weather. He was also an explorer, internationally known and greatly admired, who had seen things no other American and few white men had ever seen.

Seven years earlier, in August of 1850, while at anchor in the harbor of Valparaiso, Chile, Herndon had received notice that orders would arrive by the next steamer with instructions for him to explore the Valley of the Amazon, from the trickling headwaters of its tributaries sixteen thousand feet high in the Peruvian Andes all the way to Para Brazil, where the Amazon emptied into the Atlantic, four thousand miles away. "The route by which you may reach the Amazon River is left to your discretion," read his Navy Department order. It is not desired that you should select any route by which you and your party would be exposed to savage hostility beyond your means of defence and protection. . . . Arriving at Para, you will embark by the first opportunity for the United States, and report in person to this department.

Herndon had departed Lima on May 20, 1851, and arrived at Para nearly a year later, traveling the distance by foot, mule, canoe, and small boat. He had compiled lists, kept timetables, taken boiling points, recorded the weather, studied the flora, and measured and skinned small animals and birds. But he filed his report to the navy as a narrative, not only cataloging his scientific and commercial observations, not only presenting his studies of the meteorology, anthropology, geology, and natural history of the Amazon, but also rendering his experiences with natives and nature as colorful scenes that exposed the legends and the beauty and the curious customs of the region, creating one of the finest accounts of travel and discovery ever written. His report so far surpassed his superiors' expectations that Congress had published ten thousand copies as a book, *Exploration of the Valley of the Amazon,* which described his adventures with such insight, such compassion and wit, and such literary grace that he had come to symbolize the new spirit of exploration and discovery sweeping mid-nineteenth-century America.

AMONG THE GUESTS dining at the captain's table that evening were the newlyweds Ansel and Addie Easton. Ansel's short dark hair was swept back off of his broad forehead, a goatee covered his chin, and a glint of humor and serenity shone in his eyes. Addie had large eyes and a trim mouth; her dark hair, smooth and shiny, parted in the middle and twirled in soft buns about her ears.

"Captain Herndon had arranged to have us at his table," Addie later wrote to a friend in San Francisco, "and as he was a most delightful man, we enjoyed it very much."

That first night out of Havana, the early conversation turned to a topic popular on the steamers: shipwrecks. Scandal had arisen three years earlier when a captain and crew had rescued themselves from a sinking ship and left passengers to perish. Addie later recalled her host's charming segue to topics more pleasant. "How well I remember Captain Herndon's face as he said, 'Well, I'll never survive my ship. If she goes down, I go under her keel. But let us talk of something more cheerful.' And the captain told us some interesting and delightful experiences he had had in his remarkable Amazon expedition."

Much of Herndon's charm was his self-mocking humor. He told stories with punch lines that underscored the joke was on him. In one story, he remembered being on the river all day, beaching his craft on the shore, and preparing a typical meal of monkey meat and monkey soup. The monkey meat was tough, but the liver was tender and good, and Herndon ate all of it. "Jocko, however, had his revenge," said Herndon, "for I nearly perished of nightmare. Some devil, with arms as nervous as the monkey's had me by the throat, and, staring on me with his cold, cruel eye, expressed his determination to hold on to the death. . . . Upon making a desperate effort and shaking him off, I found that I had forgotten to take off my cravat, which was choking me within an inch of my life."

At the other tables in the saloon, the nightly card games had begun, and the sharp clink of silver coins blackened by salt air pierced the splashing of the paddle wheels and the leatherlike creaking of the timbers. Encouraged by good claret and beneath a white layer of smoke from fine Cuban cigars, the conversation at the captain's table continued late into the evening, until the Eastons retired to their stateroom and Captain Herndon excused himself to attend to ship matters.

Early in his exploration of the Amazon, not yet sixty miles from the sea, Herndon had reached the great divide, separating the waters that flow into the Pacific from the waters that flow into the Atlantic. He stood at an elevation of 16,044 feet, following with his eyes a road cut along the flank of the mountain, at whose base sat "a pretty little lake." When he got to the lake, he performed a curious ritual.

"I musingly dropped a bit of green moss plucked from the hill-side upon the placid waters of the little lake, and as it floated along I followed it, in imagination, down through the luxurious climes, the beautiful skies, and enchanting scenery of the tropics to the mouth of the great river; thence across the Caribbean Sea, through the Yucatan pass, into the Gulf of Mexico; thence along the Gulf Stream; and so out upon the ocean, off the shores of Florida."

In Herndon's imagination the green moss had floated along the same course he would take many times a few years later as captain of the SS *Central America:* across the Caribbean Sea, through the Yucatán pass, into the Gulf of Mexico, then north to catch the Gulf Stream: where she now steamed out upon the ocean, off the shores of Florida, into the dark.

Around midnight, the wind freshened perceptibly from the northeast.

WHEN SECOND OFFICER James Frazer assumed his four-hour watch at 0400 Wednesday morning, he recorded the sea conditions: a head sea and a "fresh breeze," seaman's talk for whitecaps and a twenty-knot wind. Just at daybreak, a lookout high in the rigging spied the whiteness of the Cape Florida bore fifteen miles to the west. Then the sky to the east reddened with the rising sun, blazed for minutes in vivid hues, and slowly drained of color as the sun surmounted clouds thickening on the horizon.

Passengers who had drifted in and out of sleep listening to the ship creak and the wind rattle shrouds high in the rigging awoke Wednesday morning tossing in their berths. They climbed the rocking gangways to the weather deck, where sailors confirmed their thoughts: The wind had risen after dark and then blown hard through the night. They could see the coal smoke swirling as it cleared the stack and feel the bow of the ship rise with the swell of the sea. The wind and the salt spray had cooled and freshened the air, filling the morning with a majesty that enchanted some of the passengers.

Returning to his watch at noon, the wind still fresh, the sea still head on the bow, Second Officer Frazer took his meridian observation. Steering along the western edge of the Gulf Stream, they had run 288 miles since leaving Havana twenty-six and a half hours earlier.

GARY KINDER

Now between the Florida coast and Grand Bahama Isle, the wind stiffened and the sea turned lead gray. Virginia Birch was chatting topside with several other ladies when, she reported, "a squall came up, and the wind blew like a whirlwind, and we had to go downstairs." Passengers who ventured onto the deck quickly returned to the main cabin to escape the wind and the spray. As the day passed from morning to afternoon, the wind continued to rise, and the waves lifted the steamer's bow higher and higher, before dropping her into the oncoming sea.

"In the afternoon there was a change," wrote Manlove. "It changed our feelings and drove the waves into mountains and valleys and made the old ship stagger."

Passengers unused to ocean weather and fearful at the first creaks wondered at the high waves and the rising winds; others watched the sailors methodically tend to their deck work and assumed that such weather was merely part of life at sea. "Everyone felt confident that the wind would soon abate," said one passenger, "and that there was nothing to be feared."

More immediate than the fear of storm was the nausea of seasickness. Most steamer passengers had never been on the ocean. During rough weather, the lee rail was often lined, as one contemporary put it, with "demoralized passengers paying their tribute to old Neptune." Beginning with dinner at Wednesday noon, the number of passengers desiring food had dwindled. Even the ship doctor took sick. By day's end, the sea rose above the plunging bow, flowed over the guards, and washed across the weather deck.

"When the twilight came," wrote Manlove, "if it could be called twilight, there was a raging storm such as we had never before seen. The waves and sky were crashing together." That evening, the dining saloon was almost deserted. A few steerage passengers stood and ate their meals, their legs wide to brace themselves, their elbows pinning a plate. Seasickness had confined the Birches and the Eastons to their berths. Another woman described the time as rather unpleasant, although she felt no danger. "At least my husband said he thought there was no danger, as we had so strong a ship."

Despite the weather, the nightly game of cards among the hardy souls in the main cabin went on as scheduled. At the captain's table a game of whist ensued, and across from Captain Herndon sat his partner at whist, Judge Monson. Although taking tricks in a four-hand game of cards was tame for the judge's propensity, he enjoyed a good turn of phrase and relished as well a good story, especially the telling of his own. Three times he had sailed east and back, and on an earlier voyage, he had befriended Captain Herndon. Now when he traveled, he always sat on Herndon's left at table.

The weather little bothered Monson, for on each of his voyages east his ship had steamed into an equinoctial storm. Late summer was ripe for these West Indian cyclones to arise far out at sea and rush toward land, whipping the Atlantic white. Day after next, September 11, was the mean date for storm season.

While the card games continued long after dark, some of the first- and second-class passengers who had lain in their staterooms nauseated all day abandoned their rolling berths for the sofas in the main cabin. That night, said Virginia Birch, "I lay down on a sofa with my clothes on, and passed a very uncomfortable time, the vessel careening fearfully."

Most passengers retired to their tiny staterooms or their cotlike berths in steerage, praying the weather would subside by morning so the dizziness in their heads and the nausea in their stomachs would go away and they could eat again and move about the ship without stumbling. "Down below," remembered a steerage passenger, "nothing was to be heard but the crying of children and the moans of those suffering seasickness, and rising above all the sounds that proceeded from the inside of the vessel was the continued dashing and splashing of the waves against the sides of the ship, and the howling of the storm as the wind surged through the steamer's rigging."

That night the wind continued and the rains began. As darkness fell on the second day out of Havana, even the seamen began calling it a storm.

THE CURVATURE OF the eastern shoreline fell away rapidly now as Captain Herndon angled from the mainland on a course for Cape

Hatteras. By Thursday morning the *Central America* had veered two hundred miles east of St. Augustine. High seas broke over the bow, sprayed across the decks, and splashed against the staterooms. Sometimes the steamer heeled so far that the housing over her paddle wheels rolled under water.

Trying to escape the cramped and humid below decks where scores of their fellow passengers had vomited, some passengers ventured up the rocking gangways to a weather deck constantly in sharp motion. They reminded themselves that the ocean is rarely benign and shipbuilders know this, that they build ships accordingly, that ten thousand ships had weathered a thousand storms just like this one.

At noon on Thursday the rain came in sideways, but the *Central America* remained on course, struggling against headwinds that had risen to over fifty knots. Despite the rain and the pitching deck, Second Officer Frazer shot the solar median again and calculated that since his observation the previous noon they had traveled another 215 nautical miles, steering almost due north by compass.

Two evenings before, the men had laughed at a woman for her timidity in the face of a little wind and sudden roll. "On Thursday," she said, "when I went on deck, the gentlemen kept assuring us that there could not be any necessity for fear." But by nightfall, even the men sensed that as violently as the wind blew and as high as the water around them now crested, the intensity of the storm had not peaked. That evening the inveterate card players, who the night before had indulged in whist and other amusements while the ship rocked through high seas, dispensed with the usual games to talk about the storm. "The storm was the leading topic of conference," remembered Judge Monson. "Some expressed their apprehension, particularly the ladies, as to the safety of the steamer. Most of the gentlemen, myself among others, did everything to prevent any alarm among the passengers."

About dark, the seas breaking over the steamer spilled into the staterooms, forcing some of the first- and second-class passengers to abandon their cabins. Just after the sky went black, the first officer turned over the watch to Second Officer Frazer and handed him a piece of paper. Written on the paper were the headings Frazer was to follow as he steered the ship through the storm till he left the bridge at midnight.

A GRAY DAWN broke on Friday with storm winds blowing out of the north northeast at over sixty knots, as the steamer pitched and rolled in waves whitened by wind and pelted by heavy rain. Thick foam blew across the surface of the sea in long streaks, sometimes flying whiplike into the air. Each evening had brought renewed hope from the passengers that they would awake the following morning to find the winds had lessened and the sea subsided; yet every morning the wind had blown with an even greater fury than the day before, and the sea had risen higher until the waves now towered above the ship.

The bow plunged into the oncoming sea, the deck heaving and falling away sharply. Waves exploded high into the air, salt spray mixing with rain, and the wind drove it all with a furious whistle through the bare rigging. Since late Tuesday night, the wind and the sea had slowed the progress of the *Central America,* but she had held her course. When Second Officer Frazer left his four-hour watch at eight o'clock that Friday morning, he estimated the ship's position as latitude 31° 45' N and longitude 78° 15' W, or 175 miles east of Savannah.

As Frazer departed the wheelhouse, a friend of the Eastons named Robert Brown sat near the top of a hatchway, beholding the fury of the storm. "The wind was very strong," he remembered, "but the sea was excessively high." Yet as the steamer took on the sea, he heard no creaking in her hull. "She all the time had her head to the sea and acted handsomely, and never appeared to even strain." Brown, a merchant from Sacramento, was so pleased with how she came up proud to meet the waves, he resolved that the next time he sailed for California he would delay his trip for two weeks, if necessary, to await the departure of the *Central America.*

Thomas Badger clutched his wife, Jane, and fought for footing on the pitching and rain-soaked deck. Shielding his eyes from the stinging spray, he studied the incoming waves and the bowsprit soaring to meet them. A powerfully built man, Badger had been a sailor for twenty-five years, a captain for the last ten, commanding his own three-masted bark in the burgeoning Pacific coast trade routes. He had sailed in many a storm and twice had traveled aboard the *Central America,* though he had never seen her perform in high seas. Like his bark, she carried full sail; but unlike a true sailing vessel she also carried 750 tons of iron in her engine works, and that could make her an unwieldy beast. Badger had come topside to satisfy himself that she still could match the sea in a tempest.

Badger judged the wind by reading the surface of the sea, and that morning he saw the air filled with foam and the sea completely white with driving spray, and he estimated they now had entered "a perfect hurricane." He reported that "the sea ran mountains high" and the wind was "directly ahead," but the ship's behavior impressed him as it had Robert Brown. She "came up finely, and was not strained perceptibly by the wind or the roughness of the sea." Badger could feel the enormous engines pounding, and he could see the giant paddle wheels "working regularly and slowly." As long as coal fired the boilers and the two massive engines churned the wheels with a full head of steam, he knew that Captain Herndon could lay her on the wind, let her bow take on the sea, and ride out any storm.

Working his way along the rainswept deck, Badger encountered the ship's chief engineer, George Ashby, hurrying headlong against the storm to report to Captain Herndon. Ashby had kept the furnaces hot and the steam pumping the pistons down in the engine room since the ship first went to sea as the *George Law* in October 1853. He was now on his forty-fourth voyage, and Badger knew him from previous travels on the ship.

Above the shrillness of the storm, Badger called to Ashby. As hard as it now was blowing, he yelled, it would blow harder still.

"Let it blow," shouted Ashby. "We're ready for it."

But at that moment, Ashby was less convinced than his words made him sound. Minutes earlier he had discovered something he could not tell Thomas Badger. He had just issued several orders to his men in the engine room, then rushed topside looking for Captain Herndon because the captain had to know immediately, but if word got out, Ashby's discovery would alarm the passengers: The ship had sprung a leak, water was rising in the bilge, and Ashby could not find the source.

STEAM ENGINES RAN on water converted to vapor, which cooled and condensed on the metal surfaces as water droplets, then combined and enlarged and joined with small leaks in the machinery, all of it dripping from the boilers and the massive pistons, sliding along the metal pipes, down the funnels and the flues, and finally collecting in the bilge. A steamship never ran dry. When the bilge water reached a certain level, the pumps sucked it up and spewed it back into the sea.

But Ashby had discovered that the water in the bilge was far deeper than normal; either a leak had formed somewhere in the machinery, or seawater was seeping into the hold. If the pumps worked properly, and the leak was not too great, they could control it. But the water alone was not the focus of Ashby's concern, and this was the other problem he refrained from mentioning to Badger.

The engines sat on oak timbers as thick as half a dozen railroad ties and occupied the steamer's entire midship, port to starboard: two furnaces, two boilers, and the stack, 750 tons of sweating iron, forty feet across and rising in the hold sixteen feet off the flooring above the bilge. Piled high in the bunkers aft of the engines, the *Central America* also carried several hundred tons of anthracite coal. Besides powering the ship, the coal provided ballast; but as the coal heavers wheeled coal from the bunkers to feed the voracious furnaces, and as the furnaces sent the tons of coal up the stack as smoke and ash, the *Central America* lightened and rose higher. Steamers sometimes rose so high in the water, the paddle wheels could barely scrape the surface.

Although before leaving New York the coal porters always filled the *Central America*'s bunkers with enough coal to fire the steam engines all the way down to Aspinwall and back again, on her return voyages, she often came onto the coast a high or "crank" ship, one that heeled too far to the wind. That was her reputation. In a gale, or even in a moderate blow with the wind abeam and the ship lightened, she careened considerably. Since leaving Havana three days earlier, with her hull pounding against a head sea and struggling against an ever increasing wind, the ship had burned even more coal than usual, so she was lighter and higher in the water. The tons of forty-niner gold she had picked up in Panama were hardly enough to compensate for the dwindling coal. The blow had heeled her over, and the water rapidly collecting in her bilge was settling on the starboard side, tending to keep her there.

To move the coal from the bunkers to the fire room, where the firemen shoveled it into the furnaces, the coal heavers had to push wheelbarrows filled with coal as much as a hundred feet. But the high seas coming over the bow and the hurricane winds hauling from the northeast had caused the ship to list at such an angle that the heavers had difficulty pushing the barrows of coal. The barrows slipped and spilt, and the men lost their footing. They couldn't move the coal fast enough to keep up steam.

Before he left the engine room to find Captain Herndon, Ashby had called in the off watches of firemen and heavers, ordered the barrows abandoned and the men to form a line and pass the coal in buckets hand to hand down to the furnaces. But the heavers could pass little more coal in many small buckets than they could in the coal barrows. They had difficulty even keeping their balance in the hot, dimly lit hold that rocked at their feet.

When Ashby reported the rising water to Captain Herndon, the captain immediately ordered waiters and stewards into the hold to form a second line of coal heavers. Few passengers wanted food, and the waiters had little to serve other than hard bread anyhow, because the water in the hold had risen high enough to dampen the stores of food.

IN A STORM, the ship rode most safely and easily if the captain headed into the sea and brought the wind onto the weather bow, using her engines just enough to hold her in that position. But as the careening of the ship and the inrushing waters slowed the flow of coal to the furnaces and the steam began dropping in the boilers, the paddle wheels turned more slowly. If Herndon lost his engines, the only way he could hope to hold position on the incoming sea was to use sail, and he couldn't wait for the engines to stop before he tried to hoist some of his canvas. After sending the waiters and stewards into the hold, he ordered the storm spencer run up the mizzen mast. The storm spencer was the strongest and heaviest of all the sails. With it unfurled aft, Herndon hoped to blow his stern to port and use the wind to keep the bow of the steamer headed into the oncoming sea.

The waves swelled, then rose, then tapered into sharp hills just before the wind ripped the tops off and sent them flying as spindrift into the air. The troughs had deepened enough to provide a moment's shelter from the wind, and in that moment the crew shot the storm spencer upward, and then the next wave swelled, lifting the ship again into the wind, where the howling filled the sail so suddenly and with such fury that within a minute it had been blown to pieces.

With the storm spencer blowing in tatters, Herndon ordered the third officer to spread more storm canvas low in the main and mizzen rigging to try again to bring her head to the sea. But the ship rode so high out of the water she still would not respond, and the storm canvas again blew apart. Unchecked by sail, the wind drove the rain and the salt spray like bullets across the deck, and the pitch of its whistle rose.

Amidships, DEEP INSIDE the hold, even the shrillness of the wind died in the bubble and roar of the fire room. Here in the fulcrum of the ship's pitch and roll, away from the wind, away from the rain, the temperature soared to 120 degrees.

Sidewheel steamers had no bulkheads, no watertight compartments athwartships to contain flooding in a small area should the ship take on water. Once water entered the bilge, it ran at will fore and aft and sloshed with the rocking of the ship from starboard to port. Shortly after ten o'clock that morning, with the black gang, stewards, and waiters all passing coal into the fire room, Ashby cranked up the starboard bilge pump with steam siphoned from the big boilers, and the pump started sucking water from the bilge and disgorging it back into the sea.

Ashby then inspected all of the pipes and the fittings and found them tight. He examined all of the metal plates covering the portholes and found no leak. As he searched for the source of the water, buckets of coal moved hand to hand to hand, steward to fireman to waiter, from the aft bunkers to the fire room, where the white-hot fire inhaled a bucket of coal and in seconds turned it to ash. No matter how many men joined the line, no matter how quickly their hands moved, no matter how much sweat rolled down their blackened arms, the buckets of coal came too slowly to feed the voracious furnaces. Steam pressure was dropping.

The starboard pumps sucked up the water in the bilge, but the water rose faster than the pumps could send it back into the sea. With the coal now coming only in small buckets, the furnaces no longer burned hot enough to keep the boilers bubbling at peak. If seawater seeping into the bilge reached the furnaces, the water would cool the fires, the steam would condense, the pressure would drop, and the starboard wheel would spin slowly to a halt, leaving all power to the port wheel, which already spun in the air because of the cant of the ship to starboard. Then the ship would fall into the trough and be at the mercy of a merciless sea.

With the paddle wheels turning more slowly, Thomas Badger suspected problems in the hold and descended into the fire room to check the progress of the coal brigade. He was alarmed at how high the water already had risen in the lee bilge. He heard Ashby warn the men that if they did not move quicker, every man on board soon would be bailing. Badger yelled to Ashby, "Don't wait till the ship's full of water. Start the men to work bailing now!"

At noon, both forward and aft, water overflowed the floor of the coal bunkers, rolled back and forth in the lee bilge, popped out the floor plates in the fire room and left the firemen standing waist deep in the water, some of them holding on to iron bars lashed into place to keep their balance. The water now rushed into the hold so fast and the ship met the storm at such a severe angle that the waterline had reached the starboard furnace. The coal passers could hear the hot furnace hiss as the seawater splashed at its undersides, and the furnace boiled the water swirling around it. Steam began to rise in the hold.

Ashby and the engineers discovered that in some of the lower starboard cabins just above the engine room, the waves pounding into the heeled-over ship now were pumping water through the porthole covers, until some of the staterooms were waist deep, and that much water so high in the ship increased her heel to leeward, keeping her down. They battened the shutters, then chopped through the deck in the staterooms to send the water into the bilge and help right the ship.

Searching in the hold, Badger had found an opening to the sea around the shaft that supported the starboard wheel. Every time the ship rolled to starboard, seawater pumped through the opening. He reported this to the engineers, then hurried topside to speak with Captain Herndon about organizing the passengers into a bailing gang. The engineers packed the hole around the shaft with blankets and old sails, and the pumping action of the ship rolling into the oncoming sea blew out the packing and they packed it again, but it blew out again and again, until engineers remained at the shaft, constantly replugging the leak. Still the water rose, and no one could find another source. Perhaps the pounding and twisting of the ship had worked the oakum out of her seams or even separated her planking. The hull itself seemed to be leaking, and no one could stop it.

Scores of passengers once seeking refuge down in the forward hold had long since fled the stench of vomit, the heat, and the incessant pounding of the hull. Abandoning their steerage berths, they had pulled themselves and their belongings up dark and jerking gangways to the dining saloon, where the air was less stifling and the ride less harsh. Some of the first- and second-class passengers had been driven out of their cabins by water rising along the starboard perimeter; others had left their cramped and tossing staterooms for the seemingly greater safety of the saloon and to share their mounting fears with other passengers. Outside, the morning sky remained dark; inside, the oil lamps dimly lit the growing number of passengers huddling in fear where three nights earlier they had gathered to eat and celebrate the beginning of the last leg home.

At noon, Second Officer Frazer returned to the weather deck to resume his watch, steering the ship into the storm. Thirty minutes later, Captain Herndon fought his way to the wheelhouse, and found Frazer muscling the wheel, still trying to bring the ship head to wind.

"It's no use," shouted Herndon. "I've been trying all morning."

Frazer asked the captain if he should try a different tack, keeping off before the wind.

Herndon departed the wheelhouse into the storm, still shouting, "Do whatever you can to keep her away!"

Gathering below in ever larger groups, their eyes searching upward into the dark beams of the saloon, the passengers listened: to the wind, to the waves, to the water hissing across the deck. Other than the jolts from confused seas slamming the hull, their only understanding of the storm came through their now heightened sense of sound.

Billy and Virginia Birch had left their cabin and spent Wednesday and Thursday nights in the saloon on a railroad bench. "On Friday morning," remembered Virginia, "the vessel careened over on her starboard side and we heard the beams crack. Shortly afterwards we were told that the vessel had sprung a leak."

Since Wednesday, the Eastons had lain sick in the privacy of their stateroom along the starboard side of the ship. About noon on Friday, a big sea knocked the ship down, those gathered in the saloon heard loud pops, like the beams cracking, and the Eastons' stateroom canted, just as Addie glanced at their porthole to see that it was under water.

"Ansel," she cried, "we're sinking!"

The steamer had listed so far to starboard neither could lie in a berth or even sit up. Forgetting their seasickness, they threw wrappers over their nightclothes and struggled out the cabin door to the saloon.

"I had always heard that the 'Law' was such a staunch vessel," remembered Addie, "that I did not apprehend any serious cause for alarm until we opened the door of our stateroom and saw such a scene as I shall never forget. Without one word from anyone, I saw at once from the appalled faces that we were in imminent peril. I do not remember but one woman in tears. The rest sat silent, pictures of despair."

Ansel told Addie they had to get to the upper side. He threw his arm about her, and their friend Robert Brown helped the two of them reach a settee on the far side of the cabin. They sat down, their hands clasped, and uttered not a word.

The steamer was pummeled so thoroughly by the sea, raked so furiously by the wind, and filled so high with rolling water that amid all of the clamor, most of the passengers gathered in the saloon did not know that the engines had all but stopped.

THE WATER SOON swamped the starboard fires, the flooded hold rocking with every incoming sea and sending the water in waves over to lap at the port fires. Then the furnaces turned more of the rising water to steam; vapor hissed in the hold, entering the ash pits, choking off the fresh air, extinguishing the port fires, and one by one dowsing the oil lamps until the men could hardly see, could hardly breathe, the firemen and the bucket gangs baking in a blackness that heaved all around them.

The port wheel still turned but ever slower, and with the coal now impossible to pass, Captain Herndon ordered teams of crewmen into steerage to rip out berth slats and break them into smaller pieces, which they carried into the fire room and stuffed into the port furnace. If they could keep at least one engine turning, and if their pumps could get ahead of the water in the hold, they could wait out the storm. But the kindling they tossed into the fireboxes burned with far less intensity than anthracite coal. Eventually, the fire only smoldered and the engines turned but a few more revolutions, and then the rising water extinguished the fires.

Outside, the hurricane winds beat the rigging against the masts and drove down on the *Central America* as if trying to crush her against the surface of the sea. No horizon now, only lead-colored clouds roiling above a lead-colored sea that swelled into overhanging cliffs of water.

BY EARLY AFTERNOON, Captain Herndon could no longer point the bowsprit of the *Central America* into the wind. He fought simply to maintain his direction without facing the blast, but the wind blew the steamer off course to the southeast, where her huge engines finally cooled and she fell off into the trough of the sea. Waves once split by the prow of the ship now rose nearly the size of foothills on the broadsides and came crashing down on top of her from stem to stern.

Hundreds of passengers now huddled in the central dining saloon, waiting for a report and trying to comfort each other until it came. Suddenly, they heard from above a crash so loud they thought the ship was headed to the bottom at once. The cry ran through the cabin that she was sinking. Then the splintering sound subsided, and afterward they learned that a heavy sea had ripped one of the lifeboats from its davits and sent it skidding across the deck to crash into the deckhouse above them before sweeping it overboard.

Shortly after this, Captain Herndon appeared at the door to Judge Monson's stateroom on the lower deck, a chronometer and sextant in his hands. He kept the instruments in his quarters on the weather deck, two rooms on either side of the main mast. He had a favor to ask of Monson. "He asked permission to remove his instruments there from his stateroom on the upper deck," remembered the judge. "There was a possibility, he said, that his stateroom, from its exposed position, might be swept away, and his instruments. I complied with the request of course, and the instruments were removed accordingly."

In the main cabin the few women who had experienced bad weather at sea tried to comfort others, assuring them that the *Central America* was a strong ship that had plowed its way through many a gale. Seemingly unaware of the storm and certainly not frightened by the rocking of the ship, two little girls about nine years old sat eating at one of the tables. Amid the anxious looks and pallor of sickness, they braced themselves against the table, fastened their plates with their elbows, and seemed to enjoy the confusion.

One older woman watching them recalled later, "When the dishes flew about smashing and crashing as they fell to the floor, the girls laughed merrily, thinking it was rare sport. They were decidedly jolly, little realizing the danger in which they stood."

As the little girls sat merrily at the table, the captain's boy rushed into the saloon and cried out, "All hands down below to pass buckets!"

"At this," said one woman, "the women burst out into lamentations, knowing then that the vessel was in peril."

Within seconds of the captain's boy leaving, Herndon himself appeared at the door of the dining saloon. He announced loudly but calmly, "All men prepare for bailing the ship. The engines have stopped . . ."

A murmur went round the room. The thought surprised Addie. She turned to Ansel. "What does that mean?"

". . . but we hope to reduce the water and start them again," continued Herndon. "She's a sturdy vessel and if we can keep up steam we shall weather the gale."

Though the *Central America* had leaked for hours and eight to ten feet of water now filled her hold, few of the passengers knew of the leak until that moment. Upon hearing Captain Herndon's plea, several men in the cabin stood up, pulled off their coats, and approached the gangway. The captain walked over to one passenger and remarked with a smile, "You must take off your broadcloth and go to work now."

Ansel Easton and his friend Robert Brown rose to join the other men. Before her husband could leave, Addie said to him, "Ansel, if you hadn't married me, you wouldn't be in all this trouble."

"If I knew it all beforehand," replied Ansel, "I should do the same again."

"Here in the midst of mortal peril," remembered Addie, "with death before me, with all the joys of life, so wonderfully loved, disappearing, those words made even the storm and shipwreck nothing."

The Eastons decided that if the time came, they would go down together "hand in hand."

"But," said Ansel, "until all hope is past, we must work."

Then he kissed his wife and, with Brown, joined the other men.

FROM THE GALLEY and the staterooms, crewmen had rounded up scores of wash buckets and water pitchers. Herndon explained that some of the pumps continued to work well, but that they alone could not keep ahead of the water; they needed the supreme effort of each man to dry the hold. Hundreds of volunteers rose up and made their way past Captain Badger, who directed them to one of three bailing brigades: one forward to rise from steerage, one all the way up from the engine room, and a third that wound from the aft cabin hatches through the main cabin and up the gangway. All three lines would end on the weather deck, where the last man would dump the buckets with the wind, returning the water to the sea, and send the empty pails back down to be filled again.

In the saloon, they removed the grating over the forward hatchway, and in steerage they worked in water up to their knees. They passed the heavy buckets along quickly, developing a rhythm, heavy and spilling going up, lighter and swinging coming back down, hand over hand, the water coming up out of the hold.

Most of the men were miners who had left farms, where the work was hard, to seek fortunes in the Sierra Nevada, where the work proved harder still. Many of those who went west had not survived, but these men had, and the experience had hardened not only their bodies but also their spirit. They felt themselves a special breed: unafraid to face hardship and to endure. One passenger remembered, "The voices of the workers rose merrily and powerful above the din of the storm and the lashing of the steamer's sides by the waves."

Song added a sense of camaraderie and dispelled thoughts of fear; song provided cadence and seemed to lighten the weight of the buckets. Most of the men were seasick and had had little or nothing to eat, but song and the energy of the other men helped them clear their heads and forget their hunger.

Captain Herndon seemed to be in all parts of the ship at once, assuaging the fears of the ladies in the main cabin, on deck ordering the crew, up and down the bailing lines, encouraging the men.

"Work on, m'boys," he would say, "we have hope yet."

Though one woman commented that she preferred to hear his voice on deck shouting commands, most of the women appreciated his visits to the main cabin to "cheer up the spirits of the passengers and to quiet their fears," as one woman put it. "He did not try to disguise the danger, but he made us look more cheerfully at it than some other men might have done."

Most of the women had been in bed seasick for almost three days, yet Friday afternoon, several of them demanded that they be allowed to stand shoulder to shoulder in the bailing lines with the men, but the men rebuffed them. So the women held on to the children and watched the line of men winding through the saloon. They searched the men's faces for clues of the ship's condition, waiting to hear that the men had gained on the water, waiting to feel again the rumble of the big steam engines deep in the heart of the ship. The men told the women that they were gaining on the water, that there was nothing to worry about, that the water could not gain on them. They told the women that word had come up with the buckets that the steam again had risen, and that they had the pumps working, and they offered that the storm must abate, that the wind had blown so hard, it could not continue much longer.

Annie McNeill, nineteen, remembered how the other women showed "great courage and self-composure; not a tear was shed by any of them. The men told us to be cheerful, that it would soon be all right; indeed, although we considered ourselves in imminent peril, we did not know the full extent of it. The men did all they could to keep that knowledge from us."

News filtered back to the women that the men were gaining on the water so rapidly that the engineers thought they might be able to fire up the furnaces and get some steam into the boilers to power the wheels. And soon they felt the throb of an engine coming up through the decks. The vibrations gave everyone renewed hope, and the men seemed to work even harder. But the wheels went round two, maybe three, times and again they stopped, and the water came in even faster now and rose back over the furnaces and the boilers, and the engines stopped forever.

Upon feeling the engines die, the women dispatched a small boy up the gangway to inquire the reason. When he returned to the main cabin, he reported that a man had said they stopped "because the wheels were tired and wanted to rest a while."

"He gave this answer," said one woman, "partly because it was a little boy that we sent upstairs to ask him, and partly because he wanted to prevent exciting an alarm."

But the women knew.

"The sea broke over us in avalanches," said Virginia Birch, "completely swamping the cabin and staterooms, and the vessel would be so completely buried that it was as dark as Erebus. The ladies never spoke a loud word and kept perfectly calm and collected. I never saw a calmer set of women in my life; one or two asked to be permitted to share in the labor of bailing, but were told by the gentlemen to keep quiet and all would yet be well."

Many of the men had spent their years in California handling river rock. Buckets of water about twenty pounds apiece were not an unusual load for their muscles, but working in the bailing lines without rest had knotted the muscles in their shoulders and backs and in their forearms, yet they couldn't stop. The water kept seeping into the ship, and the moment they hesitated, they lost another inch to the sea. After three hours, some of the men had grown weary of trying to keep their footing, trying to keep the buckets from spilling, trying to move the buckets fast enough to stay ahead of the rising water. No more than thirty of them had family on board the ship; most of them had no one to protect but themselves. Yet an unspoken code kept them in the bailing lines and at the pumps, as though but for their single efforts the ship would go to the bottom, taking with her all of the women and children.

The women huddled in the main cabin all afternoon, waiting and watching the men, holding and comforting the children, and feeling the creaking and pounding of the ship. The men continued to pass the buckets, but their singing had long since stopped.

With the ship in the trough of the sea, making heavy lurches to leeward, and men bailing in lines all over the ship, Captain Herndon continued trying to bring her bow back into the oncoming swell. He ordered his men to hoist just enough canvas to edge the bow to starboard and get the ship off before the wind. But the moment they set the sail, the wind blew it to pieces. Then they lashed another sail to the deck and raised it up only enough to show canvas, but as soon as the sail had cleared the bulwarks, the wind ripped it out of its bolt ropes and tore it into tatters now horizontal and almost stiff in the storm.

Herndon ordered his men to abandon all further attempts to use sail and to work instead on setting a drag, or sea anchor, overboard to bring her around. For the drag, they needed to lash a heavy anchor to a stout yard in the rigging and toss it all overboard, but the heaviest anchors hung from the bow and the ship listed so far the men couldn't reach them. Second Officer Frazer dropped the yard, bent a thick rope to it and to a smaller anchor, pushed it over, and paid out forty fathoms.

It was then approaching half past five, and the ship listed so far to starboard that no one could walk her deck. Her three heavy masts angled out over the water, nearly lancing the incoming waves as they swelled, crested, and broke over the ship. With sails and rigging no longer of use, Herndon told Badger to take an ax and cut down the foremast.

Badger, Frazer, and the bos'n, John Black, worked their way forward, leaning into the wind, grabbing hold as the ship lurched and the waves broke over her broadside and the water ran like a river at flood across her deck. First they cut away the rigging that held the foremast tight, then crouching and holding fast with one hand they hacked with axes at the tree-thick mast, until at last they heard a sharp crack and then another, and the tossing of the ship shook the mast at its splintering base, finally tumbling it over the fore rail.

But as the mast fell, the rigging they had cut free flew in the wind and snagged in an anchor rest; the mast somersaulted into the water and, ensnared in the rigging, shot under the ship and began to pound against her hull. "I do not doubt," Frazer said later, "that when the foremast went under the ship's bottom, she was injured by it, and probably the leak increased thereby. I don't know such to be the fact, but she thumped there for some time."

They now paid out another hundred fathoms on the sea anchor and made it fast around the stump that had been the foremast. But the heavy drag tugging at the starboard bow did nothing to bring up the ship's head, and later that night, the thick line holding the drag would fray, unravel, and disappear into the sea. Herndon's last attempt to right his ship was to spread bits of canvas in the aft rigging, a token gesture of desperation, for they were so small to avoid being shredded that they captured too little wind. With evening approaching, nothing that would bring the steamer's head to wind remained in his power.

A⊤ 7:00 P.M., nearly every man on the ship, over five hundred in all, was either in a bailing line or helping to run the deck pumps. Long lines of men coiled throughout the ship, up every gangway, some gangways supporting two lines, the buckets passing in both directions, the men's arms moving like the myriad legs of a centipede. The heaving of the ship sometimes sent the men slamming into one another, and buckets slipped from their hands or banged together, spilling at their feet; waves broke over the ship, and water flooded back down the gangways.

All afternoon, they had seen the water drop, and the men in the hold had had to edge down closer to scoop the buckets. But at nightfall, the water fought them to a standstill, and after two hours of darkness, it was rising again. Still, without food or sleep, the men continued steadily into the night, moving muscles long past weary, stumbling as they tried just to hold their positions in line. Outside, howling through the darkness, the wind shifted and turned back out of the northeast, blowing heavily.

Earlier in the day, the women huddling with the children in the dining saloon had listened to the sloshing of the water, thinking it was the waves dashing against the sides of the ship; then the men were called to bail, and the women realized that the water sounds came from the hold. Now the sounds grew louder as the water rose to the cabin beneath them.

"In that condition," remembered Annie McNeill, "we remained all night, the sea running very high, and breaking over us, the wind blowing a perfect hurricane, the ship rolling and beating about, everything making a most fearful noise, the rigging and spars cracking and groaning, the dishes, lamps, furniture smashing and crashing together. It was an awful night, but the women still endured it without tears or moans."

Angeline Bowley held and comforted her babies, Charles and Isabella, ages two and one. "We all appeared to grow more calm and resigned," she said. "Those who had no children to take care of, and be anxious for, were quite as brave and hopeful as the men. But as for myself, I must confess that, being sick, and weak, and with these two helpless little ones clinging to me, I became somewhat discouraged and disheartened. A few of the ladies showed no signs of fear, and kept up to the last. It was wonderful to see their composure."

Late that night, the women supplied the lines of men with what hard bread and fresh water they could find, and even a large supply of liquor and brandy made its way to the lines. A few of the men, dejected and exhausted, drank heavily, broke from the bailing lines, then hid themselves in staterooms and locked the doors. Others, very much sober, gave up in despair, and left the lines so fatigued they could hardly move. Most of the men worked until they were long past fit for it; they dropped to the deck, "as if they were dead," and they lay there for minutes until Herndon or Badger called for more recruits. Herndon told them to relieve each other like men, and not to suffer one of them to drop in place, while another man stood idle. Then he asked if they wanted water, and had it brought to them, and they found fresh courage to rise again.

"Our only comfort," said Angeline Bowley, "was that we knew the men were making every exertion in their power. They worked like horses. I never saw men work so hard in my life."

Ada Hawley asked her husband, Frederick, if he was tired, and he replied, "Yes, I am tired, but I can work forty-eight hours in the same way if necessary. I am working for your life, for you and my children." While Frederick bailed, Ada calmed their two sons, two-year-old DeForest and five-month-old Willy.

Twice, Addie Easton and two other women started toward the bailing lines, determined to work alongside the men, but they were not allowed to do so. "I never wished so much to be a man," remembered Addie. "We sat there I know not how long but until after dark, with no visible outward emotion at what seemed our inevitable fate, that a few hours only at most were between us and eternity. The bailing was continued vigorously all night, my own dear husband taking his turn and when exhausted returning to my side; and when a little rested again resuming his place. We talked with each other calmly, and mingled our prayers to him who was our only hope and refuge. He heard us and in answer gave us sweet consolation in these trying hours. . . . How little do we realize in our earthly security the preciousness when every human hope has fled, of trusting, whether we live or die in an almighty power. I could not think of anything I had ever done to merit his love but still I felt that we were in his hands and resigned to his will. All that fearful night we watched and prayed, not knowing but that every hour might be the last. My dear husband and I talked calmly of our dear, dear friends, of our brief happiness together, and hopes in the future. Life had never seemed so attractive or dear to either of us, yet I think we could both say 'thy will be done.' We resolved that when the moment came we would tie ourselves together and the same wave would engulf us both."

About eleven o'clock that night, Addie remembered the bottles of wine and the crackers and biscuits and other foodstuffs she and Ansel had received as wedding presents. The hampers filled with these items were still in their stateroom. She made her way across the cabin and down to her stateroom and brought up all of the hampers filled with the packages of food and began passing them out to the exhausted men. The men stopped only for a few minutes to sit down, eat, and drink, and then began bailing again. At intervals throughout the night, Addie made the rounds with her hampers of food and wine, until the men had consumed it all.

"Mrs. Easton," recalled Joseph Bassford, "furnished the men a large number of bottles of wine. The liberal bestowal of the wine, and the spirit which prompted its donation, won the admiration of all. Not only was increased vigor given to the men, but it roused them to work still bravely on."

All Friday night, hurricane winds ripped across the steamer's decks, the storm waters crackling with phosphorescence, and every hour the water in her hold rose another six inches. But the bailing never stopped. Hand to hand to hand, the water-filled buckets passed up the gangways, out of steerage, out of the engine room, out of the second cabin, the empty buckets traveling down again to be refilled. Too exhausted to be heard above the shrillness of the wind, and perhaps too fearful to speak, the men worked on in silence as the ship tossed in a dark and relentless sea. From midnight till four on Saturday morning, they grew wearier and wearier from incessant labor and exposure to the storm, and the water gained fast. Yet they continued, and now the women offered encouragement by repeating, "Only another hour to sunrise."

"Oh that long weary night!" wrote Addie. "How I counted the moments as they slowly dragged along! And as morning came, about three o'clock the Captain came in and said if they could keep up the ship about three or four hours longer, he thought we might be saved. The storm might cease and then perhaps they might get up steam, or when daylight came a vessel in sight might give us the blessed means of rescue. So they toiled on, and never was a daylight more gratefully welcomed than on that Saturday morning—the last that ever dawned on many a noble heart."

Aboard the
SS *Central America*

———————— ⚓ ————————

Saturday Morning, September 12, 1857

T HROUGHOUT THE SHIP, the coming of dawn fired the men's spirits. They could see through the rain to the haze hanging along the horizon, the sea not cresting as high as before and the clouds beginning to thin. The wind had shifted and dropped, now blowing in at about forty knots from the west and southwest, though higher-velocity squalls within the storm still spun through to rock the ship. Captain Herndon pointed to the thinning clouds and predicted that their breaking up portended an end to the storm. He spoke to the men at the pumps; he cheered the men in the bailing lines. He told them he thought the storm was abating, and that if they would just continue to bail until noon, the steamer might be saved. He delivered the same message to the passengers in the main cabin: They must not abandon hope.

GARY KINDER

"This announcement caused a general cheer from the men at the pumps," said Judge Monson, "and sent joy and gladness to the hearts of the lady passengers."

Though the passengers received the captain's comments with great cheer, Herndon knew his hope was false. He knew the sea would rise again and the wind would blow with even greater fury. He knew that a ship floating 750 tons of iron with water filling her hold, and more water constantly rushing in, could remain afloat but a short while longer. He also knew that every bucket of water tossed back into the sea gave the ship and her passengers a few more seconds afloat, and that in those hours gained, real hope might appear on the horizon. He was in a frequently traveled part of the ocean, and if he could keep the steamer afloat at least until the storm abated, he had a chance of saving everyone by transferring them to a passing ship.

About eight o'clock that morning, Captain Herndon went again to Judge Monson's stateroom. In privacy, he told Monson they had no hope of surviving unless the storm ended soon or a vessel came in sight.

"I presume I was the only person on board to whom he communicated that fact," said Monson. "The captain was perfectly calm, but intimated that it was only to keep up the courage of the passengers and crew until the last moment."

Herndon ordered the flag lowered and then hoisted again upside down, a signal of distress to passing ships that might be able to assist. He ordered the bos'n to rig pulleys from the mizzen stay and to run lines down each of the three aft hatchways. To the pulleys they attached pork barrels and beef barrels and lowered them to men waiting with buckets and pans to fill them with seawater from the hold. Then a gang of fifty men topside heaved on the lines to hoist the barrels to the upper deck, where they dumped the water and lowered the barrels again. Each minute now, four hundred gallons of water left the ship in barrels.

By midmorning they had nine rigs operating, in the aft hatches, over the fore hatch, between decks, and coming up from the engine room. And three gangs continued to pass buckets hand to hand. The women offered again to join the men in the bailing lines, and again the men refused. But Thomas Badger noted, "They cheered us up in our labors by their calmness in these trying times. And the men worked with such

48

increased vigor that for the next two hours the water in the hold noticeably dropped.

Captain Herndon continued to visit the bailing lines and the barrel gangs, cheering everyone to push beyond their limits and to keep hoping. But about ten o'clock Herndon was in his quarters when Badger reported to him that although the storm appeared to be abating, the water in the ship was once again gaining on the men rapidly. The engines, the boilers, the furnaces were immersed in fourteen feet of seawater, and the water had risen to within four feet of the second cabin floor.

"The vessel must go down," said Badger.

"I believe she must," agreed Herndon. "I have made up my mind to that."

As the two men talked, Chief Engineer Ashby rushed into the captain's quarters.

Badger said to him, "The ship will sink."

The remark seemed to startle Ashby. "She shan't sink!" shouted the engineer. "I'll be damned if she shall! We must all go to work and bail her out!"

Badger replied that he wished talking in that manner would make the waters recede, but he and all the rest on board had been hard at work all night bailing, and still the water was rising. No one knew when, but the ship would go down.

In front of these two men, Herndon allowed his true feelings to show. He too was dealing with his mortality and thoughts of never again seeing his wife, Francis, or his daughter, Ellen. He was tired and dejected and seemed resigned to his fate. He told Badger and Ashby that it was hard to leave his family this way, but, of course, that could not be helped; he was the captain, and as long as others could be saved, he would not leave his ship.

But outside his quarters, Herndon was the forthright commander. He might lose his ship, the mail, and millions in gold, but he still had nearly six hundred souls entrusted to his keeping, and until that final moment when the sea closed over the decks of the *Central America* and dragged them all into eternity, he still harbored a waning hope that lives might be saved. On deck and in the cabin he exuded enthusiasm and control and talked as if only a short amount of time separated the dismal and trying present from a glorious and certain rescue. The passengers caught his hope and themselves clung to little things that buoyed that hope.

Addie Easton wrote of Saturday morning. "How we thanked God for his mercy and the daylight for its cheering presence. Then commenced renewed exertions. Barrels were rigged through the skylight— three of them—and for a time they seemed to gain on the water, or at all events, it did not increase. The clouds began to disperse, and the wind to lull. Every countenance changed and brightened, and all labored more heartily and cheerfully. The steamer nearly righted. How I watched to see the lamps hang level! But no sail in sight yet. We talked of another steamer that floated eleven days water logged, and felt more hopeful."

THEIR HOPE LASTED but a few hours. By noon, the clouds had thickened again overhead, and the wind blew fiercely, and the sea swelled even higher, and as hard as they worked, more than five hundred men could not keep up with the water rushing into the hold. The *Central America* now sat so low that the sea had begun to seep through the starboard portholes, and some of the cabins were three feet deep in water.

"Alas," wrote Addie, "even with all our increased efforts, it was soon evident that the water was gaining, and to our unspeakable dismay the fury of the storm returned."

Without complaint, without demonstration, one woman solemnly gathered her children in their small stateroom, "thinking that if we went down we would all go together."

When her husband left Captain Herndon's quarters and came to comfort her, Jane Badger said to him, "I am prepared to die." But her husband was short with her; it would not help for anyone to abandon hope. Hundreds of men continued to work, and their efforts could not cease, or, yes, everyone would die. He had to help keep them going. Before his wife could finish, Badger turned away and climbed again to the hurricane deck, and though she tried not to let the other women see her, Jane Badger wept. Then after moments of struggling to control herself, she turned to a woman near her and in a strained but cheerful countenance she said, "The Lord is merciful. Perhaps some vessel will come in sight, and we may yet be saved."

As the prospect of salvation grew darker, the bailing continued throughout the ship, but time and the sea finally had beaten all hope out of the men. They showed little fear, little anguish, working methodically on, resigned to their fate, and almost too exhausted to care. A few more left the bailing lines and returned to their berths or locked themselves in their staterooms. Deeply fatigued and spiritually broken, they refused to come out.

The men still bailing no longer labored under the delusion that their efforts would allow the engineers to start the engines once again. Now every bucketful of water thrown overboard simply bought time and only a sliver of it, but most of the men preferred the exhaustion of their labor to the anxiety of sitting and awaiting whatever fate had in store.

They had been passing the buckets or manning the pumps or hauling the barrels up on pulleys with little rest and no sleep for twenty-four hours, when, nearing two o'clock on the afternoon of Saturday, September 12, they heard the cry, "Sail ho!"

ON THE TWENTY-NINTH day of August, the *Marine,* a two-masted brig, had departed Cárdenas, Cuba, bound for Boston and carrying barrels of concentrated molasses squeezed from sugar cane. She ran 120 feet from stern to bow. Captain Hiram Burt and his crew of five had left port at daybreak, sailing on a benign sea for twelve days before they encountered heavy seas off Savannah, where the vessel creaked and rolled, and water casks lashed to the deck broke loose and smashed or bounced into the sea. She took so much water over the bow, the crew had to cut away some of her bulwarks on the starboard side to drain her decks. On Friday morning, a heavy sea ripped away the flying jibboom, the foresails, and all of the rigging. Then the wind and the rain struck, carrying away the mainyard and the main topsail, and when they had cleared away the wreckage and got the bilge pumps running, they discovered that several barrels of molasses had cracked open, and the water siphoned out of the hold had turned the color of mahogany.

By late Friday afternoon, remembered Captain Burt, "it was blowing a complete hurricane." He ordered all of the canvas taken in and had the vessel lying to under bare poles, trying to hold her with the wind on her bow so the heavy seas did not break aboard. "But the gale," reported Burt, "continued with unabated violence during the night."

Just before daybreak on Saturday, the storm moderated slightly, and at 5:00 A.M., Captain Burt adjudged it best for the preservation of his ship and cargo to put her before the wind, and bore away for Norfolk to repair damages. By noon the wind had dropped to the force of a moderate gale, but the sea ran heavy, and with little sail left and the wind still fierce, Captain Burt had the *Marine* scudding under bare poles, carried onward only by the following sea.

Aboard the *Central America,* Herndon had lookouts watching the horizon. In early afternoon one spied a distant dot, which quickly grew to be another ship tossing in the storm. The lookout bellowed, "Sail ho!" shocking all of those passengers and crew within earshot, the shock spreading across the deck and along the bailing gangs and through the main cabin where the women and children huddled. News of the sighted ship cut through the passengers' grim resignation so sharply that passengers and crew alike suddenly laughed and cried and denied it was true all at the same time. No specter of good fortune had ever appeared so magnificently in the lives of so many doomed souls.

"Such a sudden hope," said one woman, "where nothing but death had stared us in the face, at once overcame our self-control; there was shrieking, crying, weeping; agonies of joy, where late was nothing but agonies of death. The severe calmness that had set on each cheek was displaced by the flush of excitement, profuse tears, and the embrace of friends, mothers and children, husbands and wives. The excitement pervaded the whole ship."

Addie Easton remembered, "We were on the verge of despair again, when about two o'clock in the afternoon was heard a joyful cry of 'a sail, a sail!' and in a few moments, it was seen bearing down towards us, and they gave it three cheers. For the first time during all the storm my eyes filled and I wept tears of joy and thankfulness. Strong men wept, women laughed and cried and it seemed for a few moments as if there would be a panic on the deck. A stern word from the captain brought order and every eye eagerly watched the vessel drawing near."

Captain Herndon stood on the quarterdeck, his glass aimed to the northeast at a speck tossing on the sea. He ordered the signal guns fired at once and a second flag of distress hoisted. The signal guns could be heard for miles across open water, but in the storm the smoke from the guns dissipated immediately in the wind, and the howling muffled their report. However, storm skies loomed so dark overhead that the brilliant flash from the muzzles caught the eye of Captain Burt. As he scudded to the southwest in early afternoon, a ship had hove into sight off his lee bow, and as he drew closer, he saw that it was a steamer flying a signal of distress. Captain Burt kept off for the steamer to render assistance if he could. For over an hour, he maneuvered his crippled ship through high waves with little more than a rudder, even that all but useless in a following sea.

At so great a distance, Captain Herndon had no idea of the size of the brig: whether she could hold merely a few or as many as all six hundred. And the sea remained so high that some questioned how they could ever board the other ship. When he was certain the captain of the brig had seen his signal, Herndon turned to Judge Monson, who stood at his side on the quarterdeck, and told him to come immediately to the captain's stateroom.

"He said he was afraid there might be a rush of passengers for the small boats," remembered Monson. "He wanted the ladies and children saved first. He desired, he said, some of the passengers to assist in preventing a rush for the boats."

Herndon had five lifeboats left, and he assumed the brig also carried boats the crew could launch. He told Monson he had to transfer all of the passengers to the brig, because he estimated the *Central America* could remain afloat no longer than another fifteen hours. As he talked with Monson, the brig approached off his weather bow.

At about 3:00 P.M., Captain Burt spoke to the steamship and found it was the *Central America* in a sinking condition. He rounded her stern less than one hundred feet out and hove to under her starboard lee. From their heaving, slippery deck, the passengers could see the faces of Captain Burt and his crew. The *Marine,* less than half the size of the *Central America,* waterlogged and partially dismasted, the jibboom snapped off, her captain with very little command over her and shipping nearly every sea, had appeared suddenly out of the storm as the salvation for which no one had dared hope. As she passed under the stern, the passengers cheered, believing now they all were safe.

Captain Herndon hailed the brig and, according to Burt, "with all the calmness of an ordinary occasion," shouted into the storm, "'We are in a sinking condition. You must lay by us until morning.'"

Captain Burt shouted back, "'I will stay by you as long as I can.'"

MEN HAVE ALWAYS pondered how they will act under fire, and mostly the reality when it comes is far more sober and sickening than imagined, and the acts are much less quick and noble. When a ship seemed destined to sink, the captain and his officers often had to hold the crew and the male passengers at gunpoint to keep them away from the lifeboats until they could safely remove all of the women and children. Sometimes not even the captain and crew acted nobly. Four nights earlier, Herndon had jokingly turned the dinner conversation from shipwrecks to topics more pleasant by declaring that if his ship ever went down he would be under her keel. It was a charming seaman's segue, and from the mouths of perhaps most men it would have remained no more than that. But the remark had been prompted by the sinking of another steamer three years back, when the captain and crew had commandeered the lifeboats, and 259 of the 282 passengers, including all of the women and children, had perished. Herndon's friends knew that for three years the story had haunted him. He was now in the middle of an even bigger disaster, but he had already determined that surviving by less than honorable means was not worth a lifetime of scorn.

Customary for the day, but grossly inadequate, the *Central America* carried six lifeboats, five of wood, one of metal. The night before, one of the wooden boats had been smashed against the wheelhouse. The remaining boats each normally held four oarsmen, a helmsman, and forty to fifty passengers, but with the strain on the oarsmen in seas such as this, Herndon could load no more than fifteen or twenty. As soon as the *Marine* rounded to, he ordered the first officer to clear away two of the lifeboats, one on the port, one on the starboard, and to get the port boat to the lee side of the ship. Then he ordered all women and children below to don life preservers.

Although the sea nearly swamped the first lifeboat under the guards of the steamer, the crew launched it and the second boat safely. Then they lowered a third lifeboat, and as soon as it hit the water, a heavy sea quickly sucked it away, then rose up and smashed it against the hull of the ship, the planks shattering and the remains of the boat hanging from its block.

Two boats were now in the water, the oarsmen trying to keep them away from the sides of the steamer. Two boats remained on the upper deck, one wood, one metal. The crew lowered the wooden lifeboat safely into the water. Chief Engineer Ashby helped launch the metal boat, riding it down into the waves, but a heavy sea caught the boat, drove it hard under the lee guard, stove it in, swamped it, and sank it immediately. Ashby disappeared with the boat, and the crew had to pull him from the water.

Below in the main cabin, the women and children gathered to prepare for the trip to the *Marine*. To give them as much freedom as possible and as little weight, the women were instructed to strip off their undergarments and layers of skirts, everything but the outside dress, then put on a life preserver. They also wrapped the older children in life preservers and the babies in blankets to be held in their arms.

Many of the women traveled with a great deal of money they had not registered with the purser. All of them were now advised not to carry more than two twenty-dollar gold pieces with them. Two women retrieved a satchel from their stateroom and upon returning to the cabin, opened the satchel, and weeping, shook eleven thousand dollars in gold onto the floor. Through tears, they said that anyone who wanted the money could take what they pleased. "That money is all we made in California," they added. "We were returning home to enjoy it."

As the women discarded their extra clothes and put on life preservers, the captain's boy appeared at the entrance to the cabin and shouted, "The captain says all the ladies must go on deck!"

The few who had managed to ready themselves and their families for the trip made their way across the room and started climbing the steps to the hatchway, their dresses long and sagging from lack of hoops and petticoats, their upper bodies covered in cork and tin life preservers, their hands holding or pushing forward small children, until they reached the deck. There the water crashed around them and the wind blew the spray over them, and they were wet through in an instant. By the time Ashby was hauled from the sea, women and children were struggling their way to the lee side of the ship, where the five-man crews fought to keep the lifeboats from being smashed against the ship or swamped under her guards. Herndon ordered Ashby and his first officer not to let a single man into the boats until all of the women and children were off.

"While they were getting into the boats," observed one man from the bailing lines, "there was the utmost coolness and self-control among the passengers; not a man attempted to get into the boats. Captain Herndon gave orders that none but the ladies and children should get into the boats, and he was obeyed to the letter."

In line to be transferred to a lifeboat, Annie McNeill glanced at Captain Herndon as he stood on the rain-soaked deck. She thought he seemed saddened. She talked with him briefly, and he said he would not try to save himself, that he would go down with his ship.

"Nevertheless," she remembered, "he did all that lay in his power to save others. He was a very kind, generous, gentlemanly man, and if he had any fault it was that he was not severe enough to his own hands."

THE ONLY WAY Captain Herndon could get the women and children into the boats was to lower each by rope one at a time from the upper deck, while the oarsmen tried to fend off, and the lifeboats crested close to the ship's hull. He and his men had fashioned a rope chair. "A noose was passed around our feet and dress," recalled one of the women. "There was nothing to support our backs, but we held a rope, which came down in front, with our hands. The boat could only approach the steamer between the waves, so we had to remain suspended sometimes while the wave passed. These waves would also drive us under the side of the steamer."

The water sucked away from the steamer's hull, then rose up and slammed against it, sending salt spray high into the air and hissing back into the sea. The oarsmen fended off, trying to keep the boat steady, close enough to catch the women and children dropped from the deck, yet far enough away to avoid the waves smashing the boat to splinters. Already the storm had claimed half of the lifeboats.

The women and children had to jump from the deck as far out over the water as they could, dangling on the rope, and then drop suddenly when the waves pushed the boat higher and toward the ship. In that instant, the men holding the ropes often just let go. Some women fell into the boat, others hit the water, and then either the hands on deck raised them hanging from the rope to try again, or the oarsmen grabbed ahold and hauled them in over the gunwale. The sea moved suddenly and with great force, seemingly in all directions at once, and a body slammed against the hull of the ship or dropped suddenly into the boat found them both hard and unforgiving. Many of the women were bruised and cut; some sprained shoulders or twisted ankles. Most of them fell into the sea at least once, several of them twice, and one disappeared beneath the surface three times. As soon as a woman or child had dropped to the boat and an oarsman could free the rope, the men on deck snapped it up and quickly began rigging the next passenger.

Some women being herded toward the rope looked around wildly for their children and called to their friends, but their voices were drowned in the confusion. Some got shuffled forward so quickly and found themselves in the sling and over the side so suddenly that they ended up in a boat without their children. Some watched after the children of other women. Captain Herndon supervised the evacuation, constantly moving from one part of the ship to another, ensuring that only women and children got into the first boats.

Jane Harris started up the gangway from the saloon to the upper deck holding her baby in her arms, but she could barely move because the steps under her feet slid sideways then suddenly dropped away or rocketed upward as the ship pounded in the trough of the waves. Herndon saw her trying to negotiate the stairs and sent one of the passengers to help her. Moments later, as she stood on deck prepared to descend to the first boat, Herndon assisted in loading her into the rope swing.

"The captain tied a rope around me," she remembered, "and I think he was one of the men that had hold of it when I was lowered down. He was a noble man, and I shall never forget him as long as I live. When I began to slide down, a great wave dashed up between me and the little boat, which threw the boat off from the ship and left me hanging in the air with the rope around my waist. I was swung hither and thither over the waves by the tossing of the ship, then I was dropped suddenly into the boat when it happened to come directly under me. As soon as I got into the boat, I looked up and saw the captain was fixing a cape around my child, and a few moments afterward he lowered her down to me."

The minstrel Billy Birch had left the bailing lines, found his wife, Virginia, and helped her with her life preserver. Then the two of them went to their cabin to find a cloak for Virginia. Amid the water and debris, Virginia saw something she could not leave behind: the canary she had carried aboard in a cage. "It was singing as merrily as it ever did," she remembered, and she hated to think of the small bird being drowned or crushed as the steamship broke apart.

"On the spur of the moment I took the little thing from its prison and placed it in the bosom of my dress. My husband remonstrated with me, hurrying me to leave the vessel, and telling me not to waste time on so trifling an object."

Together they hurried through the crowds of people belowdecks and up through the hatchway, trying to keep their footing as they made their way against the wind. They prepared Virginia quickly, and she bid her husband of three weeks good-bye. "I expected that it was the intention to transfer all the passengers to the brig, otherwise I would not have left while my husband remained behind. But he told me to go and he would soon follow, and so I went." The canary stuffed into her dress, Virginia swung over the side of the steamer and disappeared beneath the waves before she could be hauled, soaked and gasping, into the first of the small boats.

Many of the women expected their husbands to accompany them in the lifeboats, or assumed that their husbands would follow soon in another boat. But every man refused to leave with his wife until all of the women and children had been safely ferried to the *Marine*.

Mary Swan was a young wife traveling with a baby not yet two. When the order came for the women and children to prepare for lowering to the lifeboats, her husband left his place at the pumps and came to her. "About an hour before I left, he took me aside and bade me, 'Good-bye.' He said, 'I don't know that I shall ever see you again.' He was very glad to think that I could be taken off. He wanted me to go, and said that he did not care about himself, if it were possible that I could be saved, and the little child. He told me that he would try to save himself if an honorable opportunity should present itself after all the women were taken off. He had been sick for three or four days before the disaster, but notwithstanding this, he persisted in keeping his place at the pumps."

Behind Virginia Birch in the first boat came the only black woman on board, the stewardess, Lucy Dawson, known affectionately as Aunt Lucy. A stout, older woman, Aunt Lucy fell into the water three times before they could land her in the boat. On one of the dunkings, a wave lifted the lifeboat, which slammed into Lucy, then smashed her against the side of the steamer.

Three more women and five or six more children left the deck of the *Central America,* swung out over the water, and descended into the first of the lifeboats. Four crewmen manned the oars and at the helm stood one man Herndon knew he could trust above all others, the bos'n, John Black. As the oarsmen pushed off and dug in their oars, Virginia Birch heard Captain Herndon yell to Black, "Ask the captain of the brig to lay close by me all night for God's sake, as I am in a sinking state, and have five hundred souls on board, besides a million and a half dollars!"

AWAITING THE WOMEN and children, two more lifeboats now remained at the sides of the steamer, each already beginning to fill, fore and aft, as crews on deck prepared the rope slings, loaded them with passengers, and swung them out over the water.

Before boarding in San Francisco, Thomas Badger had given his wife $16,500 in twenty-dollar gold pieces, which she had sewed up in toweling, in three parcels, and laid flat in a trunk. The trunk sat in their stateroom, now in water up to Jane Badger's knees as she picked her way among the "rubbish which strewed the cabin." She found the trunk, unlocked it, took out the gold, placed it in a carpetbag, threw a crepe-shawl on top, locked the bag, then had to leave it sitting on the lower berth: It was so heavy, she couldn't lift it.

In his coat hanging in the stateroom, Badger also kept a memorandum book containing notes and other records of debts owed to him in New York, the sum of which was several thousand dollars. Wading through the water, Jane Badger retrieved the book "with all its contents," secured in a small bag about $1,500 worth of diamonds, bracelets, and rings, together with a purse containing $40, and made her way back to the deck. When she found her husband and told him what she had done, he told her to throw away the jewels, anything with any weight to it, but she declined, insisting she would keep them in her pocket.

On deck, Badger helped rig his wife for her descent to the next lifeboat. He bid her good-bye, and Jane Badger swung out over the water; a wave sucked the small boat away from the hull, and she dropped into the sea. On the second try, she landed in the boat but at first could only crouch in the bottom, her legs too unsteady and the sea too unpredictable for her to risk moving to a seat. When she finally moved to a bench, "a lady of remarkable stoutness" dropped from the deck above onto her neck and shoulders so hard she thought the blow had broken her neck. Some of the women in her boat took up buckets to help bail, though their haste, the wind, and the motion of the sea combined to keep much of the water from ever making it over the gunwale.

The men knew they were unlikely to leave the ship, but to coax their wives into the sling ready to be lowered, many allowed their wives to think they soon would be reunited on the brig. Annie McNeill, an orphan of nineteen when she married a man of thirty-three, had retrieved seventeen thousand dollars, chiefly in drafts, plus her diamonds and jewelry. "I am sure I should never have left the steamer had I known that the men were not coming," she said. "I should never have left my husband." They had been married five months, and she had no other family. "He constantly assured me that he was going with me until he got me on deck and the rope tied around me, when he said he could take care of himself and wanted me to be safe first."

In their stateroom preparing to leave, Ada Hawley asked her husband if he would go with her to the brig. He took his money out of their trunk and said not a word. She had been ill for several days, and she needed help with DeForest and little Willy. Her husband grabbed the baby, and a friend carried the older child, and they hurried on deck. Mrs. Hawley looked for the *Marine,* and it appeared to lie about a mile and a half away. She went over the side first, and there she waited in the boat.

"The little children were passed down," she recalled, "the officers lowering them by their arms, until the boat swung underneath, and they could be caught hold of by the boatmen. It was frightful to see these helpless little ones, held by their tiny arms above the waves. My babe was nearly smothered by the flying spray, as they were obliged to hold him a long time before he could be reached by the boatmen, but when I pressed him once more to my bosom, and covered him with my shawl, he soon fell asleep. I took nothing with me except a heavy shawl and my watch."

The instant the boat filled, the order came for the crew to shove off, and Herndon again shouted to the helmsman, "Tell the captain of the brig, for Heaven's sake, to lay by us all night!"

As the oarsmen set their oars to, Ada Hawley saw her husband. "He stood on the wheel-house and kissed his hand to me as the boat pulled away from the ship."

With her husband on deck working, Addie Easton went to their stateroom and put on a "dress skirt to cover my night dress and wrapper, which I afterwards took off, as I thought I might get in the water and it would cause me to sink. Then I went to my small trunk which was in the room and took my dear mother's miniature, also one of my brother James, and some money. Taking a shawl and putting on a life preserver, I started to go above."

Just as she reached the door, Ansel came in and told her to hurry. "We shall be saved," he said, "but the women and children are to be taken off first."

"I can't go without you," said Addie.

At the thought of leaving her husband on the ship, her courage vanished. Ansel told her she had to go, and that he should follow very soon. Then their friend Robert Brown came to the door of their cabin.

"Come, Easton," he said, "you must hurry. They are taking another boatload now."

Ansel quickly reached into the trunk, took out a coat, stuffed the remainder of his money, about nine hundred dollars, and some valuable papers into the pockets, and rolled it into a bundle.

When the Eastons and Brown got to the deck, the second boat was nearly full. Ansel found Captain Herndon and asked him how many boats were left. The captain replied, "Only one. We had five but two were dashed to pieces as they were being manned, so we have but three." He calculated, however, that the three boats could make several trips before dark.

"I left in the third boat," Addie later remembered. "I said however to Ansel, 'I don't want to go till you do.' He said, 'You had better go now,' or something of that kind. I then kissed him and said, 'I'll pray for you.' In a moment I was swinging from the deck, and when a swell brought the little boat underneath, the rope was lowered and I dropped in the bottom of the boat. It was a dreadful moment for we were in great danger of swamping or being stove to pieces. And just then the contents of one of the barrels they were bailing with came down on my head, completely drenching me. Ansel threw me his coat containing the money and also took off the coat he was wearing and threw it to me to put about my shoulders."

One passenger described the evacuation of the women and children as "a dangerous, heroic and almost superhuman effort" that "can scarcely have a parallel." But the passenger also noted that "through some strange and mysterious influence there were several young and unmarried men taken in the life-boats to the brig, leaving behind those men who had wives and children."

Virginia Birch had pleaded with Ashby to allow her husband, Billy, to go with her. "But he refused," she recalled, "using insulting language."

Lynthia Ellis, a woman of delicate constitution, dehydrated, and suffering acutely from four days of seasickness, asked that her husband be allowed to go with her to help care for their four children, two of whom were sick, all of whom were young. But the hands refused. "No men would be allowed to go until all women and children were safely off."

Other women, too, had pleaded with the officers to let their husbands go with them and were refused, yet somehow in the confusion single men now sat at their side. With Addie safely in the lifeboat, Ansel Easton and Robert Brown returned to the bailing gang, but before that lifeboat filled, several men got in with Addie and the other women and children, the crew unaware that three women and at least as many children remained on board the steamer.

One of the men who boarded a lifeboat early was Judge Monson. He had applied to the first officer to allow an elderly gentleman, Albert Priest, to board that third boat. Perhaps realizing the judge was a friend of the captain, the first officer agreed, and the crew lowered the old man.

"I gave Mr. Priest a message to my brother in New York," Monson said later, "in case I should not be saved myself. Mr. Priest said, 'Never mind the message, come, Judge, yourself.' The first officer said, 'Certainly, Judge, it is your turn—all right, jump in.' I immediately was lowered into the boat. A moment previous I had not the slightest idea of leaving the steamer then."

Ann Small was the last of the passengers loaded into the third boat. Mrs. Small was a new widow with a two-year-old daughter. A few weeks earlier, her sea captain husband had died at sea, and she had buried him in Panama. When she boarded the *Central America* in Aspinwall, the American consul had asked Captain Herndon to deliver her and her child to New York. Herndon replied he would personally guarantee their safety. Now as the officers placed her in the rope swing, Herndon came up to her.

"Mrs. Small," he said, "this is sad. I am sorry not to get you home safely." Then he turned away, and she saw no more of him.

The boat, tossing at the side of the steamer, waited below for its final passenger. The crew quickly wrapped Ann Small into the harness, swung her out, and paid out line on the harness. Twice she fell into the sea. When at last she made it into the boat, wet and shaken, the oarsmen pushed off to begin the long journey to the brig. Seated in the lifeboat, Ann Small now looked up: On the deck of the steamer high above her she saw her little girl. The crew had wanted the mother lowered to the lifeboat first, so she then could receive her child. But the men in the boat below did not realize that one more tiny passenger remained for them to rescue; they had pushed off immediately, and they couldn't go back now. As the small boat pulled into the high seas, Ann Small saw her little girl on deck still in the arms of a crewman.

WHILE HERNDON AND his crew worked to lower the women and children into the lifeboats, Captain Burt of the *Marine* had made sail, trying to bring his brig to windward, but with only one sail against a gale of wind and a heavy sea he could not succeed. Before the lifeboats could be lowered and filled and set off, she had drifted nearly two miles, now almost an apparition in the storm, at times disappearing altogether.

For most of the women, risking the journey through high seas and a gale wind in such a small boat seemed only slightly less frightening than remaining aboard a ship certain to sink. One woman described the waves as "running mountains high." Angeline Bowley, clutching Charles and Isabella, later recalled, "After I got safely into the little boat, and my babes with me, I had but little hope of getting to the brig. The water dashed into the boat, and we had to keep dipping it out all the time. Two high waves passed entirely over us, so that it seemed as if we were swamped and sunk; but the boat recovered from them both. The commander of this boat encouraged the sailors to keep every nerve steady, and told them that it would require the exercise of all their skill and courage to reach the brig in safety."

Almira Kittredge had gone out with three children of other women. "I put one in my lap, another between my knees, and the third I held by the collar. At length I got tired of holding the one by the collar and let him sit down in the boat, the water clear up to his neck. He sat in that condition and never spoke a word."

As soon as the oarsmen had pushed off, a huge wave rolled over the third boat, half filling it with water. With hills of sea passing under them and the oars frequently knocked from their hands, the oarsmen fought just to keep the oars in the water and the boats from capsizing. "It was a good thing just at this moment," wrote Addie Easton, "when I felt that I could not keep from breaking down at the parting of my dear husband, that I had to rouse myself for action. The men were all needed to row the boat, the other women were hysterical and so all the long miles we had to go before we reached the brig were spent in bailing the boat." While the men rowed, several women in the three boats bailed hard with large tins.

About two miles to the lee when the first of the lifeboats shoved off the *Central America,* the *Marine* had drifted another mile before the lifeboats began to arrive. The crewmen had pulled on their oars incessantly for an hour and a half.

In calm weather the *Marine*'s deck rose nearly eight feet from the waterline. In a storm, her decks were awash as she shipped sea after sea. When the first lifeboat came alongside, the oarsmen stood to fend off against the hull. The waves swelled, then crested, lifting the lifeboat higher and higher until its gunwale towered above the bulwark of the *Marine*. But Captain Burt saw a way to use this to his advantage. He stationed himself on the deck, his feet braced close to the railing. The wind blew and the sea again rose, pushing the lifeboat higher and higher, and Captain Burt yelled for the women one at a time to hold out their hands when he directed. Then, with two of his men standing by to keep the boat from crashing onto the deck, Burt watched the boat rise on the next wave, shouted the signal to the first woman, seized her slippery arms, and hauled her aboard in the moment the lifeboat seemed to hover above the deck.

"I did not have much time to stand on ceremony," said Burt. "My only object was to get them safely on my ship."

Captain Burt and the deckhands had to be quick, the oarsmen strong and alert, and still not each attempt went smoothly. The sea rose each time in a different way, coming at different angles to destroy the rhythm gained on the previous attempt. But the women and children were frightened enough to do as they were told, and some remained strong enough to try to get aboard themselves. Jane Harris watched for a chance to spring at the rigging and grab hold of a rope. "I caught the rigging with my hands, but my life-preserver under my arms was so large that I could not get between the ropes. I hung there for a few moments over the side of the ship, in almost equal peril as when I dangled at the end of a rope over the side of the steamer. I was every moment expecting to fall, when the captain caught hold of me, and pulled me in by cutting off my life-preserver."

Mary Swan did not leap for the rigging but held out her hands to the captain and his mate, who caught hold of her. "But they slipped their hold, and I fell into the water. I was got hold of again and partially lifted out, but fell into the sea three times before I was finally rescued."

"Captain Burt, with his mate, stood with open arms and a willing heart to receive us," remembered Ada Hawley. "Captain Burt took my little Willy, and the mate received DeForest, playfully saying, as he passed him over the side, 'He is all gold.'"

The *Marine* was deeply loaded and rolled badly in the high seas. When the women and children came off the lifeboat high in the air, the captain and his mate set them down again in the seawater that constantly swept her deck. Then, despite their weariness, the crews pushed off, dug in their oars, and returned through heavy seas for more of the passengers. Meanwhile, the *Marine* drifted farther and farther in the storm, and lightning flickered in a close and darkening sky.

THE GAP BETWEEN the *Central America* and the *Marine* widened, the wind pushing the smaller and much lighter brig at a faster pace. The now useless iron of the steamer's engines and her hold full of water kept her low, and the gale force winds could not move a steamer with a hold full of water as readily as they could a sailing brig only a third its size. The oarsmen in the lifeboats could fight the wind and the sea just so long, and if the *Marine* could not hold position, Herndon would have to try to maneuver closer to the brig. The wind had somewhat subsided, and the heavier sails should now withstand even the fiercer gusts. About 4:00 P.M., he ordered the main spencer set, hoping to use the wind to keep the gap between the two ships from widening even more.

With the new sail set, Captain Herndon told passenger Theodore Payne to take the first boat that returned. He had a favor to ask, but first he wanted Payne to go into his office and fetch his gold watch and chain. When Payne brought them to the captain, Herndon said, "If you are saved, deliver them to my wife. Tell her to . . ."

But then he choked and his voice quit, and he stood silently, saying nothing. When he could speak again, he asked Payne to meet with the president of the steamship company and the agents, and to explain to them what had happened. After saying this much, he walked away a few steps and sat down on a bench, his head in his hands. In that position he remained only for a few moments. Then he arose and resumed giving orders, as the first lifeboat, helmed by the bos'n, John Black, had returned from the *Marine*.

WITH THE MEN pushing to fill the first of the returning boats, the officers learned that three female steerage passengers still were on board. Theodore Payne went below to bring them up, and Ashby helped them into the sling and down to the bos'n's waiting boat. An elderly Englishwoman, Mary Ann Rudwell, was the last woman to come off the steamer. Her husband traveled with her, and she asked the captain that he be allowed to accompany her in the lifeboat to the brig. "He said he was very sorry," she recalled, "but he could not permit a man to get into the boat until the very last woman had been taken from the vessel."

In all of his moving about the ship, Herndon seemed unaware that husbands and fathers were being refused passage in the lifeboats, while unmarried men, including Judge Monson, had left the ship in earlier boats; that on this, John Black's second trip, carrying the last three women, Black would also carry more than a dozen men. Billy Birch asked Ashby to allow him on the boat so he could join Virginia, and the engineer said he would do all he could, but Birch did not get on that lifeboat. Neither could Ansel Easton secure a seat to take him to Addie. But he found a pencil and scribbled a note on a small piece of blue paper, which he folded and gave to one of the last three women to leave, asking her please to deliver it to his wife upon arriving at the brig.

When the boat was nearly filled with passengers, Ashby himself went up to the captain and insisted he be allowed to go to the brig in that boat. Captain Herndon replied that as a senior member of the crew he had no right to do that, that it was his duty to stick to his post.

"If you let me go," said Ashby, "I will bring back the two boats and all the small boats that can be spared from the brig."

Captain Herndon questioned whether he could trust the engineer. Ashby promised to do everything he could to bring the brig closer and to make sure the lifeboats returned; besides, with the engines silent and the engine room immersed in seawater and the women and children now evacuated, he had nothing to do, and the captain needed someone aboard the brig to ensure that her crew was doing everything it could to rescue those passengers remaining aboard the *Central America*. Herndon finally consented.

Ashby said, "I promise you, Captain, most solemnly, that I will come back to the steamer and not desert her."

He seized a rope and slid down to the boat. A steerage passenger, who had given an assistant engineer six hundred dollars for the promise of a seat on the first returning lifeboat, saw Ashby swing down. He saw no other lifeboats and figured this would be his last chance to exercise the

privilege he had paid for. He leaped for the same rope and quickly lowered himself into the boat, landing almost on top of Ashby.

Ashby yelled, "You son of a bitch," drew his dirk, grabbed the man
by the throat, and threatened to kill him if he did not jump overboard.
Then he looked up and threatened to take the life of any man who dared
jump for the boat. Herndon saw the incident and yelled to Ashby to put
the knife away.

The oarsmen shoved off. Joseph Bassford, holding a knife and trying to fasten around his waist a money belt containing two thousand
dollars in gold, saw that the boat had been cut loose, so he stuffed the
whole belt into the side pocket of his coat, leaped from the steamer deck,
and landed in the boat. But somehow in his haste to make the boat, he
dislodged the money belt and the belt hit the gray water as hard as he
hit the boat, and it sank instantly.

Captain Herndon walked forward along the rail of the steamer and
shouted to Ashby one more time, "I will depend on your returning with
the boats!"

To which Ashby replied, "Captain! You may depend your life on
my returning!"

THE OTHER TWO lifeboats returned, the first coming alongside forward.
Now that the women and children had been safely ferried to the brig,
the captain and his officers no longer assisted the passengers with a rope
sling. The only way down was to jump. Passengers and firemen and
stewards crowded the deck, watching for a boat and an opportunity to
leap.

When the second boat came in aft, two of its oarsmen had succumbed
to exhaustion. The helmsman had instructed them to lay on the same oar
and pull the best they could, which they managed to do. But the trip from
the *Marine* had taken two and a half hours, and when they came alongside, a wave jammed the boat against the side of the ship, knocking off
part of the gunwale and stoving in some of the timbers. The boat began
to leak, and they had to set two men to work bailing her out. Just then
five passengers and three firemen jumped from the steamer deck into the
boat, and Captain Herndon yelled to the helmsman to shove off before
more passengers could jump and swamp the boat.

Perhaps the reason even more men did not jump is because they felt safer on the big ship. The lifeboats seemed so small next to the steamer and the waves threw them at will, and the *Marine* was a long way off and drifting farther. The wind had somewhat abated, and this increased their hopes that the brig soon could maneuver back to the steamer and take them all safely on board, and they wouldn't have to make the journey across open sea.

Herndon estimated that the *Central America* could stay afloat no longer than midmorning of the following day. Saving his ship was now beyond his power, but the lives of five hundred men remained in his hands, men who had worked beyond exhaustion with little food and no sleep, while the wives and children of mostly other men could be safely evacuated. Herndon now wanted to provide these men with every means possible of saving themselves.

Even before the last of the women had been taken from the ship, he had ordered that life preservers be brought up and distributed to all of the men. He ordered the hurricane deck cut away, the doors ripped from their hinges, the hatch covers pulled off, the gratings removed, the planking lashed together, so that if the ship sank, plenty of rafts would be in the water for survivors to cling to. The once proud steamer, her boats gone, her foremast cut down, her sails in tatters, her furniture and dishes broken, her staterooms deep in seawater, her engines long silent, now had her very decks peeled away.

At four in the afternoon, with the wind still blowing heavy and the sea high, Captain Samuel Stone of the schooner *El Dorado* faintly descried a ship off his windward bow but could not discern the character of the vessel. Ravaged herself by the storm, the square-built schooner plowed into the wind and white-capped swells, her bulwarks stove, her foresail shredded, her bowsprit snapped, and her hull caked nearly up to its chains with barnacles. Waves came over her quarter rail, and seawater infested her cargo of cotton. For half an hour, Captain Stone watched the other ship through his glass. "She was a steamer, with all her colors set as signals of distress. As soon as I saw this I hauled my wind, and shaped my course for the distressed vessel; I could see that she was disabled, and was deep in the water."

Just after six o'clock, a watchman in the rigging aboard the *Central America* spotted the *El Dorado,* the small schooner plowing through the storm toward the crippled steamer.

About six-thirty that evening, Captain Stone gave orders to stand by the main sheet, to heave the vessel to, and his helmsman brought the *El Dorado* within fifty feet of the steamer, so close, remembered one passenger, he could have flung a cracker onto her deck.

Captain Stone hailed the captain of the steamer. "Can I render any assistance?"

Herndon called back, "Lie by me till morning for I am in a sinking condition."

The *El Dorado*'s first mate remembered the steamer captain's voice being "as steady as if he had the best vessel in the world under him, in a smooth sea."

"Immediately," said Captain Stone, "I gave the order to put the wheel hard down and haul aft the main sheet and hove to, directly under his lee, say about a gunshot distant. I warned him to commence at once putting his passengers on board, supposing that he had good boats, while I had but one, a small jollyboat, which would not live in the high sea then running for a moment."

Again came the captain's reply, "No, no, lie by me till morning."

Stone held as tight to the steamer as he could but within a minute or two began drifting away. He expected someone from the steamer to throw a line for him to make fast, but in the few precious seconds that passed as the schooner swept by, no one from the steamer tried to throw a line. Captain Stone assumed that the steamer captain would rather wait until daylight to begin launching his lifeboats than risk losing men overboard in the darkness. He yelled, "Set your lights," and by then had drifted out of hailing distance.

"During the time that I was talking with the captain I could hear the passengers crying and halloing, sounding like one simultaneous burst of shouting."

As they moved slowly away, the seamen of the *El Dorado* estimated the crowd on the steamer's deck at seven hundred, and even above the din of the wind and the waves they could hear a roar of voices rising. The schooner was trying to hold position as Captain Stone prepared to receive the steamer's passengers at first light the following morning. Rain fell and the captain ordered his men to catch the water running from the top of the deckhouse over the schooner's cabin, so that they might be supplied with fresh water. Then he went forward and gave the mate orders to get ready to throw over their cargo to make room for the passengers. He told the steward to be careful of every drop of fresh water and the provisions, "as we should have that whole ship's company on board in the morning."

Captain Herndon now went around his ship asking the men to stand by her till morning, that he thought he could keep her afloat if they just continued to bail. At daybreak they would see the early light wash in colors across the receding clouds and the waves lying down and the wind abating, the *Central America* still afloat under them. "We all agreed to do so," said one passenger, "and continued to bail." Herndon directed James Frazer to take charge of the arms chest and every half hour to send up a rocket.

Lights had been set on both vessels, for in a sky already darkened by storm, night was coming on.

THE BRIG *MARINE* had drifted so far to leeward that she now lay five miles off, the sea making a clean sweep over her deck. Most of the women sat or stood in the water on deck, holding on to their children and on to each other, watching for the lifeboats to arrive on their second trip from the steamer. Each expected her husband to be in the next boat.

Looking back across the distance they had come in the small lifeboats, they could see the steamer. "The fog cleared away," remembered one, "and we saw the steamer very distinctly against the sunset clouds." With the sky growing dark, the women's fear for the safety of their husbands renewed, but they had seen a sail near the steamer, a schooner, which buoyed their hopes.

Addie Easton was among those on deck, looking out toward the steamer, searching the water for the first of the lifeboats. Her only thought was for her husband, and although the water on the *Marine* deck often ran a foot and a half deep, no one could prevail upon her to go below. Surely, she thought, her husband would be in the first boat. But as she searched the faces, she could see that it was the bos'n steering the boat alongside, and that it held but three women, Mr. Payne, the engineer Ashby, and other men she did not recognize. She kept her vigil for the next boat.

Ashby leaped aboard the *Marine* and immediately relayed Captain Herndon's request that Captain Burt work his way closer to the crippled steamer. He had been trying, said Burt, but with no mainyard, no main topsail, and no jibboom, he could not work to the windward. Ashby then implored him for the use of his lifeboat. He could have the boat, said Burt, but it was only a yawl and would last but a moment in such seas. Ashby offered him five hundred dollars to get the brig "alongside" the *Central America,* and Burt repeated his statement, that in its condition his vessel could not be coaxed closer.

While Ashby pleaded with Captain Burt, John Black appealed to his oarsmen for a third trip, this one with the distance between the two ships now grown to over five miles, every pull a fight against heavy seas and strong winds. With their tacit consent, John Black shoved off once more.

Addie Easton now spied another lifeboat tossing in the waves, coming closer to the brig, filled with men. Her hopes rose, and as the boat neared, her eyes combed every man.

"With anxious heart and a choking fear," she recalled, "I saw a third boat come close to the ship and in it was not the one I longed to see."

When the men in the boats had safely transferred to the brig's deck, the oarsmen pulled in their oars and refused to embark on a third trip. The women pleaded with them to return, and Captain Burt also spoke to them, but to no avail.

Ashby jumped into one of the boats, screaming an offer of one hundred dollars for any man who would get in and help pull for the steamer. One oarsman said he would return to the steamer for no pay if she were fifty miles away, but two men could not pull a lifeboat in those seas.

One woman remembered, "Mr. Ashby called loudly from the boat, 'If you have any humanity in you, for God's sake, come back to the ship!' He and one boatman were ready, and if he could get another man, they would have been able to row back."

The oarsmen had been in the lifeboats with no rest for six hours. Their legs were cramped and strained, their back, shoulder, and arm muscles knotted from pulling against the sea. For twenty-four hours prior to launching the lifeboats they had bailed and pumped without stopping, and for nearly two days they had had no sleep and little or no food. Plus, the two boats leaked from being smashed against the sides of the two ships and having their timbers stove in. The battle back to the steamer would be against the wind and the sea for many miles, and when they arrived, they knew the scene that would be waiting for them: five hundred men facing death and looking at the twenty-passenger lifeboats as their only salvation, and now, with the women and children gone, not even chivalry to hold them back.

Ashby climbed out of the boat and accosted passengers on deck whom he claimed were deserted sailors. "He tried to make me leave the brig," said one man, "and pull the boat back to the steamer, pretending that I was one of the steamer's men who had deserted, and that he had come after me. I think he did this to cover his own desertion. The captain of the brig told me to stay where I was."

One oarsman said they could never reach the ship again, and he refused to leave in his damaged boat. Finally, Ashby gave up, and with that went his promise to Captain Herndon.

ONLY THE BOS'N Black had yet to return from his last trip to the steamer. The other crews dropped their two boats astern and bailed them out and fastened them to the brig.

"It was growing dark," remembered Addie, "and the boatmen refused to go back again to the ship. I put my face down in my hands, too wretched to speak, reproaching myself that I had not stayed with him, regretting that I had not defied Captain and all when they commanded us to leave."

Then someone touched her on the shoulder, and she heard the voice of Captain Burt.

"Here's a letter from your husband, Mrs. Easton. It was brought by someone in the last boat." He handed her a small scrap of blue paper.

"My Dear Wife," Ansel had written. "If the captain of the 'Marine' will send a boat forward for me, you can give him what he will ask. I will watch for it and be on hand. Your aff husband, A.I.E."

Addie pleaded with Captain Burt to send one of his own lifeboats to the *Central America* to rescue her husband. The captain told her, as he had Ashby, that his only boat could not survive in a sea such as this, especially now that darkness was setting in.

"But Captain," she pleaded, "they may all die before morning. Anything, ten thousand dollars if you will send another boat."

"My dear, dear lady," he replied, "if I could send it, one should go without a cent of money, but a boat such as we have would not live a moment in such a sea. I will try to take the brig nearer the steamer and she will probably float until the morning."

"Language fails me to give even a faint description of that night," wrote Addie. "If anything it was worse than the one previous. There were thirty women and twenty six children in a cabin but very little larger than a *little* parlor, and most of us sitting on the floor, while every wave dashed over the brig and most of the time, the water was several inches deep. We were all wet, and I had not one dry thread on me, but I was not conscious of my physical sufferings, my agony of mind was so much greater."

The widow Ann Small had been lowered into the third lifeboat, but it had loaded quickly and cast off before the second lifeboat. Throughout the entire trip she had suffered the uncertainty of her two-year-old daughter's fate. But on the brig they were reunited. "I afterwards learned that Captain Herndon took charge of her, and sent her to me by another boat, by a lady named Mrs. Kittredge, who handed the child to me soon after I reached the *Marine*." She now went back out on deck, her daughter safely in her arms, and sat looking out across the waves.

"It was a melancholy spectacle we were now compelled to witness," she said later. "Two staunch lifeboats floated uselessly upon the rough waves, while the wreck of the steamer, black with people, was visibly sinking before our eyes."

Some of the women had gone down to the brig's small cabin, there to comfort and feed their children and to strip off their wet clothing and don the wardrobes of pantaloons and shirts offered them by Captain Burt's sailors. Looking through a windward scuttle they could make out lights burning a few miles away on the *Central America*. In the cabin, Virginia Birch reached into the bosom of her dress and carefully retrieved the little canary. Its feathers were disarranged, but it was still alive, and placed in a cage, it began to sing.

The sea still ran high and the wind blew hard, but the storm at last seemed to be passing. Captain Burt continued trying to work his way closer to the steamer, his brig so badly disabled he had to tack in a circle of several miles.

Addie was on deck with other women, watching the only thing they could see now in the darkness that had descended over the Atlantic. "As we came back towards the steamer we watched her lights flashing," remembered Addie. "Suddenly a rocket shot out obliquely, the lights disappeared beneath the waves, and all the world grew dark for me."

Late that night, the sky black, the *Marine*'s lamps only dimly illuminating her flooded deck, the men and women aboard heard the creaking of oarlocks. Into the dim light, rising and dropping with the waves, glided the boat of the bos'n John Black, his oarsmen beyond exhaustion, his boat beaten by the sea and filled with water.

Black had left the *Marine* for his last trip to the steamer about six o'clock, when the sea was running high and confused. Still some distance from the steamer he had seen a fore-and-aft schooner round her stern and cross her bow, and then he had lost sight of the schooner. His boat made the steamer about half past seven. By then it was dark and the ship was sending up distress rockets, the uppermost part of her deck now nearly even with the waterline. Black had seen Captain Herndon and Second Officer Frazer on the wheelhouse, and Herndon had sighted the boat in the dark and hailed Black, who reported that his boat was stove and leaking. Herndon directed him to keep off at a hundred yards. Minutes later Black had seen a rocket come off the *Central America* wheelhouse at an odd angle, not high and arcing but straight out across the ocean.

Save for the exhausted oarsmen and excessive seawater, Black's lifeboat was empty. He came alongside and looked up at the men and women gathered on deck. "The steamer has gone down," he said, "and every soul on board of her lost."

TOMMY

Defiance, Ohio

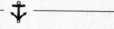

The 1960s

A T THE CONFLUENCE of the Maumee and the Auglaize rivers, the town of Defiance, Ohio, is an island of homes in a sea of tilled brown earth, with silos and black-and-white cows and brush-pile fires and big barns and white-frame farmhouses all floating in the landscape. Here dwell small-town insurance agents and fifteen-year-old boys with large hands, whose vision of the future stops on the Buckeye gridiron in Columbus, two and a half hours by car to the southeast.

In 1787, Congress ordered a fort built here, where the two rivers meet far up in the Northwest Territory, and when the fort was completed, General "Mad Anthony" Wayne told one of his colonels, "I defy the English, the Indians, and all the devils in hell to take it!" Fort Defiance it became, and later the town of Defiance, with a population in the

GARY KINDER

1960s of eighteen thousand, many of whom worked at the General Motors foundry with its tall stacks visible to the south across a little roll of hills, or the Johns-Manville factory to the east no more than a mile.

In town, another institution older than either the foundry or the factory had sat at the corner of Fifth and Clinton since before the depression: Kissner's restaurant, where Tommy Thompson and Barry Schatz and some of their friends often stopped on the way home from school for a brain sandwich.

"Got any brains today?" they would ask Bruno Kissner, and he would always reply, "If I did, I wouldn't be working here."

That was part of the reason they went to Kissner's, just to ask the question; the other part was they liked eating brains, pig or cow, Tommy for the novel idea, Barry for the taste. Barry liked his brains open-face on toasted rye. When he was seven years old, he had also loved oyster sandwiches, early glimmers of a young wanderluster and epicure. Tommy never much cared when he ate or even if he ate; like the rest of life, food was only something to wonder at and explore.

Born in 1952, Tommy and Barry shared the same birthday, April 15, tax day, and the same day forty years earlier that the *Titanic* had gone to the bottom in an icy sea. They had become friends in the seventh grade, when Tommy's family moved to Defiance from Huntington, Indiana.

Tommy's father, John, was an engineer, and his mother, Phyllis, was a nutritionist. They had met while in college at Purdue. Early in their family life, they had seen polio strike both of their daughters, Patty and Sandee. Patty had a mild case and recovered quickly; Sandee's condition was worse. Twice doctors called John at work and told him to come home immediately or he might not see her alive again. After surviving both episodes, she could move only her left elbow, and doctors said she would never again sit up. But Phyllis massaged her arms and her legs and her back at intervals, all day and all night, for weeks and then for months. And Sandee sat up. And then she stood, and then she walked, and then she ran, and then she became a cheerleader and won a college scholarship for cheerleading and singing, and for years the media and the doctors would interview John and Phyllis about her recovery. The experience gave them whatever additional perspective they needed to see the value of family. John Thompson wanted to be home from work at five. He wanted his kids in small schools, growing up in a small town surrounded by a countryside of white farmhouses. Tommy lived his first twelve years in neighborhoods where he could drive the three-wheel hot rod he and his dad had built around the block, around and around, out in the street, until all the tires wore off. Two blocks over were fields of corn.

The Thompson household was an eclectic mix of conservative values and liberal views: Make it instead of buy it; repair it instead of replace it; curfews for Patty, Sandee, John Jr., and Tommy; church every Sunday for everybody, at the church where John's father preached. But the kids could pedal their tricycles around the basement and paint their own rooms whatever color they wanted. Phyllis made sure they got the paint. They could go to a dirty movie or read a dirty book if they wanted to, as long as they told their parents, so John and Phyllis could see the same movie or read the same book and talk to them about it. When a black gospel singer came to sing in Huntington, where blacks could not stay the night in a motel or eat in a restaurant, she was welcome for dinner at the Thompson home. The kids set the table and sang with her at the piano.

Long after Tommy had gone off to college, a friend saw the family up close and remembered him even then "getting these unwavering messages from his parents that he really was good and that what he did was good, that he could do it really well. It was all mixed up with a lot of affection. He thought his mom was the best mom in the world and he would say it, and his dad was wonderful."

Of the four children, Tommy was the youngest, a child always building something or fixing something or tearing something down to build over again. Even at night, with the lights out and a plea from his parents to get to sleep, he would take a flashlight under his covers and continue building with Tinkertoys. Phyllis said he was the most sleepless child she had ever seen. Patty frequently brought her little brother a bedtime glass of water, then found herself for the next two hours listening to his ideas and answering his questions.

In high school, Tommy's older brother, John, lettered in football, basketball, and track and was president of his class for three years. But sports and politics did not interest Tommy the way working with his hands did. In the evenings and on weekends he was in the basement or the garage with his father. Together, they built young Tom's hot rod, three little wheels from a coaster wagon powered by a lawn-mower engine. When he wore that out, they built a bigger hot rod, using heavier wheels from a wheelbarrow and a little help from welders down at the plant. Then they rebuilt the engine of an old English Austin. As they worked, Tommy asked a lot of questions, like, "Why is it they call this alternating current?" His father would stop and draw him a picture and say, "Well, see, the voltage goes up here, and then it goes past the zero, and then it goes down again, and it does that sixty times a second."

John Thompson encouraged the questions. He once told his son, "The only reason the engineer is good at what he does is curiosity."

Tommy loved it all so much he never wanted to quit, even as a little boy. He would be working with his father on one of the cars or the hot rod and it would be ten o'clock at night, and his father would be tired and suggest that maybe they should hit the sack, and Tommy would say, "Oh, come on, can't we just do a little bit more here?"

When Tommy reached the third grade, his teacher discovered that he couldn't read, that he had only memorized what the children reading before him had recited. So Phyllis bought him fourth-grade-level books on science, and Patty and Sandee taught him phonics, and within weeks he was reading the science books and *Popular Mechanics,* lying on his bed, going through them page by page, combing the experiments and studying the how-to's. Before long he had built a control panel in his room, an electrified collection of old switches and wires. One switch turned on the light in the closet, another turned on the light overhead, another turned on the radio. It looked terrible, but it all worked. He had the whole room wired. "It was kind of makeshift," recalled his mother, "junk that he picked up somewhere. He never spent a lot of money on things. He never spent any money."

On trash day every week, Tommy would borrow a screwdriver and a pair of pliers from his dad's toolbox and slip them into his back pocket, and when he walked home for lunch he would pull old motors or sole-noid valves and other parts out of broken fans, radios, record players, and the occasional washing machine, freezer, or television set sitting on the curb. When his father asked him what he planned to do with all of those old parts, he'd say, "Well, we might need those." The Thompsons had a basement full of electric motors and other parts Tommy had brought home. "We thought he was going to be a junk man," said his father.

One day when Tommy was no older than eight, a man from the telephone company knocked on the front door of the Thompson home. When Phyllis answered, the man told her it was against the law to have two telephones unless she paid for both of them.

Phyllis said, "We don't have two phones."

"Yeah, you do," said the man. "Come here."

Phyllis went outside and saw a line coming off the telephone pole and looping down into the window of Tommy's room. "Wait a minute," she said, "I'll call Tom."

A little boy in the third grade came out the front door, and Phyllis said, "Tom, this man wants to ask you something about that wire going out of your window."

Tommy took the man into his room while Phyllis waited outside. When the man came out, he said, "That kid's made a telephone."

Tommy had wired the phone inside an old jewelry box Phyllis had given him, and although he couldn't dial out with it, he could open it up and listen when the phone rang. He liked to hear what Patty and Sandee said to their boyfriends.

The phone man told Phyllis, "This kid knows more about the telephone than I do. Why don't you just let him play with it."

WHEN TOMMY WAS thirteen, John and Phyllis moved their family to Defiance, where John became head of the automotive division at Johns-Manville, conceiving and designing specialized machines to stamp out fiberglass hood liners. John rode his bicycle a mile every day to the plant.

Tommy and Barry and the rest of a group of boys who had dubbed themselves the Vigilantes rode their bicycles all over town and hunted for turtles in tiny creeks that trickled through the ravines. Tommy was about as smart as Barry and his friends had seen but also downright, good-naturedly goofy. He was always working on a project, or reading a science manual, or playing with a math formula, or experimenting with an idea. One Saturday morning, his new friends found him mowing the Thompson lawn sitting on his hot rod and pulling four old rotary lawn mowers he had bolted together, two behind two and overlapped so they left no gap. Another time, they went to the country club pool and watched him sit on the bottom for ten minutes while he breathed off scuba gear he had made with a gas furnace regulator and four propane cylinders.

Tommy talked so frequently about his experiments and his projects and ideas and things he had read about, and he talked about them so fast and in such depth that Barry and his friends were never sure if Tommy was serious or if he had made it all up. When he talked, he laughed and grinned a lot, and his eyes crinkled into near slits, and they could see the wide gap between his two front teeth.

Some people thought that gap was the reason he had been nicknamed Harvey. In 1950, Jimmy Stewart had starred in a movie called *Harvey,* in which the title character is an imaginary white rabbit over six feet tall. But the audience never sees Harvey, except in a portrait, and in the portrait he has his mouth closed. The real story of how Tommy came to be known as Harvey was that after he got to Defiance, he, Barry, and the rest of the Vigilantes selected nicknames for each other; Tommy was new in town and he liked to travel around, so someone nicknamed him Harvey the Hobo, and the nickname stuck. Everyone started calling him Harvey. Later, when people discovered Harvey was not his real name, they would ask Tommy how he came to be called Harvey, and he would say, "How do you think it happened?" If he liked their version, he would say, "Yeah, that's it!"

John Thompson had two brothers who taught philosophy; Phyllis had two brothers who taught engineering and psychology; Phyllis herself taught nutrition. John's father was a minister, and his mother taught Sunday School to three hundred parishioners every Sunday. The teaching tradition in the family went back at least to Tommy's great-aunt Edna, who taught in a little schoolhouse at the turn of the century and spoke seven languages and traveled around the world with his other great-aunt, Claire, who also was a teacher. Perhaps Tommy's greatest teacher, his father, was the only one in both families not employed as a teacher.

"My whole family's a family of educators," said Tommy, "and I was steeped in all that kind of thinking, but I had developed my own views."

Tommy often talked to his parents and his uncles about education theory and the philosophy of learning. He would think out loud with them about how one might educate himself, and through this process he refined his own philosophy, all the way back to how to think. Along the way he had decided that he wanted to think like no one else thought, and that confounded people, because no one knew where it came from. It just bubbled up from somewhere deep inside him like a bottomless spring, with no outward signs until he opened his mouth, but here is the crux: When he was in grade school he could not add two plus two; he knew that the answer everybody wanted to hear was four, but he didn't care about the answer; he wanted to know why. Although he made the National Honor Society in high school, he sometimes flunked math and science tests, because he had to understand a problem conceptually before he went on. A high school math teacher once told Phyllis that Tom could get to the right answer, but it took him forever to get there. Instead of using the accepted formulas and heading in a straight line, he would go around the problem, look at it from different angles, turn it upside down, challenge the formula he was supposed to use, then give his answer. He did this with everything. Always. And he didn't care how much time it took or what grade he got or what anybody else thought. "If you're going to be educating yourself in this world," he would say, "you might as well be thinking about things that will yield value all the way along."

He extended his philosophy to his social life, another curiosity to turn upside down and observe from different angles. Trends did not matter to him, and he did nothing to try to fit in. He preferred to analyze the socialization process and then, not always unconsciously, strike out in a different direction to see what would happen. While the cheerleaders and the athletes hung out at Kuntz Drugstore and drank cherry Cokes for a dime, Tommy went to Kissner's for a brain sandwich. But his teasing manner and his wackiness usually attracted a crowd.

"There was always activity going on around him," remembered one of the Vigilantes, Mark Steiner, "and he was usually the catalyst."

One day, sick at home with mononucleosis, Tommy read an article on vitamin C by Linus Pauling. The article so intrigued him that he called Pauling's office in California, got through to Pauling himself, and told the renowned doctor he was doing a high school research project on vitamin C and mononucleosis. Pauling explained to Tommy how the vitamin C worked, how to keep it in his system at the renal threshold, and how to check his urine to see when it spilled over. Tommy then took thousands of milligrams of vitamin C every four hours, until he reached a dosage of well over 20,000 milligrams a day. Five days later, he felt so good, he snuck out of his house late at night and orchestrated after-hours bicycle races on the bottom of the drained country club pool.

As he grew older, Tommy began pushing everything to what most people thought was the limit and then went way beyond that, even physically. "He would do things," said a friend named John Radabaugh, "that showed he had no concern for the welfare of his body." On water skis, he would fly off a ski jump, sometimes doing a flip, sometimes dropping a ski. Sometimes he would hit the ramp, his skis would explode into the air, and he'd drop down on his rear end, then shoot off the ramp with no skis. "He was tough," said Radabaugh. "He wouldn't shy away from that stuff."

When Tommy turned sixteen, his uncle Jim gave him a 1948 Buick Roadmaster convertible, and he took it apart until not one piece of anything, the body or the engine or the interior, was connected to another. With help from a dozen friends, he puttied the holes, sanded the rust, fixed the hydraulic seats and windows, rebuilt the engine, painted it a royal blue, and bought a new white top for it. He loved to cruise in the Buick, to chase girls in it and get their attention; but once he had their attention, he never knew what to do next. No follow-through, said his friends.

That was the difference between Tommy and Barry. Barry was a romantic. "He'd go in for the kill, and he'd just lay it on," said Radabaugh, "where Harv didn't know how." The best line Tommy could come up with was guessing their weight in kilograms, or asking for a strand of hair so he could put it into a machine he had invented that would reveal their true measurements. Barry made them twitter; Tommy made them laugh, sometimes nervously. "He was always a jokester," said Gina Cullen, the only girl Tommy dated in high school. "But he was so smart, you thought, Maybe he can do that."

Tommy the mad scientist and Barry the romantic shared one trait: an enormous curiosity. It took little coaxing for either of them to head off into the unknown. In January of their junior year, they left Defiance in Barry's MGB-GT in the middle of the night and drove 850 miles to Winter Carnival in Quebec. There they surfed on sheets of plastic down the closed toboggan run, searched in vain for a glimpse of a legendary prostitute named Angie, ate sheep brains soaked in black butter, drank two bottles of sweet wine, and stopped the car in Ontario to throw up on the way home. Two teenage pals soaking up a little of the world outside the confines of that little patch of earth called Defiance.

IN THE FALL of 1970, Tommy enrolled at Ohio State fearing that a regimented educational system might destroy his ability to see things in ways that other people could not see them. "I had a very specific feeling," he said, "about how I wanted to be educated and how I wanted to think."

For years he had carefully cultivated a creative mind-set, worrying that if he ever stopped being different, if he ever stopped experimenting, if he ever stopped pushing and questioning and exploring and looking at life upside down, he would no longer think the thoughts that could lead him to ask the questions that no one else had asked, which is what made him unique, which is what allowed him to be what he had wanted to be since he was a small boy: an inventor. He wanted to take old ideas, turn them inside out, and apply them in new ways; he wanted to suck the world through his senses and exhale a vision.

But a fine line existed in Tommy between calculated antics and true compulsion, and sometimes it was difficult to separate the two. His dorm room at Ohio State glowed like a spaceship just landed, a place of mysticism and metaphysics and hard science mixed with an ample amount of buffoonery. Telephones hung from every wall and sat on the desk and sat on the shelves; he could be anywhere in the room and answer the phone without moving his feet. An old alarm clock turned on the TV, the sun lamp, and the radio, but the radio played through the speaker on the TV. All were perfectly coordinated to go on and off at the same times every day, except that the alarm clock had no hands, so you could never tell when the lamp would flood the room with light, the radio would go off in the television, or the television would suddenly start to glow.

He crawled through the ventilation ducts from one floor to another and rode up and down on top of the elevator, talking to people through the ceiling as they got on. He persuaded ten dorm friends to chip in ten dollars apiece to buy a 1964 Chevy convertible; no one got a key, but the ten bucks got them each a demonstration on how to start the car with the four wires in the glovebox. So they could find the car easily on the south side of one of the largest universities in the world, Tommy and a few friends decoupaged the passenger side with pictures of Jimi Hendrix and Janis Joplin and posters from Peter Max and the Beatles from the *Sergeant Pepper* album and little *National Geographic* portraits, and shellacked the whole thing purple.

Tommy reveled in so much craziness his room became a mecca for other students. "You could always count on Harvey for something off the wall," said Radabaugh. "He had that wild-eyed look and curly hair, like a nine year old ready to get into anything."

One night, Tommy shut down all the lights on his floor and formed a long string of students holding hands and winding out of his room down the hall. Then he had the first one grip the terminals of a small generator and the last one cup a neon tube, and as he cranked the generator, electricity flowed painlessly through the bodies of the other students, and the neon tube at the far end glowed brighter and brighter. Another night, he filled balloons with helium and passed them out to everyone stuffed into his room, had them suck some of the helium into their vocal cords, and directed them as a choir, which sounded like twenty-five ducks singing the Buckeye fight song. The other students liked all this stuff so much that five hundred of them elected him president of Stradley Hall.

Then there was the Tommy so deep into his own thoughts and his sense of what was important that his conscious mind made no effort to pull out and see the superficial things so important to everyone around him. At a time when socially acceptable nonconformity in campus fashion dictated that the students wear long hair and knee-popped bellbottoms, Tommy would show up with a beard and his hair poofed in natural black curls three or four inches off his head—groovy so far—wearing a weird maroon shirt and old flannel dress pants covered with little fuzz balls. "He had shirts you just wanted to burn," said Ellen Leahy. But that was the real Tommy. He could mix plaids with stripes with paisley prints and socks that didn't match each other, much less anything else he was wearing. It wasn't for effect; he just didn't care.

Ellen Leahy met Tommy in June of 1972, only a few days after she had arrived on the Ohio State campus as a freshman. Tommy was in his junior year, not in school that quarter, just dropping by Columbus to see a friend. They stayed out all night, running around in a warm sticky rain, and then Tommy took Ellen down to the lake and introduced her to his favorite mud slide. "It was so fun to run into someone who wasn't stodgy and thought at some point you should call it quits," remembered Ellen. "He never thought there was some point where you had to call it quits."

Tommy left that day, and Ellen did not hear from him again until August, when he called her from Utah as he hitchhiked across the country with a tape recorder. He had left Defiance with eighteen dollars and a friend who carried twice that amount, headed for California. A circus picked them up on the way to Flagstaff, a gigolo gave them a lift to Las Vegas, and during the whole trip, Tommy talked into his tape recorder about what he saw and what he thought, his views on science and civilization, banking and nutrition, philosophy and stars and life on other planets. When he got back, he had so many theories that Ellen and the other kids numbered them: why movie stars have serial marriages, how we got white bread, when turtles behave like humans. "He was real serious," remembered Ellen. "He would propose these theories, and you could tell he'd been thinking on them and then out they'd pop."

Cruising in the purple Chevy with Ellen one day, Tommy spied something that epitomized his very being. It was mechanical, it was silly, it was daring, it was of the water: a cream-colored, German-built, limited edition, amphibious car, the same model used in the James Bond movie *Thunderball*. "God," said Ellen, "he wanted that thing so badly." It looked a little bigger and a little boxier than an MGB. It had small fins coming off the back, license plates fore and aft, and an Ohio Boat registration on the front fenders. The owner was asking four hundred dollars, and without telling his parents, Tommy got a loan from a bank in Defiance and bought the car. From then on he and Ellen spent much of their time together looking for bodies of water to drive into and people to scare the bejeezus out of. Ellen would sit in the passenger seat, and Tommy would drive. The victims would be in the back, top down, everyone taking in the warm summer air. When Tommy got near water, he would say something like, "Isn't this a nice night?" and then he would start talking about one of his theories, and Ellen would ask the right questions so he could continue bending an increasingly outrageous line of thought until it was about to break, and then all of a sudden he'd leave the road, head across a meadow in the moonlight, still pontificating, and drive through the cattails into a pond. The victims in the backseat would freak, but the key to the act was that Tommy and Ellen had to play it straight, no smiles, no acknowledgment that the car was now afloat in the middle of a small body of water.

There was another side of Tommy that others, including Ellen, rarely saw, and when they did, they weren't sure what to think of it. Sometimes he disappeared for days, and nobody knew where he was. Sometimes an offhand remark or the look on his face as he talked about his theories made them wonder. "I think a lot of people had the same feeling," said John Radabaugh. "They weren't sure, Is he bullshitting me or are we onto something serious here?" One night he mentioned to someone in the dorm that he had to leave with one of his professors early the next morning for Detroit, because they were working together on a flywheel car. No one believed him. "He's not going to Detroit to work on a flywheel car," mimicked one of his friends, "he's just another guy on the tenth floor." They could believe the goofy Harvey, but they had a hard time with the serious, compulsive Tommy.

OHIO STATE HAD one of the largest engineering schools in the world, eight thousand students enrolled in fifteen departments and graduate schools. Like his father, Tommy had gravitated toward mechanical engineering. "I wanted to be an inventor," he said, "but there's no college training for that. As close as I could get was the mechanical engineering school." He also was determined to be an ocean engineer—out of the entire College of Engineering, the only one heading to sea.

Although Ohio State was landlocked and offered no courses in marine engineering, Tommy had intrigued his advisor, the dean of the School of Mechanical Engineering, Don Glower. Glower saw in Tommy even more than the makings of a creative machine designer; he saw an eclectic mind, as curious about social dynamics as it was determined to make things work.

Glower himself was a marine engineer, and he had emphasized to Tommy that working in the ocean was only mechanical engineering taking place in an extreme environment: In addition to all of the other problems, the ocean was wet, corrosive, and heavy. So he and Tommy designed a major in mechanical engineering with a specialty in machine design and an ocean engineering option, a five-year program, and Glower helped tailor Tommy's studies to expose him to concepts he would need to work in the ocean: courses in aquatic microbiology, corrosion sciences, marine geology. And Glower got Tommy involved in special studies projects in solar energy, a flywheel car, and pseudo-plastic.

In his third year, Tommy selected a course called Advanced Topics, a one-hour private tutorial three times a week with Dean Glower, during which the mentor and his protégé talked about engineering, mostly engineering in the ocean. During these sessions, Glower conveyed to Tommy that he didn't have to be what Glower called a "cookbook" engineer: look up the recipe, mix in the same ingredients in the same way everyone else does, and every time end up with the same bridge. Glower encouraged Tommy to take chances, to look beyond what everyone else was doing.

"He didn't have to say this to me in words," said Tommy, "but he made me feel real easy about doing things that most engineers wouldn't think about."

Glower exposed Tommy to ideas like entrepreneurship, explained to him that an entrepreneur got out of the secure atmosphere of a large corporation or university and tried something on his own; and if his ideas didn't work, the entrepreneur found other ways, but he kept working at it till he got it right. He told Tommy that inventions were nothing new, that they were just taking a number of things that already existed and putting them together in a different way. "Einstein didn't create anything new," said Glower. "Everything he did on the theory of relativity was already in the literature, but other physicists just didn't quite see how to put it all together."

During these tutorials, Glower assigned Tommy research projects, suggesting books and treatises for him to read, some on the ocean, some on engineering, some on other topics. When Tommy came to the meetings, Glower cleared off his coffee table, Tommy spread out the sources he had found, and they talked. Glower's purpose was to challenge Tommy's thinking, to help him see the problems others had faced and how they had solved those problems, or failed to solve them, and to encourage his understanding of diverse disciplines. The last was perhaps the most important: Inventing wasn't good enough; Glower had seen too many inventors who didn't know how to get beyond that, so their ideas died aborning. As Glower suggested new directions, Tommy kept an ongoing list of research projects, ideas that piqued his curiosity.

Beginning in the fall of 1972, Glower and Tommy met for two years, some quarters only once a week, other quarters three times a week, and Glower grew more and more intrigued with Tommy's approach to problem solving. The young engineer had a different way of looking at things. One day early in these private sessions, Glower asked Tommy what appeared to be a simple question, yet no one in the world had been able to answer it. "Tom," he said, "how are we going to work in the deep ocean?"

FOR CENTURIES, HUMANS had dreamed of flying through the air, traveling to the stars, and exploring the water world that covers two-thirds of our planet. We had conquered the air, and four years earlier we had put a man on the moon; much remained to learn of the universe, but already we knew far more about other galaxies than we knew about a world that began at the edge of our beaches.

Long before the birth of Christ, divers in China, India, and the Mediterranean, and the *ama* fisherwomen of Japan, had dived in the ocean to one hundred feet in search of pearls, coral, sponges, mollusks, and rare seaweeds. In the fourth century B.C., Aristotle recorded that sponge divers could remain on the bottom longer by breathing air lowered in large kettles. But another twenty-three hundred years would pass before humans could remain below for longer than a few minutes.

In 1942, a French engineer named Emile Gagnan designed a regulator to control the flow of compressed gas into the fuel injector of a car so civilian automobiles in war-torn, occupied France could burn propane instead of precious gasoline. The following year, Gagnan and a young French naval officer named Jacques-Yves Cousteau redesigned the regulator to control the flow of compressed air into the lungs of a human. Though he despised cold water, Cousteau tested the apparatus by turning somersaults in the January waters of the Marne River outside Paris. After minor adjustments, it proved successful, and Gagnan and Cousteau patented the system as the Aqua-Lung. Nine years later, Cousteau breathed off an improved Aqua-Lung while conducting the first complete underwater excavation of an ancient vessel.

National Geographic called Cousteau and his companions "fish men." But even fish men could venture no deeper than two hundred feet and remain for more than a few minutes, or during ascent the nitrogen in their blood would suddenly fizz, and the bubbles would lodge in their arteries, joints, and spine, paralyzing and often killing them. Divers called it "the bends." At those depths the compressed nitrogen also became a powerful narcotic, affecting judgment: Cousteau had once seen a diver offer his mouthpiece to a fish so the fish wouldn't drown.

Hard-hat divers typically could dive a hundred feet deeper, but they were susceptible to the same problems with nitrogen narcosis and other problems potentially worse. If their air hose broke they drowned; if the break occurred near the surface, the sudden depressurizing inside the helmet would squeeze all of their blood into their brains, bursting the blood vessels and killing the divers. The deeper ocean was no place for humans, unless they used machines to take them there.

IN HIS WRITINGS of the fifteenth century, Leonardo da Vinci mentioned a "mysterious vessel," a "method for remaining underwater for as long a time as I can remain without food." But he would not reveal his design because of "the evil nature of men" who would use it "to practice assassination at the bottom of the seas." Refusing to take part, da Vinci had seen the future of submarine technology: warfare.

The early submarine designers had three problems to solve: how to sink a watertight vessel, how to propel it underwater, and how to provide air for the crew. In 1620, a Dutch physician named Cornelis Jacobszoon Drebbel built an underwater boat funded by military-minded King James of England. To sink his boat, Drebbel filled pig skins with water, then twisted them closed when he wanted to rise. For propulsion, he used oarsmen rowing underwater. His real secret, however, was his method of ridding his cramped little vessel of carbon dioxide. Witnesses described it as a "chemical liquor" that somehow revitalized the spent air. Whatever it was, Drebbel took his invention into the Thames with four oarsmen and opened the pig skins; the vessel sank, the oarsmen rowed, and so moved the world's first submarine. Drebbel built two more underwater boats, both bigger than the first, one large enough to seat a dozen oarsmen. In this vessel, King James himself spent an hour fifteen feet below the surface of the Thames.

Before Jules Verne wrote his classic *Twenty Thousand Leagues Under the Sea* in 1869, at least twenty-five more submarine boats had dived and surfaced successfully, and the technology for virtually every one had focused on sinking, ventilating, and propelling a vessel of war underwater. The idea was to hide and plant a mine or release a torpedo or launch a missile: Yale University student David Bushnell created the *Turtle* and its ticking time bomb in 1776; the German inventor Wilhelm Bauer produced a sheet-iron vessel 50 feet long and powered by four men walking a treadmill and sold it to the Russians in the mid-nineteenth century; in 1864, the Confederate *Hunley* became the first submarine to sink an enemy warship. Within three decades of Captain Nemo's fantasy voyage twenty thousand leagues under the sea, the United States Navy had committed to a fleet of seven submarines built by the dean of American submarine technology, John P. Holland. One was 105 feet long, carried a fifteen-man crew down two hundred feet, and cruised clandestinely underwater at twenty knots.

Although submarines became a critical component of any navy's arsenal, and they played a major role in World War II, submarine technology did not advance appreciably until 1955, when Holland's Electric Boat Company built the *Nautilus,* the world's first nuclear-powered submarine. But the *Nautilus* was only the next step in a long line of underwater military technology deemed in the interest of national security; and because the military still had no reason to go deep, even nuclear submarines rarely ventured below a few hundred feet, and none could dive below a classified depth of two thousand feet. Yet the ocean at its deepest could swallow the twenty-nine thousand feet of Mount Everest and still have more than a mile to spare. Even if submarines had been built to dive deeper, the military did not equip them to observe and gather information, because the military wasn't interested in revealing the mysteries of the deep ocean; the military only wanted to hide in its vastness.

Then in 1963, eight years after the *Nautilus* sailed, the navy's most advanced nuclear submarine, the USS *Thresher,* had imploded on a routine dive off Nova Scotia and crashed into the seafloor eighty-four hundred feet below, and the navy had no way to rescue the crew or retrieve the wreckage. For months they couldn't even find it. The navy brought in over thirty-five ships and dangled lights and cameras off some of the decks, snapping the shutters at random, hoping to capture the wreckage on film; but with currents between the surface and bottom moving in different directions, no one ever knew where the cameras were. After a month of searching, one camera photographed its own anchor, and the picture was so grainy the navy mistakenly announced it had located the wreckage. After two months, they managed to get images of a few gnarled pieces of the *Thresher,* but they still had no image of the hull. With the search area narrowed, the navy called in the only vessel in the world built to withstand the pressure of the deep ocean, a bathyscaphe called the *Trieste,* a French-built steel sphere, which the navy recently had purchased and modified.

The *Trieste* was the most sophisticated underwater work technology available, but it could hardly move underwater, had only one claw that repeatedly failed to open, and came with viewing ports so small that the two-man crew could look out with only one eye at a time. It made eight dives before the crew even spotted the *Thresher* hull, five months after the submarine had gone to the bottom. After several more dives, the navy returned home with photographs and one four-foot piece of pipe secured after many frustrating attempts with the *Trieste*'s spastic claw.

Nine years later, in 1972, Tommy was meeting weekly with Dean Glower, and the deep sea had remained an inhospitable foe, susceptible to murky images and succumbing to the occasional probe but otherwise unconquerable. And that was the problem Glower presented to Tommy, the distinction between a presence and a working presence: We now could descend to the bottom of the deepest ocean and look; we had yet to figure out how to work there.

Glower and Tommy discussed the problems the navy had faced in trying to find and document the *Thresher*. Glower explained the principles behind what others had tried, and Tommy read whatever he could find on submersible technology. He talked to Glower about saturation diving and underwater habitats, until he realized they would be "enormously expensive." They talked about the new manned submersibles, but Tommy wondered about the cost and the danger of putting human life on the bottom. He broached the idea of having a mother ship at the surface and sending down robots to explore and to work, like NASA was beginning to do in outer space.

The sessions between Dean Glower and Tommy continued through the winter and spring of 1973, and into the following academic year, and Glower himself found them "stimulating." "Tommy had a lot of good questions," said Glower, "many of which I didn't have an answer for."

IN HIS SENIOR year, Tommy dropped out of school for the fall quarter, moved out of the dorm, and slept in the parking lot in the purple Chevy. For three months, he spent most of his time at the library studying ideas he had always wanted to pursue. He wanted to know how to get beyond innovation and creativity, how to arrange what he called "a good impedance match" between an idea and the environment in which that idea had to survive. "I used to have ideas after ideas after ideas after ideas," he said. "How are you going to find out which ideas are good and which ideas are bad? Nobody could tell till you did it."

He decided to extend the research projects Glower had assigned him during the tutorials and pursue at all times seven to fourteen projects, idea sparks that through research he could fan to see if they caught fire. Some would be long term, thirty years perhaps; others he might conclude in two weeks. Some required his attention for a day each week, while he spent no more than four hours a month on others. As he completed one project, he would take on another, always keeping the number between seven and fourteen, and he resolved that no matter what he was doing, where he was working, how much time he had to spend on other things, he would keep these projects alive to broaden his understanding of science, marketing, technology, business, human behavior, all of the disciplines that come together to make an idea work.

At the end of Tommy's fourth year, Dean Glower arranged for him to be the first engineering student to attend the university's Stone Laboratory at Put-in-Bay on Lake Erie. At the lab each summer, graduate students and faculty convened for a program incorporating advanced studies in marine biology and zoology. It wasn't the deep ocean, but at least Tommy could broaden his knowledge of the aquatic sciences, study corrosion, and work on a sixty-five-foot marine research vessel.

The director of the laboratory, Dr. Eddie Herdendorf, knew to be expecting an engineering student who was to work on some special projects; that was all he knew, but he envisioned the stereotype. "Real scientists," said Herdendorf, "look on engineers as just . . . kind of . . . you know . . . subhuman. It takes some skill to make the calculations, but it doesn't take original thought."

Dr. Herdendorf was at the lab registering students when he saw a cream-colored convertible about the size of an MGB coming toward the island up to its windowsills in the lake. As he and the students gathered by the window and then outside to watch, the car glided over to a small boat ramp and emerged dripping from the lake, its little propellers still spinning. Out stepped Tommy.

"That car caught everybody's attention right off the bat," remembered Dr. Walt Carey, Tommy's other advisor at the lab. "It set the style for that whole summer. Somebody said, 'If he keeps pulling stunts like that, he'll be lucky to get off here alive.' Yet he seemed absolutely fearless. He'd go places where no one else had gone and try things no one else had tried, but he always knew how far he could take something without going over the edge. It seemed to be an instinct he had."

That summer, the students measured the rate of algae growth in the lake and monitored radiation around a nuclear power station, but Tommy seemed more interested in the instruments than in the readings. He told the others how he thought the instruments worked, and as he spoke he seemed to be redesigning them in his head. He theorized as much as ever, but here, among the graduate students and professors, not so many wondered if he was bullshitting or onto something serious. When an instrument broke and no one else could figure out why, Tommy would approach it conceptually and isolate the problems. Fred Snyder, a graduate student in marine fisheries, said, "Tom was always unruffled and unstoppable on fixing something."

Snyder thought of Tommy as an "ornery character," energetic and ornery, yet always intellectual. "I think he appeared to be undisciplined to a lot of people, but he was disciplined in his own way. People at the lab expected a more quiet, reserved, scholarly type; Tom was plenty scholarly, he just wasn't quiet and reserved."

Toward the end of that summer, the students left Put-in-Bay for a field trip back on the mainland. Everyone else took the ferry, but Tommy persuaded Snyder to make the trip with him in the amphicar. "He was kind of a devilish little imp type," said Snyder. "He had the beard and everything, the wiry hair, real sparkling eyes, and an impish grin. And he had this way of talking me into everything."

The shallowest of the Great Lakes, Erie kicks up high waves in a storm. That day the sky was so dark that Tommy had on his headlights, his taillights, and his running lights. A northeaster was blowing the length of the lake, packing stiff winds and whipping the water into five-foot hills. "We shouldn't have tried this," said Snyder.

They put the top up to keep out the spray, and Tommy steered the amphicar parallel with the ferry, but the amphicar did not bob like a cork, so the waves swelled and rolled over them. "The first time I was ever in anything underwater with its windshield wipers going," said Snyder. "I was a much more pious man when I got to the other side. Tom was probably worried, too, but he never shows it. He doesn't get ruffled that easily."

* * *

As Tommy neared graduation, Dean Glower warned him that jobs in ocean engineering were scarce. The U.S. Merchant Fleet was almost nonexistent. At Scripps and Woods Hole and other oceanographic centers maybe five or six jobs opened a year. Don't go looking for employment in the field, he told Tommy. "Not yet, maybe in another thirty years." In the future, he saw a huge demand for ocean engineers, because ocean engineers would be mining the deep ocean. "We'll actually be farming down there, too," said Glower, "and there's history down there that's as valuable, or maybe even more valuable, than the digs in Greece." But that was still a long way off. Glower suggested that Tommy get a mechanical engineering job near Scripps in California and volunteer on oceanography projects, let them get to know him, and maybe they would offer him a job. Or go to Florida and work as a mechanical engineer and volunteer there. Glower even suggested that the only action might be with the treasure hunters down in Key West. Glower had just read about one named Mel Fisher who had spent years looking for a Spanish galleon and still hadn't found it.

Tommy went to routine interviews with companies for an engineer's position, and about the time the interviewer got to explaining the company's pension program, Tommy's eyes would unfocus and his brain would start reviewing the seven-to-fourteen. "He said there was no way he was going to work in the factory and pound the concrete," said Phyllis, "and John said to him, 'What are you gonna do?' and he says, 'Well, I have a pact with some guys,' one of them was Barry, 'that we're gonna meet in Key West after we graduate from college.'"

In January 1976, Tommy stored the amphicar in a friend's garage and left Defiance, driving a '63 diesel Mercedes and wondering how he could get the Mercedes's injectors to feed on french fry oil without choking.

Barry Schatz's father had a successful insurance business in Defiance, but Barry didn't like business; he wanted to study other cultures and languages and literature. He tried accounting and business courses, but his lack of interest showed in his poor grades. He dropped out of Miami University; he dropped out of Ohio University. He went to Que-

GARY KINDER

bec, then to Scotland, hitchhiking, drifting, and thinking. When he
returned from Scotland, he enrolled in Hillsdale College in Michigan,
where he buried himself in theater and literature. He got a job with
the local newspaper and went from photographer to chief photographer and picture editor to reporter, but by the middle of his senior year
at Hillsdale, he felt like he had yet to find a compass. One quarter short
of graduation, Barry had quit school and the Hillsdale paper in July
1975 and gone back to Defiance to plan his future. He wanted to team
up with Tommy on another adventure they had talked about since
high school.

When Barry was very young, his parents had taken him out of school
for a month every other winter and headed for Key West, where his
grandparents owned a co-op apartment. With its old conch houses and
fluttering palm leaves, the island was a lush and wondrous place to the
young Barry, a wild tropical colony about as far away from Defiance as
he could get. "Harvey and I had spent a lot of time," he said, "dream-
ing up this plan about going down the Mississippi and getting to Key
West by way of boat." They had even stalked the dry docks where old
boats sat on blocks down around Cincinnati, and they had found a dark
blue cabin cruiser about forty-five feet from the stem back to a beautiful stern that curved upward like an egg. When Barry returned home
that summer, he called Tommy in Columbus and told him he wanted
to find that old cabin cruiser again and fix it up, just the two of them,
like the time they had driven 850 miles to Winter Carnival in the middle
of the night. Only this time the adventure was grander, and they didn't
have to come back. They decided to weigh anchor no later than October.

But one thing about Tommy that Barry had learned long ago was
that he could focus on his projects so completely he would have no idea
what was going on around him. A friend of theirs once said, "You'd go
over to Harvey's house, thinking you were headed out someplace, and
two or three hours later you still weren't ready to go, 'cause he was still
doing whatever he was doing." And since they had first talked about
the adventure, Tommy had made a pact with himself to keep involved
in the seven-to-fourteen at all times, and he could get so deep into these
projects, he didn't care if he ate or slept.

After his August graduation, Tommy had remained in Columbus, juggling his projects, putting off the adventure. But Barry was ready to go. "My thing was, 'Look, if we're gonna do this, let's start working on a boat,' but Harvey was in limbo."

Tommy called Barry from Columbus one night, and the two chatted for a while. The next day, Barry was gone. "I decided to go, and I was so pissed off at him, I didn't even mention it. I felt bad about not saying anything, but I also thought it was justified, because he's like that all the time. That's part of the relationship."

In his old MGB-GT, Barry crossed the seven-mile bridge into the Florida Keys in October of 1975, landed a job as a dishwasher, and began eating Key West up with a cosmic spoon. He looked like a Latin expatriate: steady, dark brown eyes, heavy beard shadow, thick black mustache, thick black hair; he seemed more the romantic lead in a film about revolution in Central America than the son of an insurance salesman from Ohio. But that's what was so wonderful about Key West: movie star, famous writer, lost college kid, nobody cared.

Barry had hardly had time to wash a dish when his landlady read in the paper that the *Miami Herald* was looking for a Key West correspondent. Barry got the job, and the *Herald* paid him to explore this den of shrimpers and gays and tourists and writers and actors and drug dealers and treasure hunters all simmering in a mean temperature of seventy-seven degrees. Every morning, he rose and put on his reporter's uniform—sandals and shorts—and went out to scratch the plump underbelly of the country's southernmost city.

Here, no one cared if he flunked accounting in business school; and it was okay to be steeped in drama and literature. Here, that had value, and he found it easy to be accepted. He learned to cook Cuban dishes and eventually became fluent in Spanish and Portuguese. He celebrated Thanksgiving with friends and attended Tennessee Williams's Christmas Eve party, gathered blue crabs and cooked big feasts, and spent weekends hunting for urchins and eating them raw out on the bridges over the keys. After a while, he took a second job as a weekend disc jockey at WKWF, spinning platters by himself on the graveyard shift. He called his show "Schatz in the Dark."

By the time Tommy arrived in January 1976, Barry had already melted into Key West, his "bohemian savoir faire," as a fellow reporter called it, gaining him access to the cliques of writers and musicians and the rest of the avant-garde on the island. He had just published a story about the Keys' most renowned treasure hunter, Mel Fisher. Life was good, and Barry was not happy to see his best friend. He had finally found a toehold in life's climb, and suddenly there's Tommy, with all of his projects in a suitcase, about to yank on his lifeline. "You can't come into the play in Act Two," thought Barry.

Tommy moved in anyhow and seemed not to notice the inconvenience. "He didn't know what he wanted to do," Barry said later. "He was trying all these wild schemes, trying to make something, but he didn't really know how to do it. I felt like he was dumping bricks into my wheelbarrow, and I was having a hard enough time pushing my own load." After a week, Barry asked Tommy to leave, but during the time he was there, Tommy heard the name Mel Fisher for the second time.

MEL FISHER WAS a nearsighted chicken farmer from Indiana and perhaps the most colorful of all the characters in Key West. He had an obsession with diving and a lust for Spanish treasure, and for seven years he had been chasing an elusive galleon called the *Nuestra Señora de Atocha,* which sank off the Keys in a hurricane in 1622. Every day of those seven years, Fisher had pumped up his divers with the motto "Today's the day!" and it hadn't been the day yet. His divers had uncovered muskets, swords, and religious artifacts, plus 4,000 silver coins, 3 silver bars, a few gold chains, a long gold bar, a large gold disk, and 2 gold coins. But the *Atocha* had carried 901 silver bars, 15 tons of ingot copper, 250,000 newly minted silver coins, and 161 pieces of gold bullion, plus myriad other rare artifacts and finely smithed jewelry and ornamental pieces. The rest of that treasure still lay somewhere on the bottom of the sea, and Fisher was still looking for it.

The high point of Fisher's search had come the previous summer when Fisher's son Dirk had found nine bronze, three-thousand-pound cannons in thirty-nine feet of water. The numbers on the cannons matched entries on the *Atocha*'s armament registry in Seville, Spain. "There was no doubt in anybody's mind at that point," said one diver, "that we had finally, after all these years, found the *Atocha*." Then five days later, their dive ship, a rusty old tug at anchor off the Marquesas, had lurched in the darkness of early morning, and in seconds she was heeled over too far to right herself, and three stories of superstructure had crashed onto the surface of the sea and rolled underwater, flipping anchors, line, air tanks, buoys, dive gear, and some of the divers into the black ocean. Eight of the divers made it back up safely, but the incoming water had trapped and drowned Dirk Fisher and his wife and another young diver. Since then the divers had found nothing but a silver candlestick.

Fisher operated out of an old galleon, a replica down at the Key West dock with a sign on the bow that read "Pirates' Treasure Ship." The *Atocha* had nothing to do with pirates, but the gimmick attracted tourists at $1.50 a head, which sometimes was the only cash Fisher had to buy food for his divers.

When Tommy boarded the galleon in early 1976, he wasn't so much interested in being a diver as he was in talking to Fisher about "technology transfers," things that perhaps engineers in the oil industry or some other industry were doing that Fisher might be able to use in searching for shipwrecks. Fisher would give an audience to anyone who thought he could help him find the *Atocha* mother lode. He had talked to people who could communicate with dolphins, who could find gold with an underwater divining rod, who could speak to the spirits of the long-lost Spanish souls aboard the *Atocha* when it went down in 1622. One of the divers' favorite suggestions was from a guy they called "Mr. Bubbles," who wanted to electrify the entire work site to find the silver. So it was not unusual for some kid just out of engineering school to drop by the pirate ship to discuss technology transfers and have Fisher's ear. Tommy suggested that maybe new technology would make the search easier, and Fisher listened to Tommy's ideas, but his operation was in one of its frequent downturns and he had no money to hire anyone.

For the next six months, Tommy visited Fisher occasionally, while he pursued the seven-to-fourteen from Miami and Key West. He didn't care about making a living or having a roof over his head; he wanted to learn something. "The main thing was having enough resources to make phone calls," he said. That was the phenomenon he had discovered that would become his link to what was happening in technology all over the world: Scientists like to talk about their work, especially the arcane twists; he had learned that back in high school when he called Linus Pauling. He could get on the telephone and connect with experts in almost any field. "I just found out who knew the most about which fields I was interested in," said Tommy. But first he would analyze the papers the scientist had written, then develop questions around what was missing.

To save money for phone calls, Tommy sometimes slept in his car in the lot near the cemetery, where the graves sat above ground and everyone assumed that at night witch doctors performed voodoo. Not that Tommy's '63 Mercedes in any way detracted from the pins-through-the-ragdoll atmosphere of the place. The Mercedes had reached what Tommy called "stabilization." All the rocker panels had rusted out, all the places up under the fender where the water used to catch and sit and start to rust now had large holes in them.

"There weren't any more places that could catch water," he said, "so it quit rusting, the way I figured. It looked like hell though."

The Mercedes was part of a project, one of the seven-to-fourteen. Tommy had bought the car to experiment with alternative fuels: Instead of burning diesel to make the car run, he wanted to fill the tank with french fry oil from McDonald's. A professor at Ohio State had tinkered with the idea, and Tommy had tinkered with him. They discovered that with proper filtering and a little heating before the oil went into the injectors, they could burn the same stuff that browned and crispied all those precut, frozen fries. The Mercedes got about forty miles to the gallon, and Tommy planned on hauling two fifty-five-gallon drums full of french fry oil behind the car. He figured he could make it coast to coast on one fillup, and with the french fry oil selling for about a nickel a gallon, the whole trip would cost no more than five or six bucks. It wouldn't be as convenient as pulling in to a gas station, and he would have to talk to the people behind the fast-food counter, but he was sure they had vats of the stuff sitting out back. The only real problem was that the Mercedes smelled like skillet smoke and Tommy like a short-order cook.

That summer Tommy was visiting with Fisher one afternoon when he overheard Fisher say he needed someone to man the theodolite tower, a lifeguardlike structure that stuck up out of a reef near the quicksands. Someone had to sit on the tower and squint through a surveyor's tool and talk on the radio to keep Fisher's boat on course as it towed a magnetometer back and forth. The divers called the guys who manned the tower "fry boys," because sitting on the tower was about the same to them as crackling in skillet grease was to a chicken. The tower was constructed of scrap iron and offered not even a thumbnail's worth of shade, the sun hitting twice, once through the atmosphere, and again by ricocheting off the water. For eighteen miles around, that tower was the only spot for a bird to sit and rest a spell, so it was covered with guano that squished between the toes and sometimes dropped into the water, where it started its own little ecosystem, which ran all the way up the food chain to schools of barracuda. And the twice-baking sun simmered the guano just below boil, so that the whole thing smelled like an aviary dung heap. When Tommy heard Fisher say he needed to find someone to man the tower, Tommy told him, "I'll do it," and for the rest of the summer he worked for Mel Fisher.

After Fisher's son had found the cannons in the quicksands, Fisher had towed a former Coast Guard buoy tender called the *Arbutus* out to the site for a dive platform and a place for the divers to eat and sleep. Half a dozen young men lived out there for two to three weeks at a time, diving all day, blowing sand away with a big impeller, and searching with their face masks a foot from the bottom. At the end of the day, somebody cooked something, or they ate baloney sandwiches, and talked out under the stars.

On such nights, Tommy would discourse on the effects of calcium overdosing, or laser technology, or the latest in herd psychology, or the holistic approach to something or other, and crew captain Pat Clyne said it was like listening to someone tripping on LSD, how they start explaining their perception of things that to them makes a lot of sense but makes not one lick of sense to anyone listening. Clyne couldn't tell if Tommy was "off the wall" or "ingenious."

"I remember sitting out on the deck with him at night and talking for endless hours about everything and nothing," said Clyne. "He was a very personable guy, had a good sense of humor, and his stories were so colorful, his discoveries or things he was thinking about. That's what made him so likable. He could perfectly well have been on top of something, but we'd just kinda smile and shake our head."

The other crew captain, Tom Ford, couldn't figure him out either. There was something strange about him, not a bad strange, a good strange, a guy off marching to his own little drummer, but Ford could never figure out where he was headed. "He was always testing something," laughed Ford.

Since Tommy had a degree in engineering, Ford put him in charge of the new hydroflow, the pump and propeller housed inside fifteen feet of irrigation pipe they used to blow sand, and the first thing Tommy did was put on his gear, jump into the water, press his body up against the grating, ease on the throttle, and crank it up as it sucked him tighter and tighter against the grating, his flesh undulating, until the suction ripped the mask from his face and tore the mouthpiece out of his mouth.

"I needed to get a real live feeling what the flow was like," said Tommy. "We would study that stuff in school, but very few people have been able to feel it with their whole body."

Tommy experimented with the hydroflow at different speeds and at different distances from the bottom. He noticed that the divers had to rely on someone topside to run the hydroflow, that they couldn't signal when to rev it up or shut it down, so he designed a throttle they could control themselves from the bottom. He was a good diver, too; he could free dive to forty feet, but in the summer he was on the *Arbutus,* he brought up not one artifact. He preferred to build things and to experiment and to get his hands around the concepts he had discussed during all of those sessions with Dean Glower.

Some of his ideas were unusual, but unlike the ideas of many of the people Fisher consulted with, Tommy's were grounded in physics and engineering. On the *Arbutus,* Clyne saw him constantly reading technical manuals and journals and then talking to Fisher about more "crossover technology," something they might be able to adapt to the search. "He used to talk to Mel a lot about his different ideas," said Clyne, "and Mel eventually got all of that— side scan, magnetometers, subbottom profilers."

On rough days, the *Arbutus* acted as a break, keeping the wind and the waves away from the divers. One day not long after Tommy had joined the crew, the weather was worse than anyone had seen it in some time. Tommy was on tanks at the bottom, and at the end of his dive, he surfaced on the weather side of the ship. Ten- to fifteen-foot waves swelled and crashed against the bulwark, and Ford yelled to Tommy to dive under the boat and come up on the lee. But Tommy had already taken off his mask, and he had his tanks in his hand and his regulator under his arm, like he was going to pass it all up to them on deck. He was treading water in his swim trunks, holding on to his gear, and then suddenly he disappeared. When he went down, they saw that he had taken off his fins and was carrying those, too.

"He just tried to swim under the boat," remembered Ford, "fighting everything, and we're going, 'What is this guy doing?'"

The *Arbutus* was nearly thirty feet across the beam, and the current underneath ran strong to the weather side. The ship had been out there so long that coral encrusted the bottom, a calcified forest of tiny razors. The hull now rose and fell sharply, so a diver couldn't dive five feet down and head under or a twelve-foot wave would lift the hull, suck the diver up with it, then roll on and drop the hull onto the diver, ripping into his skull and shredding the skin on his back.

The divers on deck waited on the weather side, watching where Tommy had gone down and expecting him to surface again, pushed back by the current. But Tommy did not come back up. A minute passed. Two minutes passed.

The divers started drifting across the deck to the lee side, but they did not see him there either. As they were getting ready to jump in and look for him, Tommy surfaced on the lee side, frog kicking with all of his gear in his hand.

"What the hell're you doing?" yelled Ford.

When Tommy got up close out of the wind, he said, "I just wanted to see if I could do it without my fins."

"He didn't even get scratched up," said Clyne, "and that would be minimal compared to what could've happened. If that boat had come down on his head and shoulders just one time, he would've been somewhere between here and Cuba, shark meat."

Ford came to think of Tommy as the absentminded professor on Fisher's crew. He told wild stories about amphibious cars and staying

GARY KINDER

awake for eight days, and Ford didn't know what to believe, but when the diesel pump on the hydroflow stopped and Tommy fixed it, Ford began to think that maybe Tommy's wild stories weren't so wild. Ford, who was a decent mechanic himself, got lost in the old diesel floats and switches and had no concept of the pump's operation. He also had no schematic to consult. Tommy got his hands inside the pump and started pulling things out, and then he put some things back in and took some more out, and then he said, "Let's try this," and then he said, "Maybe it's this," and then he said, "I think it's this," and by the end of the day he had the pump running again. Impressed Ford to no end. He said to Tommy, "You're okay in my book."

Tommy solved, or almost solved, two more problems on the *Arbutus*.

When the divers had finished working an area in the quicksands, the only way they could move the floating hulk was to wait for the wind and the tide to align, then pick up all of the anchors and drift for several minutes before resetting them. The work was hard, and the divers hated it, but by resetting anchors they could move the *Arbutus* a few hundred feet every week or two. Then Tommy figured out a way to rerig the hydraulic system so they could lay the big irrigation-pipe hydroflow down on the water horizontally, making it a propeller instead of a digging tool. The first time they turned on the hydroflow in the new position, it propelled the *Arbutus* across the quicksands to the outer reef, three or four miles away, the 187-foot hulk of steel doing about four knots and leaving a little wake behind it.

But the *Arbutus*'s papers classified the ship as a stationary platform, the same as a derrick or a barge, and they were traveling for miles around the wreck site, "digging holes all over the place," remembered Clyne. Local fishermen used the *Arbutus* to locate their lobster traps, and all of a sudden that sight reference seemed to have moved, and they couldn't find their traps. They complained to the Coast Guard, and the Coast Guard came out and wanted to see the engine room, which was one huge, open, empty slime pit. When they asked Clyne how he moved the *Arbutus,* he said, "Anchor over anchor." They looked at the engine room again and left.

Perhaps the worst problem faced by the crew on the *Arbutus* was not the sun or the weather or the currents or the sharks or trying to move the *Arbutus,* but the seagulls. "It was seagull heaven," said Clyne, "guano city." Birds flying up from Havana to the Marquesas would settle onto the *Arbutus* for an extended stopover before continuing their journey. Hundreds if not thousands of seagulls constantly lined up on the rails and on the beams, occupied every perch a bird could find on the ship, frittering up and down the rail, looking at everything sideways, squawking constantly in kaffeeklatsch crescendos, and dumping on the deck.

"Our deck in the morning would look like Detroit at Christmas time," said Clyne. "White. Just white."

After a rain, the deck was slick and dangerous, and it smelled as bad as the theodolite tower. They had tried putting up scarecrows, but the birds landed on the scarecrows and dumped all over them, too. The crew swung ropes and brooms at them, and lunged at them, but only a few would move and they came back, squawking louder, almost like they were laughing.

Then Tommy had an idea: Let's electrify the rails; we can run a 220 line, send it up to the main rail and shoot a current all the way around to the other rails and control it all from the wheelhouse. At first the crew laughed, but the more they thought about it, the more they said, "Why not?" and "I wonder if it could be done." And pretty soon it sounded downright logical. Zap 'em. Just enough to tickle the bottoms of their feet and make them fly away.

Tommy wired the ship and they got everybody safely up in the wheelhouse where they had the battery. With all of the lines connected, Tommy looped one around the negative terminal, and then he touched the other line to the positive terminal. Suddenly, hundreds of birds shot into the air squawking and flapping away from the ship.

"You could hear them scream as they took off," remembered Clyne. "It was great. It worked! We couldn't believe it! Everybody was patting everybody else on the back, you know."

For almost a year they had been living with bird droppings thick all over everything, even their gear. From the time they arose in the morning until they finally retired at night, they had had to breathe the stuff and feel it squish through their toes. Now, they'd finally found a way to get rid of the problem.

Tommy disconnected the battery to make the ship safe for the crew, and the divers went back to looking for treasure. A short while later, the birds started coming back, and after a bunch had collected along the rails, the crew ran back into the wheelhouse, and Tommy hit the juice again, and again the birds shot into the air and flew away. All except one.

What happened next got the crew to wondering about a seagull's IQ, because what happened next not even some of the crew would have figured out.

Tommy hit the juice again and the current shot along the rail, and they thought that one seagull would fly away, but she didn't. She lifted one leg. They disconnected the battery, and she put her leg down. They zapped her again, and she lifted the other leg. Without both of her legs down to complete the circuit, the electricity just ran on through. Another bird landed, then another bird, and another bird, and Tommy touched that positive terminal with the wire again, and half a dozen seagulls lifted one leg. As soon as the electricity stopped, they set the leg down.

After two days, every one of the seagulls had returned, and it seemed as if many had brought friends and relatives, for the entire bow was lined with seagulls and the deck was white again. Now when Tommy threw the switch, hundreds of seagulls would lift one leg in unison, and by touching the wire to the terminal back and forth in a rhythmic way, he could make the birds dance. It reminded Clyne of *A Chorus Line*.

"That was the type of ingenuity that Harvey would come up with," said Clyne. "He would think of things that none of us would think could be useful, and he tried to make 'em work. Some did. And some didn't."

THAT SUMMER, AFTER thousands of hours with their face masks down in the sand, the divers on the *Arbutus* had found nothing much more valuable than a barrel hoop. Yet they all wanted to be on the bottom, because as one put it, "If you're not down *there*, you're not going to find *it*." But more and more Tommy stayed topside trying to solve bigger problems and observing how others searched for treasure. He was more interested in why they couldn't find the *Atocha* than he was in seeing treasure, and as he watched, he got to thinking.

For two hundred years, fleets of treasure galleons stuffed with silver and gold and emeralds had plied the Caribbean Sea, crisscrossed it every which way, from Key West down to Cartagena, from the Yucatán over to the Windward Islands, and every so often, one of those unpredictable West Indian cyclones would come spinning across the Caribbean and slam half the fleet onto shallow reefs, which ripped open the hulls and spewed that treasure all over the ocean floor.

Where were these ships and why were they so hard to find? Already, shipwrecks had become one of the seven-to-fourteen, and Tommy was asking more questions. Just how "blue-sky" are these projects? With all of those shipwrecks out there and all of that research available and the technology on line, it shouldn't be a matter of searching until you stumbled across something. Tommy had liked Mel Fisher from the day he met the man, but Fisher would blast holes into the seafloor, and within days or even hours sand would again fill the holes, and Fisher would have no record of where he had just searched.

"Amazing the way that place worked," said Tommy. "Absolutely incredible. I got to see a lot of the problems."

They had dragged a magnetometer all over the quicksands where they had found the cannons, and every time they'd get a hit, remembered Tommy, someone would shout, "Yeah, that's it! That's just where I thought it was gonna be, right in this area of the map I was thinking about! That's gotta be it—send the divers down!" Then they would mark the spot by throwing over a bleach bottle tied to a cinder block, except that often by the time they got that into the water, the boat would be a hundred yards beyond the hit. A diver then had to go down and see if what had set off the magnetometer was part of the *Atocha*. It was always something else, but they would say the same thing the next time, and they would keep thinking that way again and again and again. As soon as the weather turned rough, the bleach bottle buoys would drift.

"This had been going on for years," said Tommy, "and they had no method for knowing where they'd searched. They argued about, 'Well, we searched that last year,' you know, and somebody else would say, 'No, no, that was over there. We didn't search that.' It was incredible. After years of that they had no good records of what had happened. And Mel's operation was better than most."

Tommy figured that someone needed to study hurricanes and how they came across the Caribbean, and what they would do over the centuries to a ship already wrecked, how they would break it up and where the pieces might have moved. Everyone looking for the *Atocha* knew that two hurricanes had hit the ship, one only three weeks after the other; but there must be a way to narrow the dynamics of those two hurricanes, a way to quantify all of the possibilities.

And another thing Tommy pondered: How did Fisher know the *Atocha* had not been salvaged shortly after it sank 350 years ago? The water wasn't that deep: Dozens of people had free dived to see the *Atocha*'s cannons. If the wreck's in twelve feet of water or even forty feet, the technology was there centuries ago to salvage it. What were the odds?

As Tommy watched Fisher's operation and listened to stories about other treasure hunters, he began to see a pattern: They operated from day to day, with no long-term plan; they all were underfunded; no one kept accurate records; the turnover rate of workers was high; they raised money primarily through the media; investors were unhappy and filing lawsuits; the state claimed all treasure belonged to it; the storms scattered a ship's remains sometimes for miles across the shallow sea; they had no way of telling whether an artifact came from their target ship or from some other ship that had landed on top of it in another storm; they could never be sure that no one else had already salvaged the ship they were after.

Tommy's thoughts and observations about shipwrecks in the Caribbean began to fill notebooks.

"That summer was a very fertilizing experience for me," said Tommy, "just thinking about historic shipwrecks, where they were, and how they could be found, and what kind of technologies could be used to find them. Part of how to turn ideas into a project is being able to examine lots of different situations. The more you understand about the world, the better your perceptions are and the better decisions you can make. I was looking at wealth in terms of growth and knowledge and education as opposed to money, so it was a great experience. I didn't get paid much and there may have been all kinds of hazards, but I was the only engineer in the whole operation."

IN THE FALL of 1976, Tommy left Key West and traveled with Fisher and some of the other divers to Washington, D.C., to present Queen

Sophia of Spain with the bronze cannons from the *Atocha*. Then he drove the old Mercedes to an oceanography conference in Ottawa, stayed for a while with his parents in Defiance, where he ran up phone bills of four hundred dollars a month, then left for Columbus to work on a solar energy project.

In Columbus, Tommy visited Dean Glower and told him of his experiences working with the treasure hunters in Key West and what he had been pursuing at various libraries in Florida. Then he called one of his advisors from his summer at Stone Lab, Eddie Herdendorf, who was teaching an evening course in oceanography; Herdendorf asked Tommy to talk to the class about his work for Mel Fisher. Herdendorf recalled the presentation as "real matter-of-fact," some engineering graphs Tommy had drawn and put on an overhead. "But he emphasized the importance of not just randomly looking for something," said Herdendorf. "You need to have a plan and establish an electronic grid and systematically explore. It had to be done scientifically."

Next, Tommy left for Texas to consult with a start-up company trying to develop a flywheel transmission for ultra-fuel-efficient cars, then east to New York, and back over to Chicago, like a good shepherd tending his flock of seven-to-fourteen. He earned just enough money consulting to put diesel fuel into the Mercedes and pay off the two phone cards he used when he stayed with family and friends. But he was filling notebooks with what he learned, and he had a stingy old car that provided him a place to sleep when nothing else was available. And that was all he needed.

The winter of '77, he lived in Chicago with his sister Sandee and his brother-in-law Milt Butterworth, who supervised the lighting and photography of art objects at the Chicago Art Institute. In Chicago, Tommy pursued another of the seven-to-fourteen, a computer modeling of the commodities market. Much of his time there he spent on the phone. When Sandee and Milt asked him what he was doing, he would say he was "making contacts." "These phone calls were very important to him," remembered Sandee, "a matter of life and death."

In Chicago, two men he had met while working for Mel Fisher introduced him to a treasure hunter named John Doering. Doering was funny, unassuming, easy to be around, and he liked Tommy's quirkiness. He also admired Tommy's engineering skills. A year after they met, Doering called Tommy about his latest venture: *La Concepción*, the legendary ship wreck on Silver Shoals, about eighty miles off the coast of the Dominican Republic. The *Concepción* had gone down in 1641 with a registered consignment of 150 tons of treasure, mostly silver. Doering worked for Seaborne Ventures, and he was in Seattle trying to overhaul a 157-foot minesweeper called the *James Bay*. They had fired the previous engineer, and before he left, the engineer had done something to the engines so they couldn't advance the throttle above idle or run the ship on automatic pilot. They needed an engineer to get her running again and sail with them down to Panama, through the locks, and into the Caribbean Sea. Tommy signed on and flew to Seattle, where he and another engineer worked on the sabotaged engines. "Harvey kept coming up to the wheelhouse and playing with stuff," remembered Doering, "and somewhere along the line he figured it out. He was just invaluable."

As he had with Mel Fisher's operation, Tommy watched closely how Seaborne Ventures was run. When the *James Bay* left Seattle in late July 1978, three months after the projected departure date, Doering and his partners were still trying to raise money for the expedition. They got to Panama in August, but had to hang offshore for three more weeks, waiting for enough money to arrive to get it through the locks. Yet Seaborne was a lot more organized than other treasure hunting outfits Tommy knew of. They had a plan, and they had thought most things through. "It was pretty well run," said Tommy, "but it was still a treasure hunt. One of the problems was they didn't really have a solid project."

While the *James Bay* lay off Panama, Tommy flew back to Miami to work in the University of Miami's Rosenstiel School of Marine and Atmospheric Science. Then he went back to Texas and back to Columbus and back to Chicago, still tending his flock. With Tommy in Chicago, Doering and his partners altered their plans to head straight for Silver Shoals, instead hoping for some quick success along the banks due north of Panama: Roncador, Quita Sueño, Serranilla, Baja Nuevo, shallow water reefs that in storms could have snagged galleons from the Spanish treasure fleet leaving Cartagena. "We were stopping at all the good places," said Doering, "hoping to find something that would keep us financially alive." Shipwrecks littered the banks, but they found nothing of real value.

They traveled on to Kingston in September and up to the Inaguas, Great and Little, where they found cannons and anchors and a crashed airplane, and a sweet little brass, breach-loading cannon five feet long, but little treasure. After a brief respite in Key West, they finally left for their original destination, Silver Shoals, but on a dock in Bimini, Doering heard that a treasure hunting rival named Burt Webber had just arrived in the Dominican Republic, signed a contract with the government, found the wreck of the *Concepción*, and was now salvaging her tons of silver. "It was almost irrelevant at that point anyway," said Doering, "'cause we were out of money as usual."

With Christmas nearing, they returned to Key West to raise money and mobilize for a new project: searching the coastline of the mountainous island of Dominica. According to their research, at least a hundred shipwrecks lay covered in the black sand of Dominica's Roseau Harbor, and in Portsmouth Harbor at the north end of the island six ships from a treasure fleet had gone down in 1567 with a cargo that Doering's partner estimated to be worth $700 million.

While Seaborne prepared for the Dominica expedition, the *James Bay* sat dockside in Key West. Wary of other treasure hunters on his turf, Mel Fisher showed up at the dock one afternoon and saw hanging off the stern of the *James Bay* a pair of the biggest blasters he had ever seen. Doering had designed the blasters to blow through the bane of all treasure hunters: deep sand. Each was seven feet across. Fisher still had not found the *Atocha,* but he had some fresh capital from new investors. He hired Doering and the *James Bay* to poke around in a new area near the Bank of Spain, where Tommy had worked on the *Arbutus* three summers earlier. The sand there was fifteen feet deep.

With Tommy still in Chicago, Doering and some of his divers took the *James Bay* and those big blasters out to the new site. In fifteen minutes they had blown a hole eighty feet wide at the mouth and tapering all the way to bedrock. After a week, the divers began finding cannonballs, musket balls, and a few muskets and swords, signs of an old Spanish ship. But Fisher had enough swords and arquebuses and musket balls and cannons to fill the Smithsonian. He needed gold and silver. Some emeralds would be nice. He had been searching for ten years, and he was getting desperate.

On the 29th of June, the blasters blew another sand crater, and a diver named Rich Banko and his dive partner jumped into the water and went down. Before he hit bottom, Banko saw a gold disk five inches across. He grabbed the disk and pushed off the bottom. But when he broke the surface, waving the disk and hollering, no one heard him. After waiting a week for something to happen, a film crew from *National Geographic* had folded and left that morning, and the ship crew were forward or down below. When Banko finally got someone's attention, the crew radioed Fisher on the ship-to-shore. Fisher immediately motored out to the site, bringing with him an East Indian mystic named Baba Ram. With him, Baba Ram had brought an entourage, and from the entourage Doering learned that Baba Ram was really a rug merchant from Minneapolis, but Fisher was convinced that Baba Ram could help him find the *Atocha*.

Baba Ram was a little guy with a goatee, and he had given up speaking. He could talk as well as anybody, but for reasons that went unexplained he had given it up. He communicated by grunting, and that's what the entourage was for, to interpret the grunts.

"You'd ask Baba a question," remembered Doering, "and Baba'd say, 'Grunnt!' and his entourage would say, 'Baba says that you will find much gold today.'"

Four days after Banko found the gold disk, he lifted a pottery shard from the sand and found a gold chain seven feet long. "We came up with that," said Banko, "and things got really wild. Mel had a bottle of Crown Royal with him, and this Baba character was putting the Crown Royal away faster than anybody." Banko had pictures taken of himself and Mel and little goateed Baba Ram, all wrapped up in this seven-foot gold chain.

On his next dive, Banko was on the bottom again, fanning the sand, and right there in front of him was a bar of gold, seven inches long and about two fingers thick. It bore the royal seal of King Philip IV of Spain. When he surfaced, he showed the bar to Fisher and Baba Ram. "The guy grunts," remembered Banko, "and uses his little sign language to say, 'See, I told you, I brought you good luck.'"

A week after the *James Bay* returned to Key West, Banko and two friends snuck down to the pool at the Casa Marina one afternoon, and Banko noticed Mel Fisher talking to two men on the other side of the pool. About two hundred people were lounging around the pool deck when Banko saw Fisher pull a gold bar out of his shoe. Five minutes later Fisher pulled Banko's seven-foot gold chain out of his other shoe. The deal with Doering and the *James Bay* was over, and Fisher was trying to persuade the two men to put another boat out there over the quicksands. Banko ambled across the pool deck toward the three men, and when he got closer, Fisher yelled, "Here's the guy that found it!"

"By the time we left," Banko said later, "he had those guys eating out of his hand. He signed them to a deal right there at the pool!" Banko told Fisher he would love to have one link from the chain he had found, and Fisher said, "You got it." The next day they went to Fisher's boat, and Fisher broke open that 350-year-old Spanish gold chain and gave Banko a link.

Doering and his divers on the *James Bay* had found pottery, arquebuses, silver coins, swords, and an amethyst one and a half inches across, plus four gold bars, a two-pound gold disk, a chunk of gold the size of a quarter, a smaller gold chain about ten inches long, and Banko's seven-foot gold chain, which weighed ten pounds. For their work, Doering and his partners received $100,000 and 5 percent of the find: a small gold bar and a small gold chain. The finds kept Fisher's hopes alive, but after ten years of constant searching he had found only enough of the *Atocha* to lure him onward: parts of cargo and bits of armament, some coins and bronze cannons, a little gold, and a few fine pieces from her wealthy passengers. He would not find her treasure for another six years.

ALTHOUGH JOHN DOERING'S *Concepción* project on Silver Shoals had never materialized, and Tommy had seen enough treasure hunting in shallow water, Doering's plans to search the coastline of Dominica intrigued him: Dominica was a mountain rising from the sea, and the slope all around was precipitous; a few hundred yards off the beach, the bottom dropped quickly to two or three hundred feet. The search would be more of a challenge; they would need better magnetometers and maybe even a small submersible to search the harbors. It would give

Tommy an opportunity to try new ideas in deeper water. But before anyone could try anything, Seaborne had to solve another problem.

Because the underwater terrain of the island dropped away so quickly, every accessible shipwreck lay well within Dominica's three-mile territorial limit. If Doering and Seaborne Ventures wanted to work those waters, they had to have the government's permission, but the government changed hands so quickly, they had to be careful not to alienate one faction while schmoozing with another. Negotiating with Dominican officials took over a year, and during that time the prime minister left office amid accusations of graft and corruption; then the populace ousted the interim prime minister by electing a third prime minister. As the political winds blew one way and then another, Hurricane David ripped through the middle of the little island packing a blow of 160 miles an hour. Tin roofs sailed off into the sea, thirty-eight people died, two thousand were injured, sixty thousand were left homeless, the only road on the island became impassable, and all telephone and electrical power was out for at least six months. In a country so rural and poor, that meant no shore support, which was vital to any salvage operation.

Tommy kept abreast of the negotiations, noting how long they dragged on, and he added to his list of problems in searching for historic shipwrecks in shallow water the delays and uncertainties inherent in dealing with unstable island governments.

The *James Bay* finally arrived in Dominica on February 1, 1980. To survey Roseau Harbor quickly, Tommy and Doering experimented with a one-man, made-from-a-kit submarine about ten feet long with a clear bubble top. On one of the early dives an electrical fire filled the sub with acrid smoke while Tommy was inside trying to remove a nut with a crescent wrench, and he almost couldn't get out. On an unmanned test dive, the sub filled with water, and they spent the whole night coaxing it back to the surface. Meanwhile, the divers blasted holes six to eight feet deep in the black sand and recovered glass bottles, brass spikes and ships' fittings, silver spoons and olive jars and anchors, musket balls and clay pipes, an array of artifacts from a variety of ships flying the flags of different countries in different periods of history. They found debris from the sixteenth century, the seventeenth century, the eighteenth century, and the nineteenth century. They sucked up the black sand with an airlift and recovered copper and silver coins, British farthings and shillings, coins from France, Canada, Jamaica, the United States, and an old French colony—a lot of artifacts but nothing of value. As he had with Fisher, Tommy watched and took notes and asked a lot of questions.

One morning, Tommy awoke to a lot of yelling and discovered that a diver named Bob had fallen out of his bunk, numb on one side, suffering from nitrogen narcosis, or the bends. For several days he had been diving in deep water and had cheated on his meter. Bob's dive partner was Rich Banko. Banko described Tommy and Bob as "mortal enemies," at each other all the time, arguing constantly. "Bob would go out of his way to tell anybody that Harvey was an asshole and a dipshit," said Banko. Banko himself thought Tommy was "the most arrogant sonofabitch I've ever met, an eccentric inventor always trying to make a better metal detector." He had once threatened to beat Tommy's brains in, because Tommy had left him and two other guys at a deserted boatyard on a ninety-two-degree Sunday afternoon with no food, no water, and no shade and gone off in the Seaborne station wagon on one of his compulsive pursuits. But Banko figured he might have to stand in line, at least behind Bob, the diver who now had the bends.

Seaborne had a decompression chamber on the *James Bay,* but no one had ever used it. In his constant study of technology on the horizon, Tommy had learned that the foremost authority on the bends was a navy dive-medicine research team in Panama City, Florida. "The first thing Harvey did," said Banko, "was to call the navy to get instructions on how to run this decompression chamber exactly the right way. He ran the whole show."

Tommy called Panama City on the radio and talked to the navy experts about having Bob "blown down" to 165 feet. Within an hour, Tommy had the chamber properly adjusted, and they carried Bob inside, where he remained for almost two days, while Tommy gradually reduced the pressure to bring him slowly back up. Bob came out in better shape but still suffering from headaches, dizziness, fatigue, and the continuing paralysis of his right arm and leg. Tommy pushed the navy experts for additional ideas, and they reluctantly told him of a new theory that administering pure oxygen to the victim might arrest the progressive nerve damage. But the idea was so new not even the research team could cite results, and no one wanted to take responsibility for the method.

Some of the divers on the *James Bay* disagreed with the idea anyhow. They had been trained to treat the bends according to strict navy dive tables, and they refused to cooperate. Tommy ignored their protests; he told Doering to round up all of the oxygen on the boat and then head to the island to see if he could find more. Doering pulled the *James Bay* in front of a small island medical school, where Banko found a nurse who came aboard with enough oxygen for four or five hours of treatment. Then Banko persuaded the owner of a welding shop to part with all of his oxygen.

Tommy still had too little oxygen to fill the chamber, so he created a portable system for Bob to breathe on. The danger was that breathing pure oxygen could cause Bob to black out, so Tommy watched him through the porthole and talked to him on the intercom, trying to coax as much of the oxygen into Bob's system as he could without knocking him out. The treatment went back and forth for twenty-four hours.

"Harvey ramrodded the program," said Banko. "I give him credit for that. The nurse was there for medical advice, but as far as running the chamber and talking to the navy, Harvey handled the whole situation." The following day, they flew Bob to Miami, where he received additional treatment. When he left the hospital a week later, the doctors put his recovery rate at 98 percent. "I don't know if it would have killed him," said Banko, "but certainly he would have been crippled if it wasn't for the chamber and the way it was handled."

Throughout the spring and into the early summer of 1980, the divers searched Roseau Harbor and then Portsmouth Harbor, but they found only more "scatter," artifacts of little value. "We could have done better going up north and looking for the treasure fleet," said Doering, "but by the time we had worked ourselves up there and had evidence of the fleet, we were broke, and we got run out of town, so to speak."

Banko could tell the end was near. His bosses didn't want to run the boat, they didn't want to run the blasters, they were always skimping on fuel. Tommy left the ship and returned to Ohio. Doering left Banko and three others with the smaller boat on Dominica and took the *James Bay* to Guyana to replace rotten planking in the hull. Doering told Banko he'd be back in three weeks, but the repairs ran almost forty thousand dollars, and nearly two months passed before he called Banko again to tell him they were stuck in Guyana because they couldn't pay the bill. When they finally raised the money to get the *James Bay* out of dry dock and back to Dominica, there was nothing left to run the operation. Banko had a friend who had dived with another treasure hunter, and the friend had seen the same problems.

"This guy would have all these grandiose things going on," said Banko, "and all of a sudden there'd be no money and he's gone. It happened to all of them."

TOMMY SPENT LITTLE time in Key West now. He continued working around the country on various projects—solar energy, an infinitely variable speed transmission, an amphibious bus—earning just enough to eat and pay for his phone bills. He still had the notes from his sessions with Dean Glower and a list of the problems he would need to solve before he could perform complex and delicate work over long periods of time at the bottom of the sea. For long hours, he sat in research libraries, revisiting what he and Glower had discussed, trying to isolate the problems and get them straight in his mind. "Everything he does," said Glower, "you can see that he looks at it from all directions, and he sorts it all out so he can understand it."

As Tommy had watched the treasure hunters searching for shipwrecks around shallow water reefs and wondered at their lack of methodology, the leap had been a short one to wondering about the technology needed to recover shipwrecks from the deep ocean. For if he had the capability to descend thousands of feet, sift delicately through the remains of an ancient ship, and bring back her treasure unmarred, he had the means to do almost anything on the bottom: fish, farm, mine, drill, recover, test, save, explore, collect, all of the things he and Glower had discussed. The same forces had to be overcome.

When Tommy learned that big mining companies had banded together to share the cost of developing ocean technology, deep-ocean mining became one of the seven-to-fourteen. Deep-water submersibles already had imaged mineral rich manganese nodules scattered like baking potatoes across the ocean floor, but the mining companies needed better technology to collect the nodules cost effectively. He read whatever he could find on their work and spoke on the phone with several of the people involved.

Tommy kept looking at what others had tried and wondered what in their approach prevented them from doing more intricate work for longer periods of time. There seemed to be a barrier that no one could penetrate; Glower himself had labeled the barrier "impossible." Ever since the *Thresher* tragedy, submersibles had come out of the military and out of industry that could go deeper and deeper, but they could perform only the same crude functions: observe, photograph, film, grab, and hold.

Wherever he was, Key West or Texas or Chicago, Tommy kept in touch with Glower, and they talked about things Tommy had seen or learned, or about questions he needed to ask. When Glower answered the phone and heard a voice say, "This is Tom," he knew the conversation would be long and intriguing.

During this time of travel and research, Tommy wrote down two questions: "By the year 2000, what kind of technology will be available to find and recover historic wooden ships lost in the deep ocean?" And, "What prevents us from doing that at a reasonable cost right now?" As he talked with other scientists and engineers, he asked these questions repeatedly, until he had a clearer picture of the problems ahead. Once he had defined those, he could search for ways to solve them now.

By 1980, Tommy had read scores of esoteric treatises on the problems of working in the deep ocean, and from pay phones and friends' houses he had talked to scientists and engineers all over the country. On paper, he had sketched vague concepts that departed from different points and proceeded in different directions. He didn't have all of the answers, but finally he understood the questions, and that was the key: No one had been able to penetrate the "impossible" barrier, because they were still asking the same wrong questions.

Aboard the
Central America

⚓

Near Dark, Saturday, September 12, 1857

Captain Herndon talked in private with Thomas Badger, and the two men agreed again that the ship must go down. Herndon repeated that he would not leave his ship while there was a soul on board. As before, the two captains kept their conversation to themselves and urged all hands again to renew their efforts. Captain Herndon told one passenger that he had "strong hopes" the steamer would survive till daylight, when the storm would lie down and the brig *Marine* and the swift schooner that so recently had rounded their stern would come to rescue them all. "They will stay by us," he said, "they promised me they would." He could still see their lights in the distance, in the direction blown by the wind.

Herndon retired to his quarters and emerged in full dress uniform, the oil silk covering removed from the gold band on his cap. On the wheelhouse, he took his stand, gripping the iron railing with his left hand and striking a pose that seemed inhumanly serene, as though he had reached deep and found a strange peace.

Just after the schooner began blowing to leeward at dusk, Herndon had told his second officer Frazer to fire a distress rocket, and to fire another rocket every half hour. He also ordered Frazer to remain with him until he went from the ship; they would be the last to leave. Just as Frazer lit the first rocket, they saw off the starboard bow the lifeboat of the bos'n, John Black.

Many of the men continued bailing, but the lines had dwindled now that the women and children had all been carried safely to the *Marine*. The men could hear the water still rising in the hold. As darkness approached, they could feel the ship slipping away beneath them, and they abandoned the bailing lines to search for life preservers and pieces of the ship to keep them afloat.

Herndon ordered the men to cut off portions of the upper deck housing and stack them on deck, so they would be scattered about on the surface if the ship sank and men struggled to stay alive in the sea. One man tore the door off the wheelhouse, others ripped planks from the hurricane deck to lash together for a raft. They pulled doors off cabins and pried away the thick boards of the steerage berths and carried it all to the upper deck.

Weak from lack of food and sleep, exhausted from their labors, and discouraged by the losing hand fate was dealing, some of the men made no effort to secure a life preserver or a board, instead returning to their cabins or crawling into someone else's to await fate's final card.

One forty-niner was returning home penniless to a wife who had gone insane since he had left her eight years earlier. As the sky grew darker and the ship settled deeper into the sea, the man's bunkmate stood in steerage pleading with him to try to save himself. "When the critical moment arrived," said the bunkmate, "he refused to make any effort to escape."

In the same part of the ship with the poet Oliver Manlove were two brothers named Horn who had gone to California in 1850. Working together and working hard they had unearthed six thousand dollars' worth of gold, which they had kept in a large carpet sack that one or the other had guarded throughout the trip.

"I found Anson Horn weeping," wrote Manlove. "He said that his time had come, that he should never see his home again, which he had longed to see, praying and hoping for it. I tried to encourage him but he fully believed that his fate was sealed, that all of our fates were sealed, and that there was no use in fighting against it."

The thought that had intruded many times into the fear and exhaustion of the past two days came sharply into focus during those latter moments, for now they had to decide whether to take the gold or leave it behind. Most of the passengers were returning miners who had accumulated at least a few thousand dollars in gold, which they carried with them in treasure belts, carpetbags, and purses. But gold was dense. A red house brick weighed about four pounds; a gold brick of the same size weighed nearly fifty. Even in smaller amounts, gold could sink a weak swimmer or quickly exhaust a strong one. Yet some of the men had suffered great hardship since the summer of 1849 to accumulate the contents of that treasure belt or that carpetbag.

As if to dramatize the hysteria of such a dilemma, one man ripped open a bag containing twenty thousand dollars in gold dust and sprayed it about the main cabin as though he were a pixie and the gold were nothing more than tiny grains of sand. Others unhitched treasure belts, upended purses, and snapped open carpetbags, flinging the shiny coins and dust across the floor. "Hundreds of thousands of dollars were thus thrown away," said a passenger.

Badger himself had a satchel filled with 825 twenty-dollar, double-eagle gold pieces fresh from the San Francisco Mint; he retrieved these from his stateroom and, according to a witness, "flung them onto the floor" of the captain's cabin, telling the men to help themselves. But no one did. Purses filled with gold lay untouched. Amid the shouting and confusion, some men stood topside in a resigned daze and tossed gold coins at the wind.

Three or four hundred men waited on deck, while others remained in the cabins and corridors below. Darkness had closed in, and the *Central America* had settled so low in the water that now almost every wave broke over her. Bubbles, millions of them racing across the ship, formed and popped so fast as to make a hissing sound, and rising from below came the sound of timbers cracking and splintering.

"The scene among the passengers on deck and throughout the vessel was one of the most indescribable confusion and alarm," said one passenger. "The prayers of the pious and penitent, the curses of the maddened, and the groans and shrieks of the affrighted, were all commingled together, added to which were numerous angry contests between man and man, in many instances amounting to outright fight, for the possession of articles on which to keep themselves afloat in the water."

Badger ripped a board about six feet long and six inches wide off the front of a berth and made his way to the stern of the ship, there holding on to the stanchion that supported the after awning, ready to leap as far clear of the ship as he could when she settled. "At that time," he estimated, "there were two or three hundred on the quarter deck, breathlessly awaiting the final sinking."

Captain Herndon stood on the hurricane deck next to the wheelhouse, his trumpet in hand. The first officer, Second Officer Frazer, and Ansel Easton stood with him. When the ship had settled so deep that the water began to roll across her deck, Frazer got the captain a life preserver, and they set off more rockets. Some fizzled on the deck, others traveled upward only half as high as the smokestack. At 7:50, the captain and Frazer pulled themselves onto the deck above the wheelhouse and fired three rockets downward into the waves, a maritime distress signal to the schooner and the brig that the ship now was sinking rapidly.

Easton had secured neither a life preserver nor even a plank to float upon, but as he was standing with the captain, his friend Robert Brown came up with one of the superior cork life preservers and handed it to him. Brown had found two of them and had already donned one, and he insisted that Easton put on the other. As soon as Easton snapped the life preserver tight and threw his coat around his shoulder, buttoning it at the neck, the captain turned to him and said, "Give me your cigar, Easton, for this last rocket," and as Easton took the cigar from his mouth and handed it to the captain, a huge sea slammed into the ship, jarring every timber still in place. The ship pitched forward; Frazer looked over the side and saw the water spotted with men who had jumped to get clear of the ship before she went under. But the great mass of men remained on deck.

Then another heavy sea hit and made her pitch astern. Frazer saw a rocket go off to windward from the port paddle-box, horizontally, straight out over the ocean; an old miner yelled, "My God, we shall all perish!" and in a flash of lightning one passenger later swore he saw the entire deck and Captain Herndon standing upon the wheelhouse, his hat in hand, and within seconds a third and monstrous sea crashed across the deck of the *Central America,* breaking her, crushing her, sweeping her deck of nearly everything and everyone, lifting Frazer like a bobbin and carrying him back to midships and over the starboard side.

The stern sank below the waves, and the graceful arc of her bow aimed into the dark heavens, as she struggled, almost desperate to keep her proud head above water, and then as the hoarse screams of five hundred men rose, she began a slow watery spin, the water turning faster and faster and faster and faster, until the swirling vortex sucked the men into a suffocating darkness, deeper and deeper, cracking their ears, ripping the life vests from their bodies, tearing from their hands the planks and spars, sucking them deeper and deeper into the darkness, the pressure squeezing the air out of their lungs, salt water filling their noses and mouths and seeping into their eyes, their bodies twisting, the ship exploding all around them in the blackness, the pieces whirling, slamming into them, deeper and deeper and deeper, trapped in the vortex, entangled in the rigging, swallowing the salty water, their lungs filling, the last thoughts racing across their minds before the final darkness set in, descending with the once majestic steamer through the long column of black water, now possessed by her, and dead long before she crashed into the floor of the sea thousands of feet below.

Others shot upward through black water, bursting to the surface with a desperate gasp, struggling to breathe, coughing salt water, the night dark and the wind still fierce, the waves rolling over them, choking them, and suddenly rocketing upward from deep in the sea came the missiles from a battered ship, the spars, the hatch covers, the stateroom doors, the planking, the heavy timbers propelled up from the water and into the air before falling back with a heavy crash, to stun them, crush them, knock them unconscious, to kill many of those who had survived being sucked into the whirlpool, the surface of the sea in moments littered with torsos still wrapped tightly in tin life preservers, their arms and legs dangling from the surface, their heads pitched forward, and their hair spread across the water like seaweed.

"On rising again to the surface, the scenes presented to my view were horrifying," said one passenger. "Men, some holding planks, and others without anything, were tossed about through the sea for a great space, and appeared to me like so many corks. I could not describe my feelings at this awful moment."

Nearly every man sucked into the vortex had lost the timber in his hands, and those who found new ones upon surfacing often had them ripped away moments later by other desperate men. Many struggled in the water with bruised or broken limbs. Those who could not swim seized larger pieces of the deck, which soon were swamped, or they clung to the necks of those who could swim, pulling them under.

John Black recalled seeing "the heads of the drowning passengers like blackbirds on the water." His crew, for fear of being swamped by desperate men, refused to row closer.

One man grabbed hold of a door, but three men pulled it away from him. Then a trunk floated by and he drew it closer, but it fell apart. He then found a flour barrel and rode upon it for a while before finding a suitable board. "When I heard the waves coming," he said, "I would rise up and they would go over my shoulders."

From everywhere shrieks for help pierced the wind, as hundreds of human heads rose and fell with the waves, the cries of the men becoming an inarticulate wail. Some shouted for help from the brig *Marine,* which was far beyond hailing distance. Others screamed from their wounds or from their terror at being unable to swim. But the calls for help quickly began to subside, as the waves pulled them apart and some of the struggling forms now ceased to struggle.

"In ten minutes," said Thomas Badger, "three hundred had sunk to rise no more."

At first the waves had thrown them together, the living striking against the dead and the dying, the dying screaming for help. Every man gripped his plank so tightly as to paralyze his hands, but he was afraid to relax or the waves would rip the plank from his grasp and he soon would be as many of the others he saw around him, facedown in the water. "The scene presented can scarcely have a human parallel," said Obed Harvey, "hundreds of souls launched into the boundless sea and left at the mercy of the waves."

In the blackness, lit only by the occasional flicker of lightning, they floated, the swelling and cresting of the waves drowning calls of friendship and cries of despair, pulling them apart, then bringing them together in the trough of the sea, then separating them again at the top of the next wave. And at each successive flash of lightning, the men discovered fewer of their comrades around them, some now sinking through the black water, the waves scattering hither and yon those still afloat, until most found themselves drifting in the dark, on a slender plank, soaked, exhausted, and frightened. And alone amid the fury of a vast and indifferent sea.

"I guess I had been about four hours in the water," recalled one man, "and had floated away from the rest, when the waves ceased to make any noise, and I heard my mother say, 'Johnny, did you eat your sister's grapes?' I hadn't thought of it for twenty years at least. It had gone clean out of my mind. I had a sister that died of consumption more than thirty years ago, and when she was sick—I was a boy of eleven or so—a neighbor had sent her some early hot-house grapes. Well, those grapes were left in a room where I was, and I ought to have been skinned alive for it, little rascal that I was, I devoured them all. Mother came to me after I had gone to bed, when she couldn't find the fruit for sister to moisten her mouth with in the night, and said, 'Johnny, did you eat your sister's grapes?' I did not add to the meanness of this conduct by telling a lie. I owned up, and my mother went away in tears, but without flogging me. It occasioned me a qualm of conscience for many a year after; but as I said, for twenty years at least I had not thought of it, till when I was floating about benumbed with cold I heard it as plain as ever I heard a voice in my life, I heard my mother say, 'Johnny, did you eat your sister's grapes?'"

Far at sea, the men drifted in the night, the great swells from the storm lifting them high into the wind then dropping them again far below the next wall of water. Many rafts floated on the sea, hatch covers pulled off the ship or planking lashed together, and many men floated on each raft, clinging to ropes, their legs dangling in the water, sometimes so many men crowded onto the raft that the raft never rose higher than a foot beneath the surface. Storm waves rolled over them, immersing them time and again, and they swallowed the salty water, which caused them to choke and vomit. Often the waves twisted the rafts from their grip, flipping the boards into the air, sweeping them away, and the men had to swim in the dark to catch them again. But some of the men had muscles severely cramped, and even where the will still lived, the muscles stopped working, and the men perished with a cry for help within feet of the raft, as their comrades watched, powerless to pull them back on. Others simply quit, loosened their grip, and slipped quietly into the sea.

Those without comrades fought more than exhaustion, more than the fierce attempts of the sea to claim their souls. The waves that had thrown them together when first they surfaced, by the same motion had slowly pulled them apart, so that now as they floated in the dark, on a plank, on a door, their bodies immersed in water and shaking from the cold, they were alone. And the sum of all of their fears of deep water and storms, of high waves and darkness, could not equal their fear of loneliness, the fear of having to face all of the other fears "without seeing or hearing anything," recalled one man, "except the roaring of the dreadful storm or the faint cry of some of my companions in misfortune."

The vortex had pulled the shipboard poet Oliver Manlove down twenty feet into the black sea and ripped the life vest from his body. When he surfaced, he broke from the clutches of drowning men, and as he pulled away from the center of the wreckage, he met a friend who had two life preservers and who gave him one, and they each had grabbed pieces of the wreck, gripping tightly so the waves did not rip the boards from their hands, and as they floated on the surface of the sea, the waves began to pull them apart, until Manlove was "entirely alone drifting with the cruel and merciless waves I knew not where, often times covered up and pounded by them till I was near the end of my life. I would go down into the valley, and get nearly to the crest, when a wave would break over me and cover me up, I don't know how deep, and I would have to hold my breath till it rolled, or I could get to the surface again. It was a desperate fight for life and it seemed a hopeless one. There was no moon and I could only see a few rods. But once in a while from out of the black look of the night that shrouded the raging waves I could hear the call of a lonely voice, a wailing cry of a hopeless soul, for there was nothing to place a hope upon in such a place. I had to be ready when the waves would break over me, and hold my breath to keep from strangling, for the water was salty."

The men and the flotsam soon scattered across the ocean for a mile, and hope fled that lifeboats from the vessels they had spoken earlier would come to their rescue. About an hour after the ship sank, some of the men saw a light to leeward, that of the schooner *El Dorado,* but they knew that a ship to the lee could be of no help. The light soon disappeared, anyhow, and they turned and searched the darkness, facing into the wind, because if help were to come, it would come from that direction.

Occasionally, the wind-driven clouds parted, and they could see stars in the heavens above them, signaling perhaps that after four days, they finally were at the edge of the storm. This came seldom, but it provided for some the faintest hope that if they could survive the night, the morning might dawn fair and calm.

Besides the struggle to stay afloat, the men used a portion of their dwindling energy to cheer one another. Despite being wounded by debris in the water, the San Francisco vaudevillian Billy Birch clung to a hatch window with several other men, "as cool as a cucumber," said one of them. "To keep up their spirits he mimicked the sea monsters, told humorous stories in his own peculiar way, and on that frail bark, stretched on his back, bleeding from wounds, at midnight, tossed to and fro upon the angry waves of mid-ocean, he not only showed himself a true philosopher, but inspired courage in others."

As the night wore on, one by one more men ceased the fight and quietly succumbed to the sea, the last flicker of life beat out of them by exposure and exhaustion and the relentless pounding of the waves, their fingers suddenly going soft and spreading, their arms relaxing, their bodies slipping lower, the final moments of their lives spent unconscious, their companions watching helplessly as the sea claimed another of their own. Rafts that once held six or eight men now supported only three or four. Dead bodies wrapped in life vests floated by.

John George of England drifted in solitude, terrified by his loneliness and shouting himself hoarse to find a companion. Then in the darkness, he discerned a man wearing two life preservers and drifting toward him. He called to the other man and then paddled the best he could to meet him halfway. The other man said nothing but drifted nearer and nearer, until a wave threw them together, they touched, and suddenly George found himself staring into the face of a corpse.

Ansel Easton saw a number of men in the midst of their death agonies, and some he bade good-bye. Once he heard his name called in the dark, "Easton!"

The voice came from B. F. Parker, a merchant he knew from San Francisco, who upon coming closer said, "I cannot hold out but a few minutes longer. I am chilled, and I think we will both be in eternity in less than an hour. I hope we shall meet in heaven."

Ansel tried to encourage him and provided him a place on his board, but soon they shook hands and bade each other good-bye, and Ansel never saw the man again.

Until the last moment, Ansel had been on the hurricane deck next to Captain Herndon. When Robert Brown had brought him a cork life preserver, Ansel had removed his long coat, donned the preserver, and then thrown his coat over his shoulders, only fastening it about his neck. Then the third wave had hit the ship and a heavy arm had grabbed him round the neck, and the vortex rapidly had sucked both him and the first officer into the watery blackness. Twisting and pulling to get away from the first officer's panicked grip, he had reached to his throat and pulled open the one button keeping his coat about his neck, and as it slid away, so did the arm of the first officer. Then he had shot upward and found himself among hundreds of men floating with the debris, the subsea missiles still exploding to the surface. A large plank, once the front of a berth, floated by, and he grabbed it.

As storm waves carried him upward, he could turn his head away from the wind and see far to the lee the lights of the *Marine,* the brig that carried Addie and the other women. The thought that she would send a boat out in the morning encouraged him. In the dark he called out from time to time, and sometimes he would receive an answer to his call. He felt no alarm, or that he was going to drown, but his own thoughts seemed strange to him, as if he were delirious and at the same time aware of his delirium.

William Ede had been floating for about three hours when he became so lonely he would have been glad to feel again even the sinking deck of the steamer beneath his feet. He cried out, "Hallo! Hallo!"

Out of the blackness came a voice not far away. "Ahoy! Ahoy?"

"Who are you?" yelled Ede.

"Jack Lewis of Pine Grove, California," came the snappy answer.

Ede then was giving his name and home, when suddenly Jack Lewis of Pine Grove, California, shouted, "Where are you going to put up the night, Bill?"

Ede replied that only God knew. "How's everything with you?"

"All taut, partner," cracked Jack, and Ede heard nothing more, nor did he ever see the man again.

Dr. Obed Harvey had worked in the captain's quarters to help set the fractured arm and reduce the dislocated shoulder of a young man who had jumped from the ship and been crushed between the lifeboat and the steamer. As the water suddenly had risen throughout the ship, and the first hard wave crashed over her rail, Dr. Harvey and two others had carried the young man onto the hurricane deck. The next two waves came within seconds, and the vortex had submerged and separated the four men. Dr. Harvey never again saw the other three. When he surfaced, he found himself among hundreds of other men, struggling, and at the mercy of the waves.

"I have no words to describe the melancholy scene," Dr. Harvey reported later. "Soon after I had got to sea, I secured a small door, composed mostly of lattice work, or which appeared to me much like common window blind. This was of great service to me. When I left the ship, or rather when the ship left me, I had stripped myself of everything but a pair of pants and shirt, and although much disabled from a wound received on ship board, I firmly made up my mind that I could and would maintain myself above water till daylight. I saw and conversed with several of my acquaintances while in the water. I saw many perish, and was frequently drawn under the water by drowning men. The waves ran high, and frequently dashed over us."

It was now after midnight, and Dr. Harvey had been in the water for five hours, alone, when a man floated up alongside of him in a chair. The man was exhausted and placed his hand upon Dr. Harvey's floating door. At first, Dr. Harvey hesitated to aid, fearing that the door would not support both of them, and that if the man tried to come aboard, they both would perish. The man said his name was Frazer and that if he were lost at sea, he would leave a young family in New York destitute. Dr. Harvey then recognized the young man as the second officer.

"I told him to let his chair go, and share with me on my floating substance, and that we would sink or survive together."

In the hold of the *Marine*, hogsheads of molasses had burst and the molasses had run loose, sending thick, noxious odors wafting upward into the cabin, a room scarcely larger than an ordinary stateroom on the steamer—no more, estimated one woman, than eight feet square. Yet many of the thirty-one women and all of the twenty-six children huddled there, most of the children stuffed into the cabin's seven berths. The women had no dry clothes to cover their children, so they took sheets and made them into small garments, and their life preservers became pillows. The women themselves wore the only dry clothes they could find on the ship: red shirts and pea jackets and other garb shared by the sailors.

"Mrs. Marvin sat in a cupboard," remembered Almira Kittredge, "with a pair of gentleman's white pants on and a gentleman's coat on, looking just like a man. Mrs. Hawley had on a pair of gentleman's white drawers and socks, and a blanket having a hole cut in it, through which she put her head, wearing it a la Mexicana. Mrs. Badger got the captain's undershirt, his boots, and socks; this, with a large blanket wound round her, constituted her dress. Mrs. Easton wore the Captain's old hat."

Mrs. Kittredge had positioned herself next to the cabin door, never sleeping, the waves still breaking over the bow, still rushing across the deck and spilling through the doorway into the cabin. All night she sat in water up to her waist, while outside, "the sea broke over us, and the ship was tossed to and fro like a feather in a gale." And so they passed the night with no one but the children sleeping more than a few nods.

The men's sleeping area was a smaller space in the after hold loaded with sugar and tar, but it was so difficult to breathe there that most of the forty-one men rescued from the steamer gathered topside, using spare sails for their beds and lying cold and miserable, but content to be alive.

For Addie Easton and another woman, who chose to bed down on deck, Captain Burt spread a piece of sail across the hatch cover for them to lie on and then covered them with canvas. But Addie could not close her eyes without seeing and hearing the struggles of the drowning. "I reproached myself," she later wrote, "that I had not stayed to share my dear one's fate."

When the captain was not on duty, he would prop himself on his elbows next to her and try to quiet her fears for her husband. "The Captain's kindness I can never forget . . . ," she continued. "He is an intelligent true hearted, good soul yankee Captain, and very particularly kind to me, though he did everything he could for everybody. . . . He told us of the many wonderful rescues he had known and ended with the cheering words, 'Something tells me that you will meet your husbands when we get in port.'"

Late that night, with the sea still high but the wind beginning to drop, Captain Burt slowly worked the *Marine* back and forth, edging north to where the steamer had suddenly disappeared. But the storm had so thrashed the *Marine*'s rigging that she lacked the tautness to set before the wind. Captain Burt could only hope that the sea would push survivors and wreckage to him faster than it would force him away. But they saw no survivors and no wreckage, only the light of a schooner several miles distant.

Dawn brought light to a clearer sky and revealed a sea now rounded and rolling. The wind blew sharp but not with the fury of previous days. As the *Marine* creaked and tottered in the swells, the survivors breakfasted on hard crackers and tea passed from hand to hand in the same five cups. Captain Burt raised more sail, continuing to tack through the waters where, by the best he could calculate, they had last seen the *Cen-*

tral America the evening before. But after searching in the vicinity for hours, they saw not a trace of the carnage, no indication that the evening before five hundred souls had clung to a once proud vessel, three hundred feet long with stout masts and sturdy engines, while the sea ravaged her, beat her into splinters, crushed her, and sent her to the bottom with a roar unlike anything human ears had ever heard. As far as the horizon, gray waves rolled onward, incessantly, raking the site clean, as if only they had been there and only they would remain. About 2:00 P.M., Captain Burt decided that he had searched as long as he could, that he had a hundred passengers now who needed food and water and to be off the sea. He made all the sail he could and bore up for Norfolk.

Columbus, Ohio

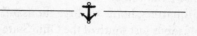

1981

Battelle Memorial Institute, a privately funded, not-for-profit facility engaged in contract research for government and private industry, had opened its doors in 1929. One of the terms of Gordon Battelle's will was that the institute would be "for the making of discoveries and inventions."

During World War II, five hundred Battelle scientists worked on the Manhattan Project to develop the atomic bomb. In the 1940s and 1950s Battelle scientists helped a nighttime inventor named Chester Carlson refine a crude process that lay concealed inside a dirty wooden box and required fourteen manual operations. Then Battelle and Carlson licensed the process to a small company called Haloid, which in 1961 changed its name to Xerox. In the mid-1960s, scientists at

Battelle created for the Treasury Department the "sandwich coin," bonding cupro-nickel layers to copper cores, eliminating the need for silver. Battelle scientists developed the insulating tiles that protect NASA space vehicles from incinerating when they reenter the earth's atmosphere.

When he was no more than twelve, Tommy Thompson had run downstairs one day and showed his mother a newspaper article about Battelle. "It's a scientific research organization," he told his mother, "and that's exactly what I want to do."

Battelle headquarters, home to three thousand scientists and engineers, was in Columbus just south of the Ohio State campus. Don Frink headed Battelle's Equipment Development Section, 60 percent of whose research and development was for the government, the majority of that work for the Department of Defense.

"Our group does oddball things," said Frink, "in space or underground or behind enemy lines or underwater. A large portion of the work is ocean engineering, which is even more unique than standard contract research."

Each year, out of two hundred applications from top engineering graduates across the country, Frink would interview about twenty and give offers to five. Four would accept. Frink was looking for the young engineer who had worked in a garage or a shop, or was raised on a farm, where a tractor busted in the afternoon had to be back out in the field by sunrise—no time to order parts, just improvise with whatever they could find in the barn. Battelle engineers didn't wallow in theory, they had to put the thing on a table, flip the switch, and have it perform.

In the spring of 1981, Tommy had finished his perambulations about the country in search of experience and knowledge and returned to Columbus. He arranged an audience with Don Frink, and Frink thought the young man fit his formula for new engineers—SWAN, he called it: Smart, Willing to work hard, Ambitious, Nice to work with. Battelle was big and had its rules and procedures, but Frink liked to provide an atmosphere where an inventor's soul could flourish, and he liked his engineers with a strong entrepreneurial bent. But he wondered if Tommy was too entrepreneurial even for him, if after he had brought Tommy along and invested a lot of money in him, Tommy suddenly would be gone, off pursuing something on his own.

"You're gonna come here a few years," said Frink, "and then the bug's gonna bite you again and you're gonna take off, 'I gave it my five-year shot and I'm done.'"

"No," said Tommy, "I have always wanted to work here."

Frink called Tommy's professors at Ohio State, and they all confirmed that Tommy fit the SWAN formula. "His IQ has gotta be sky high," said Frink, "although he might not test very well. He didn't graduate at the top of his class, but he made the Dean's List, and he impressed almost every one of his professors. They said, 'That guy's going somewhere, I don't know where, but he's going somewhere.' It was the fact that he was interested in the why behind everything."

A week after their first meeting, Tommy called again and set up another interview, and this time, Frink invited four of his engineers. One was Don Hackman, the world's foremost expert in underwater tooling. Tommy had read about Hackman and his work.

Hackman liked to phrase his questions to a new engineer so that only someone who had built things to work in the ocean could answer the question. "Have you designed underwater equipment?" he would ask. If they said yes, he wanted to know what kinds of materials they normally used, and he was waiting to hear, "I use either 316 or 304 aluminum." All the young engineers said they knew their way around a machine shop, but Hackman wanted to know if they preferred a vertical mill or a horizontal mill. "Little questions like that," said Hackman, "you can tell whether a person's done it."

Tommy was quick with his answers. When Hackman asked him about the hydroflow he had used on the *Arbutus,* Tommy not only could explain how he designed the suspension system, he could also describe what it felt like to be inside the thing with his skin undulating like a flag in the turbulence and his mask ripping off his face. He could talk firsthand about the dangers of manned submersibles and the problems with magnetometers and the latest theories on treating the bends in decompression chambers.

"If you've already done something like that at sea," said Frink, "you're going to impress the hell out of Hackman. Which is what Tommy did. Everybody voted they had to have Tom."

But Frink was still concerned. "People like him are hard to work with," worried Frink. "He's a highly creative, almost driven individual. His work hours are not normal. He'll work your buns off. He'll burn out people."

Frink's biggest worry was, Do I have 100 percent of Tommy? Because Frink needed his engineers, as he was fond of saying, "body and soul, twenty-four hours a day, seven days a week. Your subconscious has gotta be working for Battelle," said Frink. Later he admitted, "I got more out of Tommy than I thought. Four years, five years, he stayed here."

As Frink worried over how much of Tommy's brain would be engaged on Battelle matters and what portion might wander elsewhere, one thought never occurred to him: that before long he and Don Hackman both would be working for Tommy.

In one of his frequent trips from Key West back to Columbus, Tommy had met an eclectic fellow named Bob Evans, a university student majoring in geology. Bob was trained and accomplished as a classical pianist, soon to be a jazz pianist, and was a collector of odd bits of information. He seemed to remember everything he had ever read, and his mind made so many connections so quickly that listening to him talk was like watching time-lapse photography. He also was good natured and unpretentious. "There's all kinds of jokes," he once said, "double entendres about the prowess of people in different occupations, like Divers Do It Deeper. Well, Piano Players Play with Themselves. They don't put that one on coasters."

Bob lived on one side of an old Victorian double near campus, and some of Tommy's friends lived on the other side. Tommy and Bob had met through the sealed door in the basement during the blizzard of '78, as Tommy was trying to get more heat to the other side of the house. They had talked through the door for half an hour but did not see each other until the following spring. Bob told a friend he had just met "an interesting individual with no face."

Over the next three years, Tommy and Bob had visited from time to time as Tommy rolled into Columbus from Key West or half a dozen other places. Bob had graduated from Ohio State with his degree in geology and was consulting for the oil industry in Ohio. Now back in Columbus permanently, Tommy frequently putt-putted on his Zundaap scooter over to see Bob.

Bob loved to talk. He could get so cranked up, his shoulder-length blond hair would bounce and his droopy mustache would jiggle. He and Tommy would sit on Bob's front porch, Tommy oiling his tongue with tequila and Bob loosening his with a more philosophical beverage like Wild Turkey neat. So lubricated, those tongues would wag far into the night, one seemingly dueling with the other, competing crescendos rising to such levels that friends present sometimes had to press hard on the stiff little flap of skin at the front of their ears. They'd talk about crazy things, like diamond mines in Canada and treasure troves buried by Indians and Mayan temples covered by water off the Yucatán. "Bob and Harvey, when you get them together," said a friend, "become like one person in a funny way, their minds kind of meld and they know what the other is thinking and they get excited. They were always excited."

Tommy told Bob about working with Mel Fisher off the Keys and John Doering down in Dominica. He told Bob there were places in the middle of the Caribbean where hundreds of old ships lay stacked on top of each other. That was one of the problems: How do you tell the ship you're looking for from all of the other ships? They all had crashed in a shallow, high-energy environment exploding with storms and monster waves and currents that whipped the artifacts together like an egg beater. Tommy called it the "junkyard effect."

And how do you know, he asked Bob, that the wreck you're looking for hasn't already been salvaged anyhow? Treasure hunters would research the king's records in the Spanish archives, find nothing about a recovery, and assume the treasure was still at the site. But not everybody reported everything to the king, explained Tommy. There was contraband on those galleons. Or six months later, someone's sailing by and sees wreckage on an island, and they say, Let's find out what's going on, so they talk to the Indians, and they dive on the ship. "A galleon drafted about fifteen feet," Tommy told Bob, "so they generally hit reefs in about fifteen feet of water. It is not like men to leave gold lying in fifteen feet of water." Most of the artifacts Fisher had found were at twelve feet, and the only reason Spanish salvage divers had not completely stripped the *Atocha* in 1622 is because a second, far bigger storm had hit the wreck site three weeks later.

Treasure hunters had no control; that was the problem. They were subject to too many whims, of weather, of history, of government, of human nature, and that is why almost all of them failed. To succeed, they needed more control. They needed to analyze and reduce their risk. The treasure was there. During the three centuries following Columbus's voyages to the New World, much of the gold and silver on earth had been transferred from the New World to the Old World, and 25 percent of it had been lost. But don't search for it among the thousands of shallow-water shipwrecks in the Caribbean, said Tommy; the odds were too slim. Search for treasure where storms couldn't buffet the remains, where ships were not piled on top of each other, where the bottom was hard and the currents slow, and where no government could stake a claim. Tommy told Bob he wanted to recover historic shipwrecks in the deep ocean.

In the mornings Tommy reported to the Equipment Development Section at Battelle, where Don Frink quickly saw a side of him he had not guessed was there: his ability to impress sophisticated clients, like the supervisor of salvage in Washington, D.C., who was in charge of creating and developing dive and salvage equipment for the navy; or the Department of Interior, which was interested in deep-ocean mining. "Tommy was a good salesman for them," said Frink. "He meets people well, he impresses the client, and these are people that know, and they generally know more than you know. He comes across as sincere, as What can I do to help you? And he's got the knowledge to back it up. He was energetic, aggressive, active, and he made you feel good to be with him."

Frink had wanted to use Tommy as a designer, developing equipment for the navy. But he saw a better use for him up one notch: looking at an overall systems approach to solving problems. The government appeared ready to spend huge sums on deep-ocean mineral recovery, so Frink assigned Tommy to work on the concept of mining the deep ocean. Watching Tommy work, Frink noticed two things: how he carefully, almost maniacally, rationed his time, using tight priority lists; and how he amassed, assimilated, integrated, and turned out volumes of detail.

Tommy put in more hours than anyone at Battelle on the feasibility of mining the deep ocean. He talked to scientists and engineers around the country, trying to create a system that made sense economically and environmentally. He concluded that to mine the deep ocean would require "radically new technologies," and that although they could create the technology, it would cost so much to develop the equipment and to operate it, that the value of the minerals would have to increase significantly to make it pay.

Tommy worked ten- to twelve-hour days and sometimes spent weekends at Battelle, but at night he continued thinking about ships lost in the deep ocean. There, all of the other problems with trying to recover historic shipwrecks went away, and the greatest problem remaining was technology. "And if I can figure out how to solve that," he thought, "at least I'm throwing the problems back into my own hands, which I can control." His research on deep-ocean mining had given him additional ideas on how to recover a deep-water shipwreck. Although the value of minerals would have to rise dramatically to warrant the cost of developing the technology to mine the deep ocean, a single ship with a large payload might be enough to persuade investors to give him the money he would need to create the technology that could penetrate that deep-ocean barrier.

Often ideas would come to him as he was falling asleep late at night, or he would awaken suddenly, his mind flooding with what-ifs and I-wonders. "Something might occur to me, then all of a sudden I can't stop and it's three in the morning and I've got to get up at six or seven. But when it starts coming, I can't afford to disrespect it or disregard it. It'll never look the same."

He would get out his notebooks filled with new and old ideas, his notes from phone conversations, and he would make the connections and ponder the possibilities.

* * *

THE FIRST TECHNOLOGY for working on the bottom of the ocean had come in the seventeenth century with the diving bell, a large air bubble trapped inside a housing. Sir Edmund Halley, discoverer of the comet, created a diving bell with glass viewing ports and freshened the air supply by lowering barrels of air that siphoned into the bell. Three men could remain inside at sixty feet for an hour and forty-five minutes, and they could salvage the armament of a ship sunk in the harbor by securing it with ropes and chains and letting another crew haul it to the surface with ship's tackle. Three hundred years later, technology used in the deep ocean had advanced dramatically, but the work that could be accomplished there had changed little.

Prior to 1963, when the nuclear submarine *Thresher* had sunk in over eight thousand feet of water, navy scientists had already begun to lobby the Pentagon for a deep-water submersible that not only could go much deeper than submarines could go, but also could hold three people, have cameras and large viewing ports and a mechanical claw, and be able to move across the underwater landscape. Although the scientists had finally persuaded the navy to fund the project, the vessel was seen as the bastard child in the military fleet—no navy contractor would agree to build it. The navy had to go to a cereal company, General Mills, to build the tiny submersible, which would be called *Alvin*.

Two weeks after the *Thresher* tragedy, the secretary of the navy appointed a panel called the Deep Submergence Systems Review Group. Known as the Stephan Panel, the fifty-eight appointees had spent almost a year analyzing the military's deep-ocean capability and recommending technology to explore the deep ocean and to recover artifacts lying on its floor. Within three months of the report, in June 1964, the world's first maneuverable deep-water submersible, the *Alvin,* was christened in the presence of a lot of starched-white navy brass. Two years later, the navy gave the *Alvin* its first real test: to help defuse the most delicate and potentially deadly international incident in history.

An American Air Force B-52 bomber, flying a routine mission over the Mediterranean, had collided with an airborne tanker during refueling off Palomares, Spain, and the fiery wreckage had fallen thirty thousand feet, scattering across an area of land and sea ten square miles. In that wreckage were four hydrogen bombs, each with seventy times more explosive power than the bomb dropped on Hiroshima, or enough to waste much of Spain and Portugal, parts of North Africa, and perhaps most of southern France. Authorities found three of the bombs on land, but the fourth parachuted into the sea. After divers searched the shallower water and found nothing, the navy brought in two deep-water submersibles, the new *Aluminaut,* a seventy-eight-ton behemoth good to eight thousand feet, and the two-year-old *Alvin,* capable of reaching six thousand feet. The admiral in charge described the search as like trying to find "the *eye* of the needle in a *field* of haystacks. In the dark."

The *Alvin* dived ten times before the crew even spied the parachute billowing in the current at twenty-eight hundred feet. And when the mother ship dropped grappling hooks for the pilots to wrap among the parachute shrouds, the bulky and unstable *Alvin* and its awkward manipulator made the attempt seem, as one of the pilots put it, "like a drunken Swede trying to eat spaghetti with chopsticks." Twice they hooked on to the bomb and tried to winch it up, but both times they dropped it and couldn't find it again, once for nine days. After the parachute engulfed and nearly became the death shroud for the *Alvin* and her crew, the navy brought in its new ultra equipment, an underwater robot called CURV, for Cable-controlled Underwater Research Vessel, a sled equipped with lights, cameras, and a claw. But even CURV could not recover the bomb. When the claw moved, silt roiled from the bottom and blocked the cameras for fifteen minutes, then the parachute suddenly billowed and wrapped itself around CURV like a straitjacket. The robot couldn't move. Finally, the crew topside winched the robot back to the surface, and the parachute and the bomb came with it.

The navy had used the most sophisticated underwater technology in the world to defuse the most politically sensitive situation anyone could imagine. Cost was irrelevant. Yet the search and recovery in a narrowly defined space had taken almost three months, and as arduous and tense and dangerous as the recovery was, no one could explain its success without including a large dollop of luck.

After the *Thresher* went down, the navy had begun designing and building an array of underwater vehicles that could descend to the bottom of the deep ocean. Besides the *Alvin* and the *Aluminaut,* the navy built *Sea Cliff* and *Turtle,* which were bigger, slightly faster versions of *Alvin; Halibut,* a submarine with a hangar-sized bay in its belly, which could cruise undetected several hundred feet below the surface while a crew lowered cables and cameras to the seafloor; a small nuclear-powered submarine called the *NR-1,* which could remain submerged for weeks, roll along the bottom on wheels, and came with lights, cameras, viewing ports, and manipulators; and the DSRVs, or Deep Submergence Rescue Vehicles, which could descend to sixty-five hundred feet, clamp on to a downed submarine, and take on the trapped submariners and transfer them to the surface. The original cost estimate for one DSRV was $3 million; eventually, the navy built two for $220 million in the early 1970s.

During the Cold War '60s and '70s, more and more of the funding for these deep-water submersibles came not from the navy, but from a group with even deeper pockets, the intelligence community. The most notable was the 1974 ultrasecret *Glomar Explorer,* a six-hundred-foot ship created under the ruse of a Howard Hughes attempt to mine the deep ocean but really with only one mission: to send a giant mechanical claw into twenty thousand feet of water and pick up part of the hull of a downed Soviet submarine. In raising the nearly two-hundred-foot section of hull to the surface, the claw cracked, and most of the section was lost. The project cost half a billion dollars.

Twenty years after the *Thresher* had gone down, we now had submersibles that could dive much deeper and stay much longer than Halley's diving bell. They had gauged currents in the Gulf Stream, located scallop beds, recovered spent torpedoes and missiles, explored manganese deposits, studied geology, and inspected offshore drilling rigs. But if you stripped away the computerized guidance systems, the propulsion systems, the fresh air recyclers, the hydraulics, the new acrylics and seals, the sonar, and the cameras, these new submersibles could do little more than attach grappling hooks to objects and let a crew topside winch them to the surface, or grasp something in an awkward claw and hold on.

* * *

Almost any ocean engineer who wanted to recover a historic shipwreck in deep water could drop a steam-shovel clambucket on the site and indiscriminately munch it up. But things would break, specimens would be crushed, the archaeological value would be destroyed, and the treasure would be scarred and devalued, or lost again and never found. Historic shipwrecks were so complex and difficult to read, it would be easy to pile overburden on top of the treasure, burying it even deeper. Tommy wasn't thinking about clamping a big claw around a ship's hull, or draping a line around a part, or dropping a clambucket on a wreck site. He wanted to explore and document a wreck, then dismantle it piece by piece, like pick-up-sticks, moving one piece without disturbing another. He wanted to recover delicate objects and preserve them on their way to the surface, and he wanted to film and photograph it all with 35mm cameras and video cameras, black-and-white, color, and 3-D. He once said, "You have to do it smart, like a surgeon."

He envisioned an automated machine shop that he could operate from thousands of feet above; the ultimate Swiss Army knife of underwater technology, a tool for everything with but a flick of the topside wrist: saws, grabbers, backhoes, drills, blowers, pickers, and camera and light booms. Given time and money, knowledgeable scientists and engineers could design and build a robot that could do all of these things on land; they were only sophisticated technological gadgets, and that was the easy part. The secret to making them all work in the deep ocean was in the back end, away from the sophisticated technology, all the way back to the concept itself. In 1983, we had submersibles that looked like sharks, bullets, grasshoppers, tugboats, and blimps, many minds with many solutions, and not one on the bottom could do any more than Halley had done with the diving bell in the seventeenth century. The problems remained, and after ten years, Tommy at least had sorted them out, so he could see them clearly.

They began at the surface: You had to have some way of getting your submersible off the ship and into the water. But winds of only ten knots pushed the sea into three-foot waves, which slapped against the submersible as you tried to lower it through the air-sea interface. That three-foot wall of water with the weight of the ocean behind it ripped off manipulators and sometimes mashed the submersible's hull. At the first test launch of the prototype for *Trieste,* the French-built bathyscaphe used to find the *Thresher,* the waves had kicked up only a little, destroyed the gasoline-filled flotation chamber, and spewed almost nineteen thousand gallons of gasoline into the sea. You couldn't launch or recover your submersible in seas greater than three feet or you risked losing it. And not often did blue water lie calmer than that.

If you got your submersible safely into the water, your ship at the surface was rising and falling while your submersible was descending; each fall caused the cable to go slack, and each rise snapped the cable taut, like pulling a car with a chain. That load suddenly became ten times heavier than the submersible itself, and the cable often broke and you lost your submersible. That armored cable was filled with electromechanical wires that carried signals down to the sub and back again. If the snap loading didn't break it, every time that cable passed over a pulley, the wires bent and straightened with the weight of the vehicle, and often ten times the weight of the vehicle, and the wires fatigued and parted. A replacement cable took three months to manufacture, and carrying a spare cable on board meant needing more space on a bigger ship, tended by a larger crew, for much more money.

Attempting to land on the seafloor was risky and difficult for two reasons: First, the rocking of the ship would jerk the vehicle—one minute you'd be looking at the bottom, the next minute you'd see nothing, the next minute the camera would be in the mud. Second, hanging something heavy on the end of a cable twisted the cable; if you set that heavy weight on the seafloor and slackened the cable at the same time, the twisted cable tied itself in knots, like the cord on a telephone. When an armored cable with several thousand pounds on the end kinked up, and the bouncing of the ship topside jerked on those kinks, the cable again often broke, which meant you left your vehicle on the bottom and headed back to the beach for the rest of the season.

One way around the problems with impact loading and snap loading and cable fatigue and twisting and breaking was to pack all of your power on board the sub, forgo communication, and put humans inside, let them drive around at will, like the *Alvin* or the newer *Trieste.* But this put lives at risk, so every system had to have a backup; 90 percent of the engineering would go to designing redundancies, and the vehicle would have to be much heavier. Often, an entire mother ship had to be built around the submersible, driving the cost into the tens of millions of dollars. Tommy's attempt would not be a government project, national security would not be at stake, and he would not have an unlimited budget. Whatever technology he created would have to be done as cheaply as possible, a few million for the equipment, a few million more for the rest of the project.

Others in the deep-ocean community already had seen the limited future of manned submersibles. The navy was experimenting with "autonomous vehicles," because they were much lighter and far less dangerous than the manned submersibles, and they could be programmed ahead of time to go to the bottom and perform simple tasks. The French-built *Epaulard* had been the first, but it could only work on a flat, known bottom; it couldn't react to the terrain; it could shoot film and take photographs, but it couldn't send back real-time information, so the results would not be known until the film was developed at the surface. If the operators then saw something in the photos, trying to find their way back to that point would be nearly impossible. Tommy wanted something that could stay, that he could control, that would tell him what was happening as it happened, so he could make intelligent decisions from the surface.

The real problem with every system, manned or unmanned, tethered or untethered, was that it couldn't perform significant work tasks on the bottom. And that problem arose because no one had been able to overcome this fact: Submersibles were unstable. To allow them to "float" underwater, they had a narrow, unchangeable center of buoyancy, and their manipulators had to be extremely short; extending them shifted the center of buoyancy and tipped the vehicle. When you sit at the dinner table and reach out to pick up a bowl of peas, you unconsciously tense certain muscles and shift your center of gravity. Submersibles couldn't do that. If you were a submersible, as soon as your hand tried to lift the bowl, your face would drop into your plate. Even if a manipulator had a short reach that would not tip the sub, it couldn't do anything that required force, or an equal force transferred to the vehicle and turned it upside down. The *Alvin* weighed seventeen metric tons, but at the bottom it had no muscle.

Tommy already had eliminated systems that required the presence of humans on the bottom, anyhow. They were too expensive, too dangerous, too limited. "I figured that the secret was to build a stable unmanned system that could work on the bottom for days at a time with as many mechanical functions as possible." A robot, an underwater Remote Operated Vehicle, or ROV. The oil industry was starting to use them to replace divers, and the military had used them underwater for years. In 1982 only ten existed, and they still presented all of the cable problems with launch and recovery and trying to land on the bottom, and none that Tommy knew of could really work down there. But those were problems Tommy now thought he understood and could solve, because the secret to working in the deep ocean was not in the technology, where everyone else had been looking. Most of the technological pieces were there; they just hadn't been put together properly. The secret was in the concept behind the system and in the interrelation of all of the subsystems, and the keys that would reveal the secret lay in Tommy's innate, insatiable, sometimes irritating need to know why two plus two equals four. What had driven teachers and friends to apoplexy when he was a boy would be at the core of what enabled him as a man to begin dismantling a series of barriers to working in the deep ocean, to examine the pieces, to understand them, and to proceed toward what others had thought impossible. The secret was in his naive, at times arrogant insistence on the absolute simplicity of the quest.

BEFORE HE COULD do anything on the bottom, Tommy first had to find the ship; that was the other difficult technology: the ability to image things lost at sea under thousands of feet of water. Historical documents for any deep-water ship would only reconstruct an approximate location of the sinking; the error could be fifty miles in any direction. To be sure he could locate the site, Tommy would have to sweep an area of ocean so large that traditional sonar would require years of summer weather, dragging a tow fish back and forth.

In 1977, engineers working for the mining consortium that had surveyed the deep ocean had also developed a vacuum that could suck up potato-sized nodules of manganese from the fields they imaged. After dragging their vacuum along the ocean floor, the techs had returned for a second look with their imaging system, and they had seen a little white band in the middle of the manganese field: a stripe created by the vacuum sucking up the nodules. The water there was eighteen thousand feet deep, and that white stripe was only six feet wide. And the signal bouncing back from below was much stronger than it needed to be. They could open it up and search over three miles of ocean floor in one track.

Three years later, Columbia University's Lamont-Doherty Geological Institute had received funds from a benefactor to build the Sea-MARC I, a sonar prototype by the same engineers who had conceived the mining consortium's high-speed exploration system. Lamont-Doherty wanted the sonar to survey underwater mountain ranges and other large geological formations in deep water. Tommy understood the technology, and he thought that with some adjustments he could use it to find shipwrecks in the deep ocean, but the Lamont SeaMARC was already under contract with various organizations for the next two years during the summer weather window. Tommy also discovered that since Lamont-Doherty was a public institution, any information on deepwater shipwrecks he collected while using its SeaMARC automatically entered the public domain.

In five years of traveling and talking on the phone, Tommy had built a large and diverse network of scientists, engineers, and oceanographers. In 1983, one of those contacts introduced him to Mike Williamson, the geophysicist who had sailed with the mining consortium tech team to find manganese. Williamson still remembered the day six years earlier, when they had seen that thin white stripe cutting through a manganese field. He couldn't believe they could image something that small, that deep. And they could do it in swaths three miles wide. "You could really start herding up the real estate," thought Williamson. If he had one of those imaging systems, he could quickly search large areas of deep ocean, not for underwater mountains or manganese fields, but for downed aircraft, flight recorders, bombs, missile parts, and finer geology for the oil companies. The only thing holding him back was the million-dollar price tag.

By 1983, Williamson had started an ocean technology company and sold it and started a new one, and he still wanted that million-dollar sonar. "I was going to get a million bucks," said Williamson, "and we'd get one built and go off and do all sorts of things." About this time, Tommy called him and encouraged him to go ahead with the project, even to the point of helping him find financing.

To Williamson, Tommy Thompson was a treasure hunter, and Williamson ran with a different crowd, what a friend called the "black community," the intelligence people. "Williamson actually has a lot of clout," said the friend. "He's a pretty big genius in that community." And Williamson did not work with treasure hunters.

"They're generally a flash in the pan," he said, "a lot of talk and no dollars, and we considered Harvey the same way. We were interested in the project but certainly not willing to jump in and share his enthusiasm without seeing a little long green."

But Tommy pursued Williamson the same way he had serenaded other suppliers who consulted for the government and large corporations: He kept in touch, and he asked intelligent questions. And he knew that Williamson wanted to get a new SeaMARC, the IA, built just as much as he, Tommy, wanted to use it to find deep-water shipwrecks.

By the fall of 1983, Tommy had talked with Williamson frequently by phone and had met with him three times in Seattle to talk about deep-water side-scan sonar. Williamson explained his ideas on how to turn the current SeaMARC technology into a more efficient side scan, and Tommy knew the ideas were sound. It was the new-generation technology he had been looking for, and suddenly he could see all of the pieces. "I started to realize that the technology wasn't going to be in the year 2000, that with a lot of effort and the right group we could make it happen. I finally decided that the time was now."

TOMMY NOW ALLOCATED more and more of his free hours each month to studying ships that had sunk in deep water, like the *Titanic,* the *Republic,* the *Andrea Doria,* the *San José.* For each ship he wanted to know: Was there enough historical documentation to determine that the ship carried a cargo of substantial value when it sank, and that it sank in a roughly definable location? If the answer to these questions was yes, he then studied the ocean environment at that location. Environment was critical. If the ocean floor was deep sediment, the ship would be buried, and the sediment would fill in the site faster than he could dig it out; if the currents ran swift, he would not be able to study the site carefully and later position an ROV with cameras. By subjecting potential sites to such scrutiny, he could determine which ship had the greatest probability of being recovered.

One of the ships he researched was a sidewheel steamer called the SS *Central America*. For years in treasure hunting lore, the story had persisted that the *Central America* lay off Cape Hatteras a hundred feet deep. Treasure hunters with no more than scuba tanks and huka rigs found it a convenient legend to keep alive, because if that legend died so did their source of funding. They had raised millions, showing investors a World War II navy map on which was charted all of the debris near the cape, and it had arbitrarily labeled one of the sites "the *Central America*." One group had found wreckage in ninety feet of water, stanchions, a spokesman claimed, that could be traced to mid-nineteenth-century steamships. Not far away, a different outfit had found another site, and the UPI reported that artifacts found at that site "are being tested by a Florida company to determine if they are from a steamship that sank in 1857 with a cargo of gold." In the spring of 1983, a magazine piece claimed that although a few treasure hunters speculated that the *Central America* had sunk on the eastern edge of the Gulf Stream in about 105 fathoms or 900 feet of water, the bigger school of thought pegged the sinking at the western edge and hence down no more than 20 fathoms or 120 feet. "The latter argument may be valid," concluded the article. "A treasure-hunting concern recently announced that they found the *Central America* in 10 fathoms of water near Cape Hatteras."

But Tommy had read many of the newspaper articles written in 1857, then plotted the information from one officer's account, and estimated that the ship had lost power and foundered at a coordinate roughly one hundred miles east and north of Cape Romain on the Carolina coast.

One evening after work, he stopped by Bob's house on the way home. If the *Central America* lay one hundred miles from Cape Romain, it would be in the middle of a geographical feature called the Blake Ridge. Tommy wanted to know what the bottom was like out there, because a soft bottom of rapidly accumulating sediment would complicate his search. Bob didn't know the answer, but he told Tommy he would find out what he could at the geology library on campus.

What Bob read about the Blake Ridge so excited him he called Tommy at work, and the two of them met at Bob's house that night. "Harvey, this is not a problem at all," he said. On the Blake Ridge, currents moved by at one-tenth of a knot, and for miles the bottom was hard flat sediment. Bob told Tommy, "I mean the sedimentation collects out there no more than one centimeter every thousand years!"

OUTSIDE BATTELLE, TOMMY was now amassing voluminous notes on underwater technology, beginning to formulate relationships with suppliers, and corresponding with historical archives at several libraries on the East Coast. For years he had collected information on deep-water, historic shipwrecks, and the list had grown to forty. He and Bob met more frequently, together refining what they called the Historic Shipwreck Selection Process and narrowing the targets to a project Tommy could present to investors. "We developed the language as we went along," said Bob, "the selection criteria for projects in general, and then we analyzed the risks involved with each ship."

They divided risk into intrinsic and extrinsic. Intrinsic risks were those inherent to the site: probability of previous recovery, accuracy of historical documentation, and the environment around the site. All deep-water shipwrecks scored high in the first category; most of them scored high in the second category; few of them did well in the third. Shipwrecks with a high total score then advanced to form a universe of "Feasibly Recoverable Shipwrecks with Low Intrinsic Risk."

Next, they assessed the extrinsic risks, those that had to do with recovery: Favorable Operational Factors, Positive Site Security, Legal Rights Obtainable. Is the technology available to access that site, can we guarantee site security in that area of the world, and do we have legal protection?

Once they had eliminated all ships but those with low intrinsic and low extrinsic risks, each ship had to pass a final test: Was there anything on board worth recovering?

The *Titanic* was a hunk of steel seven hundred feet long that would burn a hole through a sonar chart; even if it rested in mountainous territory, they could probably find it, and the abundant historical documentation would help them narrow the search area. But the *Titanic* presented two insurmountable risks: Her steel hull would be impossible to penetrate even with the technology Tommy saw on the horizon. And if they could get inside, she carried nothing worth recovering; some loose jewelry perhaps, rings and bracelets and necklaces scattered in various small cubicles, but no treasure centrally stored, nothing they could use to make the payoff attractive to investors.

"In terms of financial risk," said Bob, "the *Titanic* was not a good project."

Other deep-water ships presented similar problems. Myths had arisen around some of them that tons of gold lay stored in secure compartments. But no historical data supported the myths. In 1909, the British White Star luxury liner *Republic* had gone down fifty miles off Nantucket, and for decades, rumors had circulated that it had taken millions in gold coins with it. But no official records existed. "Sure, there were a lot of rich people on board," said Bob, "but how much was in the purser's safe? Nobody knows."

The *Andrea Doria,* an Italian liner hailed by her owners as the "Grande Dame of the Sea," collided with another ship in dense fog in 1956 and also went down just off Nantucket. She was a glistening seven-hundred-foot floating museum of murals, rare wood panels, and ceramics designed by Italian artists, and her passengers also were wealthy, but once again myth about the treasure on board sprouted from rumor with no documentation.

Tommy and Bob were convinced that the *San José* had carried more than a billion dollars in treasure to the bottom when British warships landed a cannonball in her munitions cache and sank her in 1708. But the *San José* was off the coast of Colombia in murky, turbulent waters.

After many deep-water shipwrecks were run through the selection process, the sidewheel steamer SS *Central America* rose to the top in every category. It had sunk in an era of accurate record keeping and reliable navigation instruments. Dozens of witnesses had testified to the sinking, and five ship captains had given coordinates that placed the ship in an area where sediment collected no faster than a centimeter every thousand years. The extrinsic risks looked as favorable: She had a wooden hull, which would be easier to get into, and massive iron works in her steam engines and boilers that would provide a good target for sonar, even if much of the iron had corroded and disappeared. And it was off the coast of the United States, so they wouldn't have to negotiate with a foreign government and they could more easily provide site security.

One other thing appealed to Tommy and Bob: the ship was American and its treasure symbolized one of the most defining periods in American history, that narrow window running from the California Gold Rush through the Civil War. If they could find it, they would open a time capsule representing an entire nation during a crucial period in its formation.

"The *Central America,*" said Bob, "scored much, much higher than any other project when subjected to this selection process."

And her gold shipment was documented: With gold valued at $20 an ounce in 1857, the publicly reported commercial shipment totaled between $1.210 and $1.6 million. Although many of the *Central America*'s records, including her cargo manifest, had been destroyed in the Great San Francisco Fire of 1906, some accounts estimated that the gold carried by the passengers at least equaled the commercial shipment. And the Department of the Army recently had confirmed a story approaching myth that had circulated for years: that the *Central America* carried an official secret shipment of gold destined to shore up the faltering northern industrial economy. The letter, dated April 2, 1971, acknowledged that the information about the shipment had been declassified, and it verified that secreted in her hold the *Central America* had also carried six hundred fifty-pound bar boxes, or another thirty thousand pounds of gold.

In his historical research, Tommy had uncovered two coordinates that helped pinpoint the sinking. One came from the *Ellen,* a bark that had sailed into the wreckage the following day. The other had come from the *El Dorado,* the schooner that had rounded the stern within a cracker's throw of the *Central America* only ninety minutes before she sank. The only problem was that the *El Dorado* coordinate approximated the site of the sinking, and the *Ellen* coordinate marked the area to which the wreckage had drifted over the next twelve hours, but the two positions were nearly sixty miles apart, and the wreckage could not have drifted that far in that time, even in the Gulf Stream. The search area would be too big, unless Tommy could find another coordinate to confirm either the coordinate of the *El Dorado* or the coordinate of the *Ellen*.

In collecting newspaper accounts from every major port along the East Coast, Tommy had discovered that on the day after the sinking, the *Ellen* had spoke another ship, the *Saxony,* bound for Savannah.

"Just on gut instinct I thought, God, somebody, some reporter, somewhere must have found out where the *Saxony* was and got those coordinates."

Tommy called Savannah's Regional Library and found a librarian willing to do the research. He sent his request in a letter, enclosed a five-dollar check to pay for copies, and explained that she should search the Savannah papers on the 18th and 19th of September 1857. The librarian found an article that covered the arrival of the *Saxony,* but the piece was short and contained no coordinates. Tommy knew from his research that if the *Saxony* had sailed into Savannah and the captain had not given coordinates to the ship reporter, it would be the only ship that had sailed into a port on the East Coast after the hurricane without reporting a coordinate. Tommy called the librarian again and suggested that perhaps more than one article on the *Saxony*'s arrival appeared in the paper, or that maybe the one she had found continued on another page inside.

A few days later, Tommy received a letter dated October 18, 1983:

Dear Mr. Thompson:
   Your perseverance has been rewarded. When I was checking September 19, 1857 for the "Shipping Record" column, I looked at the entry I had cited to you earlier. Contrary to my previous findings, the abstract of that article was just that. Enclosed is the entire entry. . . ."

When Bob saw Tommy's copy of the article, he figured the librarian must have had nothing else to do. It was nearly impossible to read, a scratched white-on-black microfilm copy of an obscure notice in the Shipping Register of the *Savannah Daily Morning News,* but they could read the coordinate, "lat. 3140, long. 7620," which meant that the *Saxony* had spoke the *Ellen* no more than fifteen miles from where the *Ellen* captain reported he had been tacking through the wreckage earlier that day.

"We now had a third coordinate from the day after the disaster that was in the same area of the ocean," said Bob. "It gave the *Ellen* coordinate credence. And the existence of those two coordinates within a ballpark of each other was enough to convince us we've got something scientific to play with now—it's not just hearsay." But he still couldn't explain why those two coordinates were sixty miles from the *El Dorado*'s.

TOMMY NOW HAD acquired dozens of survivors' accounts and other documents, and he could see there were probably hundreds more. He had even created a matrix, breaking the sinking into time segments, so he could understand the chronology of events. But studying technology was requiring more and more of his time and he also had to pursue financing, so he gave Bob all of the articles and letters he had collected and the matrix he had started putting together, and asked Bob to take over as historian. "He was into all kinds of subjects in different areas and a real trivia freak," said Tommy, "which is perfect for a project like this." By the fall of 1983, Bob's geology consulting practice had dwindled, anyhow. The oil industry was pulling in, drying up, and rumors persisted that the market was about to collapse. Bob put it euphemistically, "I had more time on my hands."

The sinking of the *Central America* was one of the biggest news events of nineteenth-century America. It was the worst disaster in American shipping, and several pundits of the day attributed the Panic of 1857 largely to the loss of the gold shipment aboard the steamer. Passengers and crew on the *Central America* hailed from all thirty-one states in the young union and many foreign countries. The telegraph only recently had sprouted up and down the East Coast, so that news of the arrival of the first survivors in Savannah shot straight up the wires all the way to Boston and over to New Orleans. Reporters waited dockside to get the first stories from survivors arriving in ports along the East Coast, and for the first time reporters relied heavily on the accounts of women. For weeks, dozens of newspapers ran vivid front-page accounts in the survivors' own words, the articles sometimes filling the page and several more pages inside. As survivors returned to their hometowns, and as official bodies inquired into the sinking, the story lived for months in over two hundred newspapers.

Two or three days a week, several hours a day, Bob sat on the main floor of the William Oxley Thompson Library at Ohio State, amid the bare concrete columns and the file cabinets filled with rolls of microfilm, his head in a reader, his blue eyes scanning the white on black film of articles from the front pages of old newspapers: *Frank Leslie's Illustrated,* the *New York News,* the *New York Times,* the *New York Journal of Commerce.* When he saw something pertinent, he made a copy and took it home, where he studied it and pulled information to place on the matrix.

He found a dictionary of marine terms from the 1920s, which he referred to constantly, for the ship's logs and the seamen's accounts were filled with an argot that sometimes read like a foreign language: top gallant and jib-boom, spanker and bitt, fresh breeze, mizzen, hawser, and drag. Bob needed to interpret to understand the damage done to the vessel, and how the wind might push the ship, and what the storm had ripped from the deck.

Besides reading scores of lengthy accounts, he researched early American hurricanes, and the Great Storm of 1857. He studied the Blake Ridge. He read the work of Matthew Fontaine Maury, the father of oceanography and coincidentally Captain Herndon's brother-in-law. He included in his inquiry the California Gold Rush, the country of Panama, the rise of the Panama route, the political climate of mid-nineteenth-century America, and the Panic of 1857.

Before long, Bob could recite from memory long passages of the survivors' accounts. Inside his head began to live the voyage of the *Central America* as she steamed north from Havana bearing six hundred souls bound for New York: the storm winds rising on the second night, the waves swelling, the engine room leaking, the water deepening, the engines failing, the men passing buckets, the women taking courage, the *Marine* hoving to, the crewmen rowing the women and children in high seas, Captain Herndon refusing to leave his ship, the steamer succumbing to the storm, many men going to the bottom, and many more set adrift on the flotsam of the wreck on a dark night in a black sea. He thought about the Eastons and Virginia Birch and George Ashby and

Second Officer Frazer and Captain Badger and Captain Burt and Judge Monson. He tried to picture them and know them as he would a friend. He tried to crawl inside the head of Captain Herndon.

The matrix became the Data Correlation Matrix, and it soon expanded to take up one entire wall, floor to ceiling, in the study of Bob's house, each entry a specific reference to the storm, the condition of the ship, or a statement of location. When he found a critical piece of information, he walked over to the matrix and wrote it in pencil with a simple reference code next to it. NYT stood for the *New York Times,* PB was the *Philadelphia Bulletin,* CC the *Charleston Courier;* then came the date, page, and column numbers and the name of the witness. To disguise where the data had come from, he and Tommy later created a cryptic reference code, a series of nine letters and numbers, so they could trace the information back to its source, but no one else could.

"That way we would not be giving away our sources," said Bob. "We were very, very jealous of our sources at this point, because we knew there were other people out there thinking about the *Central America.*"

In 1979, Tommy had received a phone call from a lawyer practicing in a small Columbus firm. Robbie Hoffman, a short, funny, friendly, fast-talking man not much older than Tommy, had a client interested in salvaging the *Andrea Doria,* the Italian luxury liner that had sunk in 1956 off the coast of Nantucket. Tommy had told Robbie that the *Andrea Doria* was a poor target because it had a steel hull and no records supported the wild claims of jewels and currency stashed on board. Robbie dropped the client, but he liked working with Tommy. "He's brilliant," said Robbie, "he just is. He exudes it. He knows what he's talking about, and he's extremely disciplined, and he has tons of energy. Tons of energy."

Tommy and Robbie had talked then about deep-ocean recovery one day being a viable business. "When it happens and you're doing it," Robbie had said, "you call me, 'cause I want to be involved."

By 1983, Robbie had left his firm and gone out on his own. Tommy was now convinced that one day was here, that the technology for finding a ship in the deep ocean and then salvaging that ship was possible and that he had a good target to pursue. His next step was to broaden his contacts with suppliers and contractors and professionals who would provide him with pieces of the right technology, and to explore ways to finance the project. He called Robbie and reminded him of what he had said four years earlier about being involved.

Tommy and Robbie renewed their relationship over occasional beers late at night at Robbie's house, Tommy talking to Robbie as he had to Bob about his theories on deep-water shipwrecks. They wondered out loud how with no money Tommy might secure professional services and promises of technology, and how to approach investors with an idea both costly and seemingly impossible. "I was a cheerleader," said Robbie, "someone to be there and to spend time with him, to hear the talk about how to do it."

Robbie compared Tommy to Don Quixote, always chasing windmills, and he saw himself as the "schlepper" walking behind the donkey. Before long, Tommy was at Robbie's house two or three nights a week, every week, sometimes on Saturdays and Sundays, working on the project. He seemed not to notice the indiscretion of showing up nigh on to midnight, night after night, with a gap-tooth grin and a six-pack of beer at the home of a newly married friend. Robbie thought it was outrageous, but he accepted the behavior. "Harvey was so driven that it made it difficult at times to put up with him," said Robbie, "but that's the nature of the beast, the beast being the brilliant person. You have to be able to accept a lot about them."

With Robbie's new wife often asleep in the other bedroom, Tommy and Robbie worked on Robbie's PC, drafting letters to research libraries, suppliers, contractors, and potential investors. Tommy would outline what he wanted to say, Robbie would put it into words, then Tommy would review it sentence by sentence to make sure the words not only carried the message, but also conveyed the right tone. Tone was important. Then Tommy would try to imagine every response the recipient could have to the language they had used, and he would look for ways to reword the letter either to avoid a negative response or enhance a positive one, to be a little more ambiguous or a little more clear, to pique interest but not reveal too much.

"I sat in front of this green screen," remembered Robbie, "Harvey standing there at my shoulder, the letters dancing across, eighty-five proofreads per letter. He had incredible discipline. Working over there at Battelle, going home every night, and then coming over here at eleven o'clock, work till two in the morning, three in the morning. Get these things knocked out."

With Bob Evans as his sounding board on the research and Robbie Hoffman as his business advisor, Tommy began to shape his ideas about deep-water shipwrecks into a project. Robbie likened it all to the principle of inertia: A body in motion tends to stay in motion. "It just continued to build, and as it built, it seemed to move faster."

ONE OF TOMMY's late-night letters went to the publishing branch of the Church of Scientology in Hollywood, California, another to a billionaire from South Africa, and another to the heir of the Miller Brewing fortune in Milwaukee, all contacts Tommy had established through friends and acquaintances, and all capable of providing him the millions he needed to search for the *Central America*. All three were interested. The first in-person pitch went to the Scientologists.

"Harv, in his inimitable, incredible style," said Robbie, "somehow had convinced the Scientology people to send us airfare."

Tommy could take off from Battelle no more than one day, so he planned for them to leave Columbus early in the morning, fly to Chicago, change planes and head to Los Angeles, meet with representatives of the L. Ron Hubbard publishing branch of the Church of Scientology, then fly back through Chicago and be home in Columbus not too long after midnight, so he could be at work at Battelle early the next morning.

"It's something," said Robbie, "that in my entire life I will never forget or forgive Harvey for. Ever."

What flabbergasted Robbie as much as Tommy's insistence that they make the trip in one day was that after Tommy had asked three guys wearing powder blue button-down shirts and striped ties to sign nondisclosure statements, after he had pitched the deal to them for three hours without a bite of lunch, after he had piqued their interest with his theories and his research on the *Central America,* after he had asked them to cover the search phase of the project for a million-six, he had the audacity in the foyer as they were saying good-bye to remind them they owed him cab fare back to the airport.

But that was Tommy's way of testing their sincerity and their interest. "If they tell you they're going to do something and it doesn't happen that way," said Tommy, "you have to think about that."

Tommy handed the first cab receipt to the man who had set up the meeting; the man doubled the figure and handed Tommy sixty dollars in cash, but Tommy decided later not to pursue the relationship.

Next, they flew to New York by way of Newark to meet with the billionaire, who sat in the drawing room of his suite, surrounded by dark antiques, and never asked a question. "There weren't any questions left to ask," said Robbie. "Harv was absolutely complete and totally prepared at all times." The billionaire thought the idea could work, but because of a lack of liquidity in some of his companies, he did not want to invest, and that was the end of the meeting.

When they left the hotel, Tommy told Robbie he didn't have enough money to take a taxi to the Newark airport, that they would have to walk fifteen blocks to the Port Authority and catch a bus. "It'll be a slice of life," said Tommy. But the bus died in the Lincoln Tunnel, and the driver had to find a semi to push them through to daylight, where Tommy and Robbie got out, walked to the next exit, and found themselves in a neighborhood where no one spoke English and taxis wouldn't stop. "I was so aggravated when we finally got to the airport," said Robbie, "all I wanted to do was go home and never have anything to do with this project again."

THE HEIR TO the Miller Brewing fortune was Harry John, who lived with Capuchin monks in Milwaukee, where he ran the De Rance Foundation, a nonprofit corporation organized to devote funds to religious, charitable, and educational causes. It was the largest Catholic charity organization in the world, with resources exceeding $100 million. Mr. John, then in his sixties, had created the foundation thirty years earlier with inherited stock and made himself the trustee, chairman of the board, president and chief executive officer, treasurer, and chief financial officer of the foundation. Of late, as a federal judge would soon find, the eccentric Mr. John had been surreptitiously siphoning off large sums from the foundation to indulge his propensity for looking for shipwrecks. In pursuit of two wrecks that year, he had pulled out just under three and a half million dollars, and he'd thrown in another two and a half million of his own money. Now he was after the *Central America*.

Already John had paid treasure hunters three hundred thousand dollars to look for the sidewheel steamer in shallow water off Cape Hatteras. Over the phone, Tommy told John, "Don't spend any more money at Hatteras." Anyone could read a few articles in old New York newspapers at any library in the country and determine that the ship foundered over the Blake Plateau one hundred miles out and probably sank on the Blake Ridge, as far as two hundred miles out. In October 1983, Tommy flew to Milwaukee and had lunch with John and about ten monks, all sitting cross-legged on the floor, everyone eating a bowl of grains, John at the head of the ceremonial rug dressed in a robe and skullcap.

"Well, how can we find it?" John opened the conversation.

Tommy said, "You need to do an analysis of the whole problem. You need a computer and you need to figure out the physics and develop a probability map. It isn't like X-marks-the-spot. You need to do more in-depth research."

Before he said any more, he had a nondisclosure agreement for John to sign, but John dismissed it with, "Well, I'll have somebody look at this."

Tommy proceeded but even more cautiously than he had intended. He told John that the first thing he would need was a new generation of side-scan sonar. "Columbia University has the first one built, the SeaMARC I. It's prototypical, but with the right adjustments it might work." He showed John a chart with enough information on it to prove that the *Central America* lay in deep water. They talked a while longer, then Tommy got up to use the bathroom, and when he returned, the monks had his briefcase open and the copy machine running. Harry John said, "We just thought we would xerox this stuff, and we didn't really want to bother you with it."

Tommy grabbed the papers and stuffed them back into his briefcase, muttering something about their relationship having to develop a little further before he could give them that information. He had left all of the revealing information, the important coordinates, back in Columbus, anyhow.

After the meeting, John remained interested, and although Tommy felt he could not trust the man, he refused to close the door because of one bad experience. It was one half of his philosophy: Always keep your options open. If watched and brought along properly, John could provide the cash to create the technology Tommy foresaw he would need to find and recover the *Central America,* and that included helping Mike Williamson build the SeaMARC IA.

Tommy arranged to meet with John and Williamson in Seattle to discuss the new sonar. There, Williamson explained to John how the SeaMARC evolved and how he thought it could be modified and improved to do the sort of survey Tommy wanted to do; then he introduced them both to the engineers who had built the other SeaMARC. By the end of the visit, John was talking to Williamson's accountant. "Things looked fairly encouraging at that point," remembered Williamson.

After they left Seattle, Tommy talked more with John by phone, John pumping him for more information about the SeaMARC and how it could help them find the *Central America*. Why did they have to build a new one, he wanted to know, when one already existed? Tommy explained that the other SeaMARC was owned by a university, and by law all sonar data recorded by them became public information after one year.

Three months later, Tommy heard through Williamson and the deep-ocean grapevine that Harry John had contracted with Columbia University to use its SeaMARC I for a search in the Atlantic Bight off the Carolinas. When Tommy got the story sorted out, he realized that John had pumped him for information all fall and into the winter, right up to the day John left the dock. John then had rushed to sea in February, a time of year no one, given a choice, would work in the Atlantic. The weather was predictably bad, and when the ocean was rough, the SeaMARC jerked, and John was out for ten days trying to sweep huge areas in rough seas, and getting data that Tommy was certain would tell him nothing. John had left the dock apparently hell-bent for a single coordinate, a set of numbers that had puzzled Tommy almost two years earlier: 31° 25' latitude, 77° 10' longitude. That was the *El Dorado* coordinate included in the official report of the disaster issued in 1857 by the New York Board of Underwriters. The *El Dorado* coordinate was the one sixty miles from the other two coordinates Tommy had discovered, and he and Bob had yet to reconcile the discrepancy. Tommy had told Harry John he had to approach the search systematically, to use computers, create probability maps, generate search grids, cover a wide area and gradually narrow the field, not to look for an X marking the spot. Why was that coordinate so far from the other three? John hadn't taken the time to find out, but Tommy wouldn't sail until he did.

With John no longer a possible backer, Tommy continued his conversations with Williamson, encouraging him to get the SeaMARC IA built any way he could. Tommy wouldn't lease the other SeaMARC even if the frequencies could be modified, because he wanted everything confidential, so he needed a private sonar operator with his own Sea-MARC. He couldn't commit to a contract yet, but he told Williamson that as soon as he got financing for the *Central America* project, he would be Williamson's first client.

The Partnership

1984

IF ONE HALF of Tommy's philosophy was Always keep your options open, the other half was Acquire as many options as you can. Before he had talked to the Scientologists or the billionaire, before he had met with Harry John, Tommy had already discussed financing with his old mentor Dean Glower.

Glower was now dean of Ohio State's College of Engineering, and since Tommy had returned to Columbus to work for Battelle, he had talked periodically with the dean about the progress of his ideas for working on the bottom of the deep ocean. As Tommy refined his ideas, he had become more and more convinced that finding and recovering a deep-water shipwreck was the ideal project to try them on, and he had told Glower about some of the possible targets. Then one day in early 1984, he called Glower and said, "I think I found our ship."

Tommy made an appointment to meet with Glower at the dean's office. There he told Glower the story of the *Central America* and how the new SeaMARC technology might enable him to find it, and how he would design a vehicle that could go to the bottom and document and carefully dismantle the ship and recover the gold. But who can I get to fund the project? he asked Glower. He had enough ideas; now he needed backers.

Glower listened, then called the office of Herb Lape, fifty feet down the hall. "Herb," he said, "come up here a second. I want you to hear a story."

The person in charge of fund-raising for the College of Engineering, Herb Lape was an old crony of Glower's who had connections Glower didn't have. He had grown up in a Columbus suburb called Bexley, home to the community's ultra wealthy. "I'm the dean of engineering, which is a credible job," said Glower later, "but Herb is their buddy. That's why I got Herb with Tommy."

Glower told Lape that what he was about to hear might sound crazy, but that he, Glower, knew enough about engineering and about the ocean to realize that the theories were sound, that someday before long someone would be doing exactly what this young man was talking about doing now. "It's a very exciting idea," said Glower, "and Tommy is the kind of guy who can maybe pull it off. Maybe not."

Then Tommy told Lape about deep-water historic shipwrecks and why they were preferable to shallow-water shipwrecks, and how he thought the technology was available if someone would just pull it all together. Lape, who had to exude a certain amount of enthusiasm in the fund-raising business anyhow, listened with growing disbelief, and when Tommy had finished explaining his idea, thought, "My God, I've got to get some friends of mine to listen to this! This is the biggest crap shoot I've ever heard of!"

Glower suggested that Lape set up a luncheon, get some of the Bexley people to come, and he would host it out of his own pocket. Lape started calling his friends, most of whom groaned. "Hey, it's a free lunch," said Lape. "You gotta hear this!" He told them that the idea might be wild, but that the young man with the idea was an engineer at Battelle who came highly recommended by Don Glower, who also would be at the luncheon.

Lape reserved a small conference room at the university's Fawcett Center for Tomorrow. Prior to the luncheon, he told Tommy that for two reasons the key person there would be a CPA named Wayne Ashby: One, Ashby knew how to put the whole thing together, how to structure the offering; two, he had all kinds of contacts.

The room held twelve people, and it was full: Glower, Lape, and Tommy, plus nine potential investors. By Glower's estimate, the nine investors had a combined wealth of more than a billion dollars. He introduced Tommy, giving a brief testimonial: He had known Tommy for years, and he knew Tommy wasn't going to disappear. "You guys may not know that, but one thing you do know is that I'm not going to disappear. Tommy's honest, he's smart, and he'll give his damnedest. We just want you to hear the story, and you can take it from there. We guarantee nothing."

Then Tommy got up and talked about the history of treasure hunting, how the vast majority of shipwrecks had occurred in shallow water, and how the recovery of these shipwrecks presented many problems. He showed slides he had taken when he worked for Mel Fisher in Key West, and he noted the primitive methods Fisher used in his search. Fisher had been searching for the *Atocha* for fourteen years and still had found only pieces scattered across miles of ocean. In deeper water, the environment was far more stable, adulteration of a wreck far less likely to occur, and you could be certain that no one else had recovered the wreck. The only thing preventing someone from recovering these deep-water wrecks was technology, and even that finally appeared possible.

As Tommy was talking, one of the men excused himself early and said on his way out, "It's all very interesting, but Wayne's our accountant. Before we do anything, we would want his blessing."

As Lape had told Tommy, Wayne Ashby would be the key man there. He was the managing partner of 150 CPAs in the Columbus office of the national accounting firm of Deloitte, Haskins & Sells. He or his firm did the books and the tax work for just about everybody in the room and a whole lot of other people in the Columbus area whose net worth was figured perhaps not in the hundreds of millions, but certainly in the millions. He knew virtually everybody who might be interested in a deal like the one Tommy proposed. But Ashby sat through the whole presentation, which lasted an hour, and left less than intrigued.

"My impression was just a nice young man, and I didn't think much more about it beyond that. Everybody was pretty busy, and we rushed out of the meeting, and I'm sure we all thought that was interesting and informative and that was the end of it."

Tommy waited for two weeks before he called Ashby. When he reminded Ashby's secretary about the luncheon and who he was, Ashby picked up, and they talked for a short while, Tommy asking some questions about business and financing and giving Ashby an update on the progress he had made since the luncheon. "I figured if I could just keep the idea in front of him until he could see the logic of it the way I could," said Tommy, "then he might help turn it into a 'proper business'." Tommy had thought through the conversation as carefully as possible, considering every reaction Ashby could have to each approach he could take, and he knew he had to be brief and professional. He kept the conversation informative and slightly inquisitive, but his approach, no matter how professional, could not change one thing: "I just didn't have much time for him," said Ashby. Tommy closed by saying he would call again in two weeks to give Ashby another update. And he did.

In that conversation, Tommy asked Ashby two specific questions about financing, and at the end of the call he said he would be in touch again in a few weeks. Whenever they spoke, Tommy closed the conversation by saying he would call back in a week or two weeks or a month, and, as busy as he was, back in New York about half the time, Ashby always said that would be fine.

Though he had little time to think about it, what Ashby had locked on back at the investors' luncheon, and the tiny spark he just could not extinguish, was the ratio of risk to reward inherent in what Tommy proposed. It was a long shot; the odds were great; it would probably never happen. But if it did, look at the payoff. You could realize your investment a hundred times, a 10,000 percent return. That thought kept rewinding and playing itself out again in his head.

"It was a bet you could not ordinarily make anywhere," said Ashby. "And it was an investment that just doesn't come along in Columbus, Ohio." That was the other intriguing thought: Tommy was not propos-

ing another strip mall northwest of town or an apartment complex or a
pharmaceutical start-up; he was offering an opportunity to be part of
an adventure to a new frontier.

Each time Tommy called, Ashby found himself thinking about the
possibilities, and then those thoughts gradually would subside, and he
would have just about forgotten the last conversation when Tommy
would call again. After several phone calls, Tommy met Ashby for
lunch; after that, Tommy continued to call every week or two, and occasionally they met, and Tommy gradually slipped into a comfortable
rhythm of calling Ashby with updates on his progress, and how he was
proceeding with some of his ideas for financing the project.

The more Tommy talked, the more Ashby began to appreciate his
intelligence. Tommy just seemed to know what he was talking about,
and he was talking about doing something no one had ever done before. Ashby would have dismissed the whole idea as a fantasy, a shot
far too long, a return on investment too good to be true, if the greatest
of his intrigues had not been with Tommy.

"My confidence level in him just grew as I spent time with him, and
I became increasingly intrigued by his knowledge. Sometimes I didn't
know what he was talking about, but he just came across with a lot of
credibility. He could talk fluently about his subject, and his small-town
Midwest background came across, good character."

Months went by, months of talking with Tommy, answering questions, giving advice, and meeting with him occasionally. In that time,
with Ashby's knowledge and input, Tommy had pursued financing
with the Scientologists, the billionaire, and Harry John, none of which
had worked out. But Ashby could see that eventually he would find a
way to finance his project. As Ashby watched him, saw his commitment,
realized the enormous amount of energy he put into everything he did,
his confidence in Tommy grew even more until the idea that he could
work on the bottom of the deep ocean, where no one had ever worked
before, seemed no longer wild, but only highly speculative.

Several months later, Ashby had talked to Tommy so many times
he knew everything he wanted to know about him and his project. He
could learn nothing more from talking, and he was a busy man. On the
phone one day, he said, "We've been talking about this a long time, and

I'm to the point where I feel I have to decide whether I'm in this project or out of this project, and I want to think about that." A few days later, he called Tommy back, and he said, "I want to be in this project. What can we do?"

By now, it was early summer of 1984. Tommy had worked three years at Battelle, and his hours at the institute had begun to taper down, from over sixty to around fifty, and then to forty. Meeting with Bob on the historical research and phoning experts about technology and talking to Ashby and pursuing investors and trying to figure out how to organize the project required more hours than he was now able to squeeze into his weekends and his evenings at Robbie's. He was certain that Harry John had not found the *Central America,* and Mike Williamson seemed more committed than ever to building the SeaMARC IA. As much as he loved Battelle, Tommy's thoughts were elsewhere.

Tommy made an appointment with Don Frink and explained to him that he had a project he had been pursuing outside Battelle, that he was attempting a recovery operation in the deep ocean. He couldn't keep working ten or twelve hours a day as he had been. He told Frink he wanted to go on a partial leave without pay, cutting back to half time or three-quarter time.

Frink called it "a shared arrangement" and agreed to let Tommy divide his loyalties as long as he worked a set number of hours each month at Battelle.

But over the next few months, Frink saw less and less of Tommy, and gradually, his hours fell to the minimum for him to remain with Battelle. "Sometimes," remembered Frink, "he came in at four in the morning, and I got a note on my desk: 'Did such and so. This'll take care of it for this month.'"

Wayne Ashby told Tommy that the first thing they had to do to give the project a businesslike structure was to get a big downtown law firm involved. To raise a large sum of money from investors in a public offering, Tommy would have to comply with strict regulations from the Securities and Exchange Commission and the Ohio Securities Division, and all of the papers had to be filed properly. Robbie Hoffman had gotten him this far with friendly legal advice, but if people of the calibre Ashby wanted to pursue were going to invest, Tommy needed a name those people knew.

One of the prominent lawyers in Columbus was Art Vorys, of Vorys, Sater, Seymour and Pease, a firm of 265 lawyers. Vorys had attended the investors' luncheon with Wayne Ashby and, like Ashby, had come away thinking that the previous hour had been interesting. Since that day he had thought little about shipwrecks or Tommy, but when Ashby called, he agreed to assemble some of his venture-capital lawyers in one of the firm's conference rooms, have Ashby bring over a few of his CPAs, and let them all mull over how they might structure a project to search for and recover a shipwreck.

Tommy's presentation went well until Vorys asked, "How far down is this?"

"Well," said Tommy, "it's at least eight thousand feet."

Vorys's jaw dropped. "Eight thousand feet!" He thought about the time he had scuba dived in the Bahamas and blew out both eardrums at thirty feet. "What has ever been done at eight thousand feet?"

"Nothing," said Tommy.

"It'll never work!" said Vorys. "That's too far, too deep."

"That's when he lost me," Vorys said later. Vorys thought that eight thousand feet was "an absolutely ridiculous depth to find anything over a hundred years later." Just the technicalities bothered him. How do you direct something eight thousand feet down to move some kind of hand four or five inches this way and then four or five inches that way?

"Tommy didn't connive you," said Vorys. "He was always very serious and very constrained and not prone to exaggeration. But I couldn't get over that eight thousand feet two hundred miles offshore!"

One of the CPAs Ashby had brought to the Vorys meeting was Fred Dauterman, the tax partner at Deloitte for twenty-five years. When Deloitte clients wanted to put their money into tax-sheltered investments, Ashby brought the deal to Dauterman. For years Dauterman had written articles and presented programs for the New York Tax Institute on tax shelters, particularly for high-risk investments like oil and gas. As tax partner in the firm, he had analyzed roughly five thousand deals, and out of those five thousand he could count on two hands the ones that turned out to be good investments.

During Tommy's presentation at Vorys, Dauterman had sat across the table from a CPA friend, and when Tommy began his talk, their eyes met. "He and I kind of looked at each other and thought, Hmmmm. This sounds like another wild one." But by the time Tommy had finished, Dauterman thought it might not be so wild after all. After the meeting, he and Ashby went back to Ashby's office.

"Ashby," said Dauterman, "this is the craziest deal I've ever heard in the world, but it's crazy enough that it might have something to it."

Dauterman thought at first that Tommy came across as a "country bumpkin," but as he listened to Tommy speak and field questions from the others, he found himself impressed with the young man's knowledge. With a steady voice and overwhelming detail, Tommy could wither a roomful of crisp, professional cynicism. "The more contact you have with Tommy," said Dauterman, "the more you're convinced that the man's a genius. He thinks on a different level than most people." That was important, because, for a deal like this to work, two things had to come together in the point man, and the first was that he had to know unequivocally what he was talking about.

The second was that he had to be trustworthy. On this, they had the recommendation of Dean Glower and contacts at Battelle; they also had Ashby's own confidence in Tommy, which had only grown after hours of listening to him talk about going to the bottom of the sea. Glower had assured Ashby that Tommy's ideas were sound, and that someday someone would do it. Slowly, Ashby had come to this thought: There was a reasonable chance that the only thing that stood between Tommy Thompson and the floor of the deep ocean was money, and the only thing that stood between Tommy Thompson and money was credibility with big investors, and that was the one thing that Ashby could give him: credibility.

Over the next few weeks, Ashby and Dauterman tossed figures back and forth, talked more to Tommy, and finally decided that the project had one chance in ten of succeeding. But if Tommy got all of the technology together to find the ship and then do the wonderful things on the bottom of the sea that he wanted to do, the payoff could be a hundred to one.

Ashby told Dauterman, "If you go out to Las Vegas and find ten-to-one odds on an even bet, you're not going to play it because you'll lose. But if the payoff is a hundred to one, you take that bet every time."

"You won't let go of this, will you?" said Dauterman.

"If I don't get it put together," said Ashby, "I won't get to make my bet. The only way I can make my bet is for Tommy to go to sea."

Art Vorys had turned them down, and he probably would not be the last to do so; Ashby understood, and in different circumstances he would agree. But he had grown to believe so deeply that this young man could do what he said he could do that something inside told him to throw his own integrity behind the project, and in Columbus that was a lot of integrity. "They all trust Wayne," said Dean Glower, "because Wayne is a very good, honest guy."

But before Ashby went to friends asking for money, he still wanted a big law firm to structure the deal and walk Tommy through the securities regulation maze, and to help lend credence to the idea. Art Vorys had been at the luncheon, so Ashby had gone to him first. He now went to the other big firm in town, Porter, Wright, Morris & Arthur, specifically to Bill Arthur. Around Columbus, Arthur was known as Mr. Venture Capital, a lawyer who had syndicated real estate deals and oil and gas exploration and myriad start-ups. He had heard more crazy pitches than anyone else in the city, so Tommy's proposal was not out of the ordinary. What was out of the ordinary was that Wayne Ashby had made the call, and that got Arthur excited about the deal before he even met Tommy.

"Here is this absolutely flabbergasting experience," said Arthur. "Wayne Ashby having people buy equity interests in one of these high-risk deals? Wayne is an accountant! He's never promoted anything in his life! He's never even hinted that somebody might invest in something he might know about!"

When Ashby brought Tommy over to meet with Arthur, Arthur didn't even grill him; he didn't need to. The whole thing sounded plausible. His firm represented Battelle, and he thought of the institute as "a bunch of crazies up there doing all kinds of wonderful, exciting things." Tommy was just one of a breed, and he had calculated as carefully as humanly possible that this could be done. "With a guy like that," said Arthur, "coupled with the thrill of finding a hundred times your money, why in the hell wouldn't ya?"

Arthur told Ashby, Hell, yes, his firm would structure the deal, and he turned the project over to one of his younger partners, Curt Loveland, to see if Loveland could put it all into a framework that would encourage people to invest in it. Loveland was not so enthusiastic. His job as a lawyer in the securities field was to be cynical, and after years of experience he could not have been much more so. He was surprised Bill Arthur was even dragging him to a meeting to talk about the deal, but an hour with Tommy changed his mind. "I came away from the first meeting, and I'm thinking, God, I think this guy can do it! And then at the second meeting I was convinced this guy can do it."

Loveland brought in another colleague, Bill Kelly, to help him structure the deal, and Kelly had the same reaction. He assumed Tommy was a crazed inventor until he heard his methodical speech, the carefully selected words, and saw in him what Don Frink at Battelle had seen: someone sincere, knowledgeable, and intelligent, who made you feel good to be around him. When Tommy spoke, somehow the ideas didn't sound so crazy.

Tommy, the CPAs, and the lawyers now got together and tried to structure the idea as an organization that would make sense to an investor. But no one had ever seen a deal like this. It wasn't like real estate, where they could look at it and touch it and smell it. Dauterman had done so many of those deals he had a checklist: If the numbers added up, the investment was good; if they didn't, it wasn't. And if he made a mistake, over time inflation might rescue him.

At the first meeting in Ashby's office, they all sat around tapping pencils on blank legal pads and asking each other, What should this thing look like? "What really set it off from other offerings," said Kelly, "was that it was so speculative." Someone looking for venture capital usually had invented a new computer screen or a new piece of software, and they could tell you the size of the market and the competition and what their product did that the competition's didn't and their production costs and the number of units they could sell in the first year and how many they expected to sell in year two. That's what venture-capital investors were accustomed to seeing. They analyzed spreadsheets, the market survey, the pro forma financial statements and decided. Whether it was computers or medical technology didn't matter. "But this one," said Kelly, "was unique, to say the least."

Bill Arthur had suggested to Ashby that they try to raise only a little money at first, some seed money, and use that to improve the presentation materials and allow Tommy a few months to pursue the real money he would need for the search phase. Loveland reiterated Arthur's point. "You've got to take it in pieces and not try to raise $5 million the first day, because you're going to fail. People want to give you a little bit and see how you use that." But they emphasized to Tommy not to budget that seed phase at $100,000 if he really needed $200,000. Too many start-up companies failed because they went in underfunded.

The meetings ran through the winter and into the spring of 1985, the lawyers and the CPAs impressed with Tommy's quick grasp of legal and accounting concepts and his determination to keep within budget. They decided it would be a limited partnership offered privately to persons of substantial means, and that nobody involved with the project would be paid up front. "We had to make sure," said Dauterman, "that when we went to people we knew, we could say, 'Look, we're not getting a nickel out of this. We're putting our money in the same as you are, so you can't bitch.' If you look at every other deal in the world, there is somebody taking some kind of a cut for promoting it. Here, there were no commissions. Every dollar went into the project."

They set up the funding in three phases: the Seed Phase, the Search Phase, and the Recovery Phase. To the Seed Phase they allocated 10 percent of the partnership, twenty units at $10,000 per unit, trying to raise $200,000, so Tommy could organize, line up contractors, complete the research, reimburse himself for past expenses, and cover the expenses of trying to raise funds for the next phase. It would give Tommy a chance to show his investors not only that he was scientifically capable, but also that he could manage a company and use their money wisely.

For the Search Phase they allocated 25 percent of the partnership, fifty units at $28,000 per unit to raise $1.4 million to complete a probability map and a search map, to mobilize a ship, rent the SeaMARC IA that Mike Williamson had just had built, pay Williamson and his tech crew for forty to sixty days at sea, search about fourteen hundred

square miles of ocean, find targets that could be the *Central America,* and
verify that one of them was the ship of gold.

In the final phase, they would send a robot to the bottom of the ocean
to study the site and recover the treasure, and for that they allocated
another 25 percent of the partnership, fifty units at $72,000 per unit to
raise $3.6 million.

That totaled 60 percent of the project. The remaining 40 percent
went to Tommy and his associates. Tommy wound up with a rich package for two reasons: First, the potential rewards were substantially
higher than in an ordinary oil and gas lease or a real estate deal, so the
investors were willing to take less. Second, Tommy was key to the success of the business. If the investors were purchasing a patent, they could
hire anyone to run the business. Here, Tommy was everything, and they
had to satisfy themselves that he was sufficiently motivated. The lawyers and the CPAs and Tommy talked about the split for months, and
in the end everyone was satisfied.

THEY PUT OUT the first offering in March of 1985: twenty units at ten
thousand dollars apiece, a half-unit for five thousand dollars. Ashby and
Dauterman couldn't raise the money as partners in an accounting firm,
but as individuals they could talk to friends and say, "Here's a deal you
should look at. We've talked to the principals, and we think they're
honest. And by the way, we're going to put our own money into it."

They drew up a list of people who could afford to lose the money,
and when they started down the list, Dauterman's conversations often
went like this: "Bill, this is Fred. Are you sitting down [chuckle]?" Then
Fred would give it to Bill in three sentences. Bill would say, "You want
me to do what?"

"It took a lot of guts to call people," said Dauterman, "and you got
some crazy reactions." He told everyone, "If you put your money in, kiss
it good-bye, that's the only way to do it." Then he would say, "But the
rewards, if it's what Thompson says it is, could be astronomical."

When Ashby saw friends socially or called a few close clients he
thought might be interested, he would tell them, "There's a little extra-
speculative opportunity that I'm sure you'll want to be part of." He said

he had a young man with an idea they might find intriguing, and would they meet with him and consider investing a small sum? Sure the idea was a little crazy, and he had never seen anything quite like it before, either, but, yes, he had already anted up his five thousand dollars. As had Fred Dauterman, Bill Arthur, and various other familiar names around Columbus.

One of the first people Dauterman called was Art Cullman, a professor of marketing at Ohio State who consulted all over the country. Dauterman told Cullman the story in the usual three sentences with all of the chuckles and the words like "wild" and "crazy." Then he added, "Wayne and I think the guy's pretty good. He comes from Battelle, he's a scientist, and he's got an idea that really sounds exciting. He thinks there's a lot of gold on this ship."

Cullman signed up for a meeting, and what impressed him most about that meeting was how carefully Tommy spoke. He told Cullman, "We have an idea where it is, but that's not factual yet. We're still gathering the material from newspapers to get a better idea." Cullman asked how he would locate the wreck site, and Tommy told him about the SeaMARC and Mike Williamson's crew, and Cullman wanted to know if he had these sonar experts already under contract. Tommy said only that he was pretty sure he could get them.

Although Cullman asked a lot of questions, he knew nothing about technology and rarely understood Tommy's answers. When he asked what happened after they found the wreck, Tommy patiently explained deep-ocean robotics, using far more facts than Cullman wanted to hear. "The idea sounded good," said Cullman, "but I didn't know too much about it."

Cullman called friends at Battelle, who told him that Tommy was solid. He called Bill Arthur, who told him the guy had a hell of an idea. He talked to his own wife, who told him, "Arthur Cullman, you can't do that kind of thing. You can't afford that." Then she met Tommy herself, and she said, "Arthur, you can't afford not to do that. He's too reliable." She told her husband she had never met anyone who seemed so reliable.

Cullman even called friends in New York and Chicago who annually blew more than ten thousand dollars on the Kentucky Derby. "I was amazed at the number of people who could easily afford to put ten thousand dollars down who had absolutely no interest in it." He would even remind them how often they threw that much money away, and they would say, "I know, but I don't want to have anything to do with something that wild."

But Cullman had talked to Tommy, and he didn't think the idea was wild. He liked the thought of backing someone qualified, someone determined to conquer a new frontier. He spoke with the other members of his family, and they agreed. Cullman set up a family investment partnership, one-sixth for Cullman and one-sixth for his wife and one-sixth for each of the four kids.

The Cullman family was in for two units.

When Tommy arrived at Buck Patton's office, Patton had several pages of detailed questions for him. The first page concerned Tommy's assessment of risk in five primary areas: search, historical research, security, legal, and operational. Patton was particularly concerned about the legal issues and the operational budget. They talked about Tommy's research methods, the SeaMARC, information leaks, problems with weather. Patton asked questions like, "What happens if a hurricane comes up and you lose your recovery vehicle?"

Patton was also skeptical about having a scientist try to guide a business along a demanding critical path. He had built a small concrete and gravel operation into a multimillion-dollar enterprise, and he had also started other companies. Start-ups were tough. Patton had seen leaders weaken themselves by trying to do too much, or, he said, "Their ego says, 'Well, shit, I can figure this out,' and they drop right off the edge of the world into Chapter 11."

For four hours he grilled Tommy on every fine point his honed business acumen could conjure, and when he finally ran out of questions, Tommy still had answers. He raised issues Patton hadn't asked about or even thought of. Patton was impressed. He found Tommy to be "refined, accurate, tough, determined, factual. There was a strong will that showed through." At the end of the four hours, he asked Tommy to give him the probability of being able to locate the wreck, overcome the legal issues, and secure the gold for the investors.

"He thought for a while," recalled Patton. "Then he looked me square in the eye and said, 'Sixty percent.' Nothing could have been more credible."

Two weeks later, Buck Patton agreed to buy two units and sent a check to the partnership for $20,000.

Many of the people Tommy met with thought he was a poor salesman. They had seen the slick approach, something colorful by someone glib, something almost a sure thing by a packager, a peddler, a promoter. They expected it. By contrast, Tommy was so precise and so careful to avoid overstatement that he made the project seem almost boring. "You could almost see him figuring out how much fuel it was going to take," said Jim Turner, who bought half a unit for $5,000. Sometimes potential investors even prompted him to sound more optimistic. Others advised him to spice up the presentation, put more pizzazz into it, more flash, print up a full-color, glossy brochure, but Tommy refused.

At only $5,000 a half unit, sometimes the sell was easy; the investors just asked Ashby, "Are you going to do it?" and he said "Yes," and they said, "Okay," and a check arrived a short while later.

Although many prominent citizens declined the invitation, and others talked to Tommy and said no thanks, Ashby was amazed at Tommy's record: Three out of every four people who met with him wrote a check for $5,000 or $10,000 or $20,000. "It was kind of interesting for me to watch Tommy grow," said Ashby. "He was thirty-three years old when I met him, and one of the first things I did was introduce him to some pretty prominent businessmen one on one or one on two. Those were frightening experiences for him in the beginning. He was nervous beforehand and nervous after, but he always knew what he was talking about, so he was always credible." Ashby remembered sitting in on one meeting with a particularly important investor, and when they were finished, Ashby said, "Tommy, you really did a great job." Tommy stood up indignantly, and said, "Well, I hope I never become a salesman," to which Ashby replied, "Tommy, you are one."

He just had a way; he could sit face to face with you in landlocked Columbus, Ohio, and never take his eyes off of your eyes and tell you that he had an idea for working on the bottom of the deep ocean, where, by the way, no one has ever worked before, and recovering a gold shipment from a mid-nineteenth-century sidewheel steamer, and he somehow convinced you he could do it by telling you all of the things that could go wrong.

Don't know if we can find the ship. Think we can, but don't know. Don't know what the site will be like. Think we do, but don't know. Don't know if we have the technology to recover the gold. Think we have, but don't know. We do know the ship sank, that it carried at least three tons of gold, that it rests in about eight thousand feet of water, that no one else has recovered the treasure. And I think I can put together a project to do what no one else has done before: find the remains of a three-hundred-foot, wooden-hulled ship in a stretch of ocean bigger than the state of Rhode Island and go down to the bottom and pluck the treasure from the debris with a robot. Don't know if we can keep the treasure. Think we can, but don't know. Don't know how to market the treasure. We'll figure that out.

Tommy's greatest fear now was that word of his project to find the *Central America* might leak to someone in the deep-ocean community. Outside of his small group and the partners now stepping up to invest, he wanted not one person to know the project even existed. That was critical. "The ocean community is a very exclusive group of people," said Tommy, "a very closed community, and it's hard to keep secrets in that community unless you really put a lot of effort into it."

At Battelle, Tommy had received his secret clearances early, and he had worked on classified programs for the Department of Defense. Eventually, he had gotten top-secret clearance and then had special access clearances above top secret. He had seen tight security up close, and he understood how it worked and why it was necessary. Part of the reason he had pursued the idea of having a single large investor back the entire project was to reduce the possibility of word leaking out to the deep-ocean community that he was launching a major effort to find the fabled treasure of the *Central America*. Security was his highest priority, "almost to the point of being paranoid," remembered Fred Dauterman, "but I think it was justified."

To show to potential investors, Tommy and Bob had drafted a rough document of concepts. Robbie Hoffman had helped them write it, and although it bogged down in detail, it was Tommy's first attempt to distill between covers his theories on deep-water shipwreck recovery. It didn't show the reader where to find the *Central America,* but it told the history of the sinking, documented the gold shipment, and in Tommy's convoluted, technical language outlined the steps anyone would need to find and recover the ship. Stamped all over it were the words "confidential" and "proprietary."

Investors never even saw the document until they signed a nondisclosure statement, agreeing not to mention what they were about to see and hear to anyone outside the organization. At meeting after meeting, Tommy admonished investors not to talk about the project. When an investor became a partner, he sent them a letter and reiterated the importance of keeping quiet. "The conditions of the offering were, I can answer any questions you want, but we're not going to release this material, like why this project can work, to anybody, any outside people, anywhere."

Toward the end of March, one interested investor met with Tommy, signed the nondisclosure, listened to the presentation, and took a copy of Tommy's concept document. He had served in the navy, and he liked the sea. A week later, he called Ashby and said, "I've got Searle coming in with a guy named Kutzleb." He wanted another meeting with Tommy, and he wanted this Searle and Kutzleb to listen to what Tommy had to say about the *Central America* project. Ashby had never heard of Searle or Kutzleb and thought little of the phone call, but he called Tommy to set up the meeting. Tommy was outraged.

Robert Kutzleb owned Steadfast Oceaneering, located outside Washington, D.C., and was one of the top underwater contractors for the navy. Captain W. F. "Bill" Searle was the former supervisor of salvage for the United States Navy and now consulted all over the world for the deep-ocean community. After the *Thresher* had imploded and crashed into the seafloor at eighty-four hundred feet, he had been put in charge of finding things in the deep ocean. He had been off Palomares, Spain, with the H-bomb and southwest of the Azores looking for the nuclear submarine *Scorpion* when it exploded and went down in two miles of water. He knew everyone who worked in blue water, and he knew the details of every deep-water recovery ever attempted. He and the investor had been friends since kindergarten, and before the investor gave money to Tommy he wanted his good friend's opinion on whether the project was feasible.

After Tommy had finally persuaded a small group of tough, cynical, self-made investors to back him, a consultant from the outside who had been involved in every major deep-water recovery since man began recovering things in deep water was coming to Columbus, and Tommy knew what he was going to say.

"Anybody brought in from the outside was going to say that it was impossible. I knew that going into the meeting. And I knew who it was, one of the top consultants in the ocean community, a guru in the ocean field, a very prominent guy, a former supervisor of salvage for the navy, top position in the salvage field. The top position. So I had to figure out how to talk about it, how not to overtalk about it, how to counter everything he did. I had to explore his background. I had to have a whole database search done on exactly everything he'd done. All his experiences. What he knew about. What he didn't know about."

If Bill Searle and Bob Kutzleb cast doubt on everything Tommy had proposed so far, the partnership that had just started to blossom might suddenly wilt and die.

SEARLE AND KUTZLEB first had lunch with the investor at the Columbus Club downtown. Then their meeting with Tommy, Wayne Ashby, and Robbie Hoffman began about 2:00 P.M. in a small room upstairs in the club. The meeting did not conclude until a quarter past five.

When Searle, Kutzleb, and the investor walked into the room, Ashby glanced quickly at Tommy, because tight in the fist of one of the consultants was a copy of Tommy's concept paper. Tommy's face went from flush to the color of lead. The investor had violated the nondisclosure agreement not only by inviting two deep-ocean consultants to come question Tommy, but also by sending them and asking them to review Tommy's bible on the *Central America*.

"These guys walked in with Tommy's treasured secret book," remembered Ashby, "and Tommy just about had a heart attack."

The investor introduced Searle and Kutzleb by saying, "I just happened to have these two good friends of mine who know something about the deep ocean, and I figured you wouldn't mind if . . ."

Tommy did mind. He reminded the investor that he had violated his nondisclosure agreement by even mentioning the project to anyone outside the partnership, and that he had further violated that agreement by revealing the contents of the concept paper to consultants in the deep-ocean community. He said he would not talk to the consultants until they had signed nondisclosure agreements, which they refused to do. But Tommy reiterated that he would not talk to them about his theories until they did, so Searle and Kutzleb took a few minutes to read the two-page nondisclosure, and with an attitude that Robbie described as, Well-I-suppose-we-can-do-this-since-we-don't-believe-you-in-the-first-place, they agreed to sign, but only after they had made a few changes. Then Tommy watched them cross out nearly half of the nondisclosure.

When Tommy objected, they said, "Hey, we're in the business of salvage. What if somebody asks us to do a salvage of this ship? We won't do it?"

As bad as he thought the meeting would be, Tommy never envisioned it would get as bad as this, and it had hardly begun. "They had all our information, access to the ocean community, and they marked out the two sections that really said the important stuff, like they weren't going to compete, and signed it. It was like, 'Okay, go on.' It was terrible, but I couldn't walk out now, that really would have been bad."

With the nondisclosures marked up and signed, Searle showed a movie about the deep ocean; then he and Kutzleb talked about their experience and gave what Searle described as "a general tutorial on underwater search at these great depths." They said that if this young man had good equipment and good people—"And we can lead him to good equipment and we're good people," Searle interjected—he might locate the ship. But locating the ship was "only scratching the surface on cost," said Searle. Next he had to recover the treasure, and that couldn't be done. "First," said Searle, "it would be hellishly expensive," and that was not the worst part. Searle was an expert not only on deep-water recovery but also on the *Central America*. The ship and the aftermath of its sinking had long intrigued him, and for years he had researched and filed articles on the story. At his home in Virginia, he kept a huge box of information on the *Central America,* and he had concluded that without the federal government behind it and national security at stake, no one could get to it. The technology did not exist to do what this young man said he was going to do. "It's never been done," said Searle. He and Kutzleb were aware of every recovery ever attempted at those depths. "This includes H-bombs, airplanes, missiles, nuclear generators that NASA put up and nobody ever knew about," he explained. "Between us, we have a handle on everything that has been looked for and either not found or found and left lay. Some of it is still classified."

Tommy now had to try to refute their assertions without revealing his research on the SeaMARC's capability, or what he had learned about the bottom characteristics of the Blake Ridge, or how he had rethought the whole concept of working on the floor of the deep ocean. Robbie noted that the conversation began to deteriorate about the time Tommy opened his mouth. He got out a sentence or two, and either Searle or Kutzleb said, "That's not the way it is."

"He was immediately attacked," said Robbie, "as to location, his thoughts about recovery, everything, and it was done in kind of a sarcastic, you're-just-a-young-kid type of way. Searle was probably a little more sarcastic about it, speaking about the age difference and all their experience picking up salvage off the bottom for the navy. How could Harv do it if they couldn't?"

No matter what Tommy said, they countered with, "No, that won't work because. . . ." They pointed out that the *Central America* was somewhere under the Gulf Stream, and that working in the Gulf Stream was extremely difficult because of the currents. They reiterated that even if he could find it, he could not recover it. "Nobody's ever gone that deep," they said. "What makes you think you can do it?"

Tommy had thought about it all, and he had answers to everything, but he couldn't give them. "All the stuff it took me all those years to figure out, like about the bottom characteristics and what kind of side scan could be used. These guys didn't even know about the SeaMARC."

So he had to phrase his answers to counter their accusations without revealing too much. But sometimes he got so irritated and angry that he had been forced into this confrontation, he became emphatic. "You're just thinking about the problem in general, you're not thinking about. . . ." And then he had to catch himself and carefully consider what he was going to say next and whether he should say it. Wayne Ashby was listening, and the investor was listening, and Wayne had had so much faith in him up to now, he couldn't disappoint him, and who knew how many other people the investor would tell about what happened at this meeting. "But I was really getting into the nitty-gritty of what I knew at the time, and I didn't want to tell them how to do it."

"I'll never forget it," said Robbie. "One of the guys would interrupt and deliver his theory, and Harvey's sitting down at the end of the table, 'Nope, nope, nope, nope, nope,' shaking his head, 'nope, nope,' just the whole time. 'You old guys have no idea what's going on!'"

They would say, "Well, how're you going to do it then?"

And Tommy could hear the investor thinking, "Yeah, how're you going to do it?"

He had to answer, so then he would stop and say, "You sign those non-disclosure agreements the way they're originally drafted, and I'll tell you."

But he couldn't say that all the time, because he would appear to be avoiding difficult questions. He tried to explain the SeaMARC without saying too much about how it could be adapted to image a wooden shipwreck, and Searle interrupted, "The ship's in the rocks on the cliffs, so this piece of equipment won't work there." Tommy had the answer, but he didn't want to divulge such a critical piece of information, even if it did help him win the argument. So he countered with, "You're just making assumptions that are different than mine. What you're objecting to doesn't apply to our situation."

But sometimes they would push him until he had to say, "Well, I can't . . . I can't tell you that."

As the confrontation continued into its second hour and then the third, Ashby and Robbie noticed that Kutzleb seemed to be agreeing with Tommy on some of the issues. More than once, when Searle said, "Young man, that won't work," Kutzleb interjected, "Yeah, Bill, they're doing that now."

"Not that Kutzleb was agreeable," said Robbie. "Harvey just caught the guy with the facts, and he couldn't walk away from that."

Although Kutzleb seemed to understand more of what Tommy was saying, he still challenged most of it, especially the amount of money Tommy was proposing and the time frame he was proposing: About $5 million and three years. Kutzleb never said how much or how long, just that if it were possible, it would take longer and require more money than what Tommy had proposed.

Ashby knew little about the technical side, but after listening to the debate, he concluded, "I had more belief in Tommy than I did in these experts." Kutzleb implied that sooner or later the deep ocean would be accessible, which is what Tommy had been saying all along. And the debate reinforced that the *Central America* really did exist, there really was gold on it, and it was in the area where Tommy was looking. "I was encouraged," said Ashby.

But while Ashby was hearing that at least the project was viable, Robbie was hearing that the experts had decided that Tommy was not the one to do it, and that anyone who tried was going to need tens, maybe hundreds of millions of dollars behind them.

Searle was adamant that even if someone could develop the equipment, it was going to cost "one pocketful of money." He had been there. He had developed equipment to do fancy things on the *Scorpion* recovery. His advice to the investor was, "Don't be taken in by the odds of success towards the search. You've got to think in terms of the whole goddamn operation, search and recovery, and the critical item of that pair becomes the recovery. So when somebody says, 'I'm raising five million to go and look for the wreck,' ask if the five million includes recovery or does that just get me a photograph or some television of the wreck on the bottom, which then becomes just more enticement? That amount was not near enough to fund a recovery. If the search costs five million, you're talking about at least another order of magnitude, that is to say fifty million, to even begin to be in the ballpark for success of recovery. Maybe orders of magnitude. I've been in situations where it's almost like this but the object was not gold. The object was some weapon or some object of great national interest and value."

Searle compared a project to recover the *Central America* with the *Glomar Explorer* project, the top-secret attempt to recover the Russian submarine in 1974, when the government poured hundreds of millions of dollars into the project, and he estimated that the *Central America* project would be more difficult, "because you don't have any idea where to look." Then he put the whole thing into perspective. "Say this had been proposed to me: Do I think he can do it for ten million? The odds of that happening are maybe one percent. It's remote. Man, I have been on several operations and we were trying to do things like that, but they didn't work that smoothly, and we had the whole goddamn United States treasury behind us!"

Despite the concerns of Searle and Kutzleb, Searle's friend invested in the project, although less than he had planned. Within three months, Tommy had thirty-eight partners who had invested a total of $200,000 in him. Their vote of confidence said, We like you, we trust you, your idea seems sound; here is enough money to mold that idea into a project we all can see and hear and touch. Now get your concepts down, finish your historical research, arrange for contracts with your suppliers, consult with your expert on search theory, start looking for a vessel, and line up the SeaMARC. Then come back to us and show us what you've done, and if we like what we see, we'll give you more.

Off the
Carolina Coast

⚓

Midnight, Saturday, September 12, 1857

About six that evening, a man-of-war hawk had tumbled from the grayness, then swooped across the quarterdeck, grazing Captain Anders Johnsen's shoulder as he stood with the helmsman and two of his crew. Seamen's superstitions long had held that an encounter with birds far at sea was a harbinger of danger ahead, but Johnsen ignored the bird. The bird flitted into the rigging, beating its wings incessantly, and then it dived again and circled the captain's head in what he later described as "an extraordinary maneuver." He ducked away, the bird darted over the deck, banked, and for the third time came at him, aiming for his face. When it neared, wings flailing, the captain grabbed it by the throat.

The bird was like none Johnsen had ever seen. Its feathers were iron gray, its body a foot and a half long, and its wings tip to tip more than twice its length. The beak was a weapon, eight inches long and lined with teeth like a hacksaw. They ripped into Johnsen's right thumb, then the bird bit two of the crew as they tied its legs. The bird lunged at everyone who came near, until Johnsen told one of his men to cut off its head and throw the body overboard.

The bow of the *Ellen* heaved in a sea pocked with hills of water blown white by a wind still harsh. On the 17th of August the Norwegian bark had departed Belize, her hold laden with mahogany logs, her course aimed across the Caribbean Sea for the Straits of Florida to catch the Gulf Stream, which she would ride most of the way to Falmouth, England. Caught in the storm, she had made considerable water, lost most of her shrouds, and had her foremast ripped from the deck.

As the storm's intensity lessened late Saturday, Captain Johnsen tried to follow an eastern heading, but the wind had forced him to alter his course to just north of northeast. He had kept to the new course for a short while until the man-of-war hawk appeared in his rigging, and when the bird fluttered about his head for the third time, he resolved to reset his course to the original. As he explained later, "I regarded the appearance of the bird as an omen and an indication to me that I must change my course. I accordingly headed to the eastward direct. I should not have deviated from my course had not the bird visited my ship."

The storm continued on the wane, and for the next few hours under full sail the ship followed her original course through heavy seas and a strong breeze, making another twenty miles due east. Around 1:00 A.M., with the helmsman on deck and the rest of the crew asleep below, strange cries near the vessel startled them awake, distinct cries, obviously not natural, but muffled by the wind and the splashing of the sea. Johnsen rushed topside. "In a moment the agonizing shrieks, as it seemed of a hundred human voices, were plainly distinguishable. I at once knew that we must be in the vicinity of a wreck, and immediately roused every man on board. In less than a minute I found that we were surrounded with persons floating in the water. The darkness of the night made it impossible to see them, but the voices calling for aid rang in my ears from every direction."

Johnsen and his men secured ropes and tossed them over the side, and in a few minutes they had raised four men from the sea, not one of whom could speak. A few minutes later they picked up another man, who cried out for his wife and something to eat. Johnsen reported later, "I could not learn from any of them from what vessel they were wrecked or what disaster had happened." They heard more voices coming at them out of the dark, and Johnsen ordered his men to tie off the three life buoys and pitch them into the water, hang as many ropes from the sides of the ship as possible, and set out additional lanterns, so the ship could be seen from all sides.

They raised more men from the water ten feet to the deck, one of whom was a husky fellow they found floating on a board. He was the only one who could speak. He told Johnsen that about five hundred men had gone down in the sinking of a big steamer and that most of those men, he was certain, had perished, though many still remained alive floating on pieces of the wreck.

Johnsen had four small boats on his vessel, and he ordered his men to cut the lashes on one of them and lower it into the water. The moment it touched the waves, six men turned it keel up, but Johnsen rescued all six. He and his crew now stood alert at the gunwales port and starboard, but the heavy sea that continued to roll and the noise on board and the whistling of the wind through the rigging made the voices of exhausted men difficult to hear.

John George floated alone in the dark, when suddenly he heard a score of voices around him crying, "Ship ahoy!" When he raised his head, he saw lanterns hung from a ship no more than half a mile away. He began to shout with the others, but the lights glided silently by and soon receded in the distance. Then the lights turned round and headed back his way. With the ship approaching, he could make out six men clinging to a log of wood and also drifting toward the ship. The lights drew closer and closer, until he could make out her hull and saw one of her masts, and when she was but a short distance away, he saw two of the men on the log lose their grip and slip quietly into the sea. As he floated nearer and nearer to the ship and then shouted and felt himself being raised from the water, the other four men on the log must have shared the same fate as the first two, for he never saw them again. When he got on deck, he could not stand up.

"I do not know whether I cried or not," he said, "but I know I was astonished to hear my own laughter ringing in my ears. I do not know why I laughed. That verse, 'God moves in a mysterious way,' kept passing in and out of me—through me, rather—as if I had been the pipe of an organ."

Most of the men were so exhausted they had no strength left to assist in their own rescue. Seventeen-year-old Henry O'Connor had been in the water for seven hours when he floated near the bark and the sailors threw him a rope. He grabbed it, and as he did, two other men got hold. He asked them to let go, and they did so, and he twisted the rope around his waist, and held it to, until the crew hauled him up. He did not know if the crew also rescued the other two men. "When I got aboard, they helped me down into the cabin, as I was not able to walk. I went to sleep immediately."

Dr. Harvey and Second Officer Frazer still floated together, Frazer beyond exhaustion, chilled, and sleepy. But when they saw a lighted ship in the distance, they paddled together against the current, and Frazer still had enough strength in his hands to hold to a rope while they drew him out of the water. The sailors threw another rope to Dr. Harvey and raised him just even with the deck of the vessel when his cramped hands loosened and he fell. Three times he fell back into the water, twice sucking under the ship, until the crew threw over a ladder. "To which," remembered the doctor, "as the last desperate effort, I tangled myself into in some way, and was taken on board at about three o'clock A.M. We were insensible for some time."

The poet Oliver Manlove was picked up about an hour later. His first sight of the ship seemed an illusion, for all he saw was what looked like a star suddenly rise from the waves and in a moment disappear. Then he saw it rise again, and he realized it was a ship's light, and that the bow of the ship was headed for him. He yelled and someone threw him a rope, and he grabbed hold and felt himself rising from the water, and when he was about six feet above the waves, his fingers unraveled from their grip and he fell and went under the bark. His life vest popped him up again and the sailors got him on deck, where he could hardly stand.

For hours, Captain Johnsen continued to tack among the driftwood, crossing the wind on one leg after another, keeping the bow of the *Ellen* working its way to windward, back to where the steamer had gone down. By four in the morning, he had rescued forty-four of the *Central America*'s passengers and crew. Out of the total, only two could stand up, talk, and help with the rescue of fellow survivors: One was the husky fellow on the board who had enlightened Johnsen on the shipwreck, merchant sea captain Thomas Badger; the other was Ansel Easton.

Easton had not seen the *Ellen* until she was suddenly before him as if dropped from the clouds. He grasped the rope thrown to him and ascended to the deck, where the crew offered him dry clothing, and he stayed on deck to help the other survivors. At times he went to the gunwale and shouted into the dark the names of his friends who had gone down. Mostly he yelled "Brown!" for aboard the steamer Robert Brown was his closest friend, the one who had brought him a good cork life vest on the hurricane deck moments before the ship went down. He told Captain Johnsen he had a friend he wanted to save, and he would pay the captain whatever he wanted if he would stay out there and try to find his friend. Johnsen refused compensation but said he would stay as long as there appeared to be someone left to save.

At daybreak they were still tacking back and forth through the debris, when they found one man alone and then two more together. A few more hours passed, and they found no one. The wind and the sea had steadily dropped throughout the night, the sun was approaching midmorning, and already the day was turning hot. Johnsen said he thought they had rescued all of the survivors they could and now should head for the nearest port. Easton persuaded him to tack just once more, and if they found no one, then they could sail for land. Johnsen came about for one last leg through the area with Easton at the rail shouting for Brown.

Around three in the morning, Captain Johnsen had seen two men on a hatchway, but before he could tack to get closer, they had disappeared. Now, six hours later, they saw the same hatchway, and Johnsen got close enough for a sailor to throw the two men a rope, which they caught, but they could not hold on. Johnsen had to tack twice more, and on the second, the hatchway drifted alongside, and they pulled the two men on deck. One was a man named John Dement. The other was Robert Brown.

Until noon on Sunday they continued looking for survivors, but the wreckage was gone now, no sign of a hurricane or a tragedy in the gentle swells and breezes. Johnsen later reported, "As the morning advanced, and we found ourselves at a considerable distance from the place where the steamer sank, I bore down for it again. It was in longitude 76° 13', latitude 31° 55'. I first took an observation at eight o'clock in the morning and again at noon. I saw the schooner spoken by the *Central America* before she went down, and do not think she picked up any passengers. After picking up all the persons I could discover, I sailed for Norfolk."

THE STORM WAS one of the fiercest in recent memory. A smaller steamer, the *Southerner,* put into Charleston in distress, her stack, her paddle-boxes, and her lifeboats washed away, part of her cargo thrown overboard, and six feet of water in her hold. The storm had destroyed so much of the *Southerner* that a crew patched her together only enough to sail her back to New York without passengers or cargo and dismantle her. A mate on another ship testified that "the gale was one of the severest we ever experienced." The *Columbia,* bound from New York to Charleston, hit the storm about Frying Pan Shoals, forty miles out from the mouth of the Cape Fear River, and one member of the crew called it "the most driving hurricane we have ever seen or conceived. It prostrated the awful seas which had come from the broad ocean and appeared to sweep its surface along in spray and foam with lightning power and velocity." The editor of the *Charleston Daily Courier* called it a storm "of almost unprecedented fury and violence."

Onshore, the steamship company and the citizenry knew nothing of the tragedy at sea, nothing of the rescue or the final moments; they had only the news telegraphed from New Orleans to New York that the *Central America* had arrived safely in Havana and had departed under full steam on the 8th of September.

An hour after the *Central America* departed Havana, another smaller steamer, the *Empire City,* had also left the harbor bound for New York. At sunset of the first day, the *Empire City*'s Captain John McGowan could still see smoke from the stack of the bigger steamer to the north. But the next day and the next, the *Empire City* had fallen farther behind. When the storm reached its height, the sea had broken over the decks of the *Empire City,* sweeping away the forward houses on her paddle guards and her entire starboard paddle-box. The wind had blown to pieces every one of her sails, water had swirled through the cabins, she had run out of coal, Captain McGowan had burned every chair, every table, everything movable, even the wheelhouses, to keep up steam, and the ship had pitched and rolled so violently that the bolts anchoring her boilers had snapped in two. But she had not foundered and fallen into the trough of the sea. Still afloat, she had limped into Norfolk. During the ensuing week, she was repaired and outfitted, and McGowan was directed to return to the steamship routes in search of the *Central America.*

Before departing Norfolk early Friday morning, September 18, McGowan was asleep in his quarters aboard the *Empire City* when a loud rap at his door awakened him.

"Who is it?" he yelled.

A voice said, "Easton."

Captain McGowan knew Ansel Easton, because Ansel supplied the steamship lines with fixtures and furnishings and frequently traveled aboard the steamers back east. The morning that both ships lay at anchor in Havana, McGowan had boarded to congratulate the Eastons on their marriage.

McGowan leaped out of bed and swung open the door and saw Ansel there with four other men, all of them dirty and haggard, their clothes stiff from hours in the salt water and days of baking in the sun.

"My God, man," he shouted, "where did you come from?"

The men crowded into McGowan's quarters without sitting and told him of the sinking and their rescue by the crew of the bark *Ellen* and their arrival off Cape Henry only the day before. In Chesapeake Bay, Easton had hailed a pilot boat and paid the captain to ferry them into Norfolk from the *Ellen,* which still was under tow in the bay. When Ansel had spied the *Empire City* at anchor, he had stopped the pilot boat to find McGowan and ask after the *Marine.* Ansel had not seen the brig since he floated free of the steamer and thought he descried her lights far to the lee. When McGowan assured him that news of her fate had not reached Norfolk, Ansel and the other men returned to the pilot boat to resume their trip into Norfolk.

W HILE OTHER SHIPS weathered the storm and reported back to ports along the East Coast, the *Marine* sat becalmed from Tuesday through Thursday. The winds that had driven down the *Central America* and ripped her sails into tatters had dropped to a strong breeze and then a fresh breeze and then dwindled and then disappeared altogether, leaving a hot, insidious stillness.

As the *Marine* sat becalmed, the stewardess Lucy Dawson died. She had fallen three times trying to get into the lifeboat and was pinched between the boat and the side of the steamer. Some said she died of fright; some said fever. She was the only woman who did not survive.

On Wednesday they spoke the *Euphrasia,* a brig bound for New Orleans, and her captain, refusing payment, supplied them with two barrels of sea biscuits, two barrels of potatoes, three hams, six chickens, cheese, water, and a quantity of coffee, tea, and sugar. He then offered to take any passengers who desired on to New Orleans.

On the morning of the sixth day, now Friday, they awoke to find themselves drifting within sight of the Cape Henry lighthouse at the mouth of the Chesapeake Bay outside Norfolk. But they still remained sixteen miles away, and the brig was weakened and listless in the absence of wind. A tug captain, upon hearing of their misfortune, announced that they could remain at the mercy of the wind, unless they paid his five-hundred-dollar fee for a tow into Norfolk. Captain Burt explained that in the storm and the confusion of loading the lifeboats, everyone had left what money they had on the sinking steamer. Most had lost everything they owned, and Captain Burt was down to the last day he could provide food and water for them. But the tug captain demanded his fee before he would tie on. Eventually, the passengers scraped together three hundred dollars, which the captain accepted.

Nearly a week had passed since the lights of the *Central America* had suddenly disappeared in the distant dark and the bos'n had returned with the tidings that she had gone down and taken all hands with her. The passengers rescued by the *Marine* had seen no one else since the sinking, and they knew nothing of the fate of the men they had left behind.

As the *Marine* rounded Cape Henry and entered Chesapeake Bay under tow, a harbor pilot boarded to guide them safely across the bay and around to Hampton Roads to the docks outside Norfolk. He brought with him the first news they had heard: Forty-nine survivors from the *Central America* had arrived in Norfolk aboard a bark that very morning. That meant that for every ten men who had remained with the ship, only one had survived, and the harbor pilot knew none of the names.

"I felt very sad and downcast and scarcely spoke," wrote Addie Easton, "for I knew my hopes were soon to be realized or I must yield to despair."

CAPTAIN McGOWAN HAD the firemen and coal passers stoke the fires of the *Empire City* and ordered her into the bay. He knew the fate of the *Central America,* but no one knew what had happened to the brig *Marine.* Within an hour he encountered the *Ellen* and the other forty-five men who had survived the sinking. After transferring most of them to his ship to continue their voyage to New York, he steamed across the Chesapeake.

About noon, as he approached the mouth at Cape Henry, the men spied the *Marine* in tow behind the steam tug and recognized her storm-torn profile. Many of the women on the *Marine* also recognized the *Empire City,* and as the two ships came closer and closer, the women realized that the men lining the rail were the survivors of those left behind on the *Central America.* The purser for the *Empire City* looked down on the *Marine*'s low decks and saw them "swarming with wretched-looking objects, many of them women and children, wringing their hands, and weeping and laughing, by turns, hysterically."

A ship reporter watched how frantically the women looked up and searched the faces of the men who now lined the rail of the *Empire City.* "The eager scanning of each face in agonizing fear and expectation, the joy or grief manifested as recognition or disappointment awaited the gazer, was touching in the extreme, straining the heart-strings." Two dozen women had left their husbands behind, but only two recognized a face among the bearded men. Mary Segur found her husband, Benjamin, and another woman saw her seventeen-year-old son, Henry O'Connor.

McGowan boarded the *Marine,* where he was "caressed, embraced, and indeed half strangled by the poor women, who threw themselves upon him as he reached the deck." His first inquiry was for Mrs. Easton.

"Her husband is awaiting her arrival in Norfolk!" he sang out.

"I scarcely knew what I did for a few moments," Addie later wrote. "A number of ladies threw their arms around me and kissed me while the captain and other gentlemen, and even the rough sailors, shook me heartily by the hand, and congratulated me for the safety of my dear husband.

"As the captain came aboard he took me by the hand and we both felt too deeply to speak for some minutes. Then he said: 'Let us sit down here, for I must tell you all about it. He is safe and as hale and hearty as ever, only very anxious about you.'"

McGowan transferred to the *Empire City* most of the passengers rescued by the *Marine,* and after raising eight hundred dollars to present to Captain Burt and his crew, they continued the last leg of their journey to New York. The remaining passengers, including Addie Easton, remained on board the *Marine* and spent the afternoon crossing the Chesapeake behind the pilot boat. They reached the quarantine ground at Hampton Roads just after dark, and crewmen in small boats then rowed them several miles from the docks to the city. From there, "the forlorn little procession," as Addie described it, walked up to Norfolk's principal establishment, the National Hotel. Addie still wore the nightdress and wrapper she had worn as Ansel and Robert Brown had helped lower her to the lifeboat for the trip to the *Marine* almost a week before. News of the disaster by now had reached the city, and by the light of the gas lamps, the townsfolk recognized the shipwreck survivors trudging through the streets. "A perfect crowd," wrote Addie, "followed us to the hotel."

When she reached the hotel, Addie looked quickly from face to face, expecting to see Ansel, but he was not among those there to greet the survivors. He had discovered that the *Marine* was at quarantine, and in his impatience to find Addie, he immediately had left the hotel with Captain Johnsen, and in a small boat the two men had rowed out to the brig, unwittingly gliding by Addie in the dark. Another hour passed before Ansel returned to the hotel, and there he found his bride of four weeks in the immense parlor, she and the other women surrounded by the proprietor and maids ministering to their needs and by citizens aghast at the stories they had to tell. They embraced and did not speak, and whatever thoughts ran through their minds during those moments they either could not remember or chose not to reveal.

"Our meeting, I will pass over," Addie wrote. "We wept together as well as rejoiced and for several nights after we could neither of us sleep, so vivid were the scenes before us that we had passed through. My watch, my beautiful ring, wedding presents and many other things I valued from their associations were all lost. Though I shall never behold them again I still have the blessed privilege of preserving them in memory and I have my darling husband, the most precious jewel of all."

On Friday morning via telegraph, news of the disaster had hit the streets of New York. The *New York Times* announced in headlines as big as it had ever printed:

CENTRAL AMERICA FOUNDERED

FIVE HUNDRED AND SIXTY-FIVE LIVES LOST

ONLY SIXTY SAVED

When the *Empire City* arrived in New York on Sunday morning, nearly one hundred survivors disembarked. For two days, the citizens of New York had read about the shipwreck in articles telegraphed up from Savannah, Charleston, and Norfolk. Hundreds of people had waited all night Friday and Saturday at the telegraph office for more news of the disaster. Speaking in the parlor of a hotel, survivor Jane Harris reminded reporters of what had happened. "The ladies that you see around here were all passengers on the *Central America*. We do not appear in such good condition now as when we started. We have all suffered much, and the sufferings of some are not yet at an end. Many of these ladies have been made widows, and many of these little children have been made orphans by the loss of that steamer."

Of the nearly 600 souls aboard the *Central America,* 149 were saved: 30 women, 26 children, and 44 men taken on board the *Marine,* and 49 rescued by the bark *Ellen.* Besides the Eastons and Mary and Benjamin Segur, Thomas Badger and his wife, Jane, met at a train stop in Baltimore; and Billy and Virginia Birch found each other in New York.

But the stories of happy reunions were few. Mary Swan, whom Captain Herndon had promised to see safely to New York, arrived with a baby less than two and her husband recently lost to the sea. When the *Empire City* docked in New York, she burst into tears. "Where shall I go after I go ashore? I have no friends in New York, nor in all the world, now that my husband is lost."

Winifred Fallon, seventeen, had gone to California with her father and younger brother in April, after her mother had died. Four months later, they returned east on the steamer. "We saved nothing but our lives," she told a reporter. "I have not heard from my father since I left him on Saturday. I saved not a cent—nothing but one shawl and a dress."

Several of those interviewed said, "I have lost everything but my life."

Ansel and Addie Easton repaired to the Metropolitan Hotel, where Addie wrote a detailed letter of everything that had happened since they set sail from San Francisco. Family had awaited their arrival in New York for a week, one day mourning their loss, the next in receipt of a dispatch from Ansel in Norfolk announcing they both had survived.

"You never saw such a rejoicing as there was when we arrived," wrote Addie. "For one week after we did nothing but receive calls and congratulations. A perfect crowd all the time, and we were real live curiosities. I can't go in a store without hearing a whisper 'There goes Mrs. Easton.' I must now close hastily, without giving you particulars since our arrival. I have had so many interruptions, and spent so much time on this letter, that I shall not be able to write even to brother Edgar. So please after reading this to sister Fannie enclose it as soon as you can to my dear brother. It has been a severe task to me to recall those trying times, and I feel that I cannot even set down to write this all again."

With those words, Addie Easton concluded her account of the disaster in a letter to her friend Jenny Page of San Francisco. The Eastons later returned to their home in California, where after the Civil War Addie gave birth to a son and then a daughter. Eleven years after the sinking, Ansel was thrown from his beloved racehorse, Black Hawk, and died of his injuries. He was forty-nine. When their daughter Jenny died giving birth to her third child, Addie raised the three grandchildren and lived to be eighty-six. Still residing in the collection of the San Mateo County Historical Association is the little blue note Ansel penned to Addie, safe aboard the brig *Marine,* as the *Central America* lay sinking beneath him.

SURVIVORS AND VICTIMS of the *Central America* tragedy hailed from twelve foreign countries and every one of the thirty-one states. Within hours after the news arrived in Charleston, disaster headlines appeared on the front page of virtually every major newspaper in the country, from New Orleans to Boston, and as far inland as Dubuque, Iowa. For the next three days, reporters met rescue ships landing in Savannah, Norfolk, and New York. Newspapers could hardly satisfy the public's appetite for the lurid details of the sinking. Many ran over ten thousand words on the incident in a single issue, rife with poignance, some featuring woodcuts depicting disaster scenes as described by the survivors. Nearly sixty survivors gave statements to reporters, and many of them spoke more than once with the press. In all, 212 newspapers ran over fifteen hundred articles on the disaster from its first reporting to the official investigations into the causes of the tragedy.

The *New York Times* later reported, "The full horror of their position is not unparalleled indeed in the desolate annals of the ocean, but . . . no story so clear and so appalling has ever before been brought to the firesides of the land." The praise for Captain Burt of the brig *Marine* and Captain Johnsen of the bark *Ellen* was effusive. "The conduct of Captain Burt was noble in the extreme," wrote a reporter for the *Philadelphia Daily Evening Bulletin*. "Like a noble-hearted sailor, and a true and gallant man, he bore down with his half-wrecked brig to the aid of those whose dangers and necessities were greater than his own." For his efforts, Captain Johnsen received a gold pocket chronometer from President James Buchanan.

But the praise was greatest for the captain who did not return. A survivor reported that when the final moment of the sinking came, he saw Captain Herndon standing upon the wheelhouse, his trumpet in hand. Then Herndon had disappeared into the waves, and no one had seen him again. Everywhere, his loss was mourned. The *New York Times* wrote of the "calm, deliberate, enduring courage of a truly brave man."

Frank Leslie's Illustrated said, "There cannot be a doubt about the fact that the name of Captain Herndon will ever be held in grateful remembrance among all the heroes who have achieved triumphs upon the sea." In England, the *Liverpool Post* noted that in Captain Herndon "the finest part of chivalry appeared"; indeed, the conduct of every man, woman, and child aboard the *Central America* testified "irresistibly in favor of the high tone of the American mind. Their country ought to be proud of them."

Captain Herndon's cousin and brother-in-law, the father of modern oceanography, Matthew Fontaine Maury, wrote Herndon's eulogy to the secretary of the navy. He wanted the world to know that a man had stood flatfooted at his post and lived an oath taken on a calm sunny day that he would not leave his ship regardless of the danger. "A cry arose from the sea," wrote Maury, "but not from his lips. The waves had closed about him, and the curtain of night was drawn over one of the most sublime moral spectacles that the sea ever saw. . . . Forgetful of self, mindful of others, his life was beautiful to the last; and in his death he has added a new glory to the annals of the sea."

The last sentence is chiseled on a granite monument, twenty-one feet tall, which the officers of the United States Navy erected at Annapolis in 1860, to honor Captain Herndon for heroism.

Columbus, Ohio

⚓

Fall, 1984

THE DATA CORRELATION Matrix on the wall at Bob Evans's house became a Lotus spreadsheet twelve feet wide, twelve feet high. Fifty-nine survivors and ship captains had given their stories to the press in 1857, and thirty-three had made comments that would help locate the site. The names of those thirty-three ran across the top of the spreadsheet, four names to a page. Heading down, each of fifteen pages contained three time blocks, each block representing three hours, beginning Tuesday at noon, just after the *Central America* had left Havana Harbor, and running all the way through Sunday midnight, when the *Ellen* already had been underway for Norfolk half a day.

Besides an intimate understanding of the story, Bob was looking for three things: coordinates or distances and their time; the velocity and direction of winds and the height and direction of seas; the condition of the vessel. Whenever a witness mentioned one of these, he placed it in the appropriate time slot under the witness's name. The idea was to reduce as much of the oral information as possible to physical data.

After exhausting the resources at the Ohio State libraries, Bob next went to the Library of Congress, then the National Archives, and the Smithsonian, and the Army Office of Military History. He visited and corresponded with librarians at state and local libraries in New York, Norfolk, Savannah, Boston, Philadelphia, Charleston, San Francisco, and Sacramento. He contacted the Webb Institute of Naval Architecture and the Mariner's Museum and the museum in Mystic Seaport. Everything that could affect the position of the steamer's sinking went onto the Data Correlation Matrix. The rest went into Bob's head, and a movie of the storm continued to evolve; as new bits of information came to the script, he rewrote the scenes, until he could close his eyes and see the ship tossing in the storm, the women and children huddled in the saloon, the men passing out in the bailing lines, and the water rising in the hold.

"I immersed myself in this stuff, because we needed every piece of data that could have bearing on the sinking location. We pulled all the human interest in as well, but the stuff that went up on the matrix were things about the sails being ripped to shreds at ten o'clock on Friday morning, things from which I could deduce wind strengths and wave heights and how far down the vessel had sunk at a certain point, when the water reached the lower deck and when it flooded the boilers."

One witness might say the same thing as another witness, but there would be a difference of two hours, and Bob would ponder the discrepancy. Which account was more likely to have occurred when the witness said it occurred? Did either witness have a reason for remembering the time? Have both witnesses consistently reported other events accurately?

On the matrix in the Friday morning slot appeared, "Vessel careened over on her starboard side and we heard the beams crack."

"That's an interesting comment from Virginia Birch," noted Bob. "None of the men put it quite as colorfully as that. She has a moment, 'heard the beams crack,' or at least as far as she was concerned that's what had happened. Friday morning is all she says, so I put it in the 9:00 slot just because that's a nice midmorning slot to aim at. Right next to that you've got J. A. Foster reporting at 11:00, 'Ship over on beam end.' He commented about it once it was over on its beam end, he didn't actually talk about it happening. So you see, there's a different kind of information there. But it's interesting that Virginia talked about the actual careening over; she probably had a colorful experience associated with that—who knows, maybe she slipped and fell or something happened that was a very kinetic act for her."

Another passenger had reported that "the storm was raging with unabated fury" the minute they left Havana, but Virginia Birch said, "The nine o'clock weather was fair." She continued, "On Wednesday at three o'clock the winds commenced to blow with increasing fury."

"That's why the Data Correlation Matrix became so important," said Bob, "because whereas one man may remember that 'The storm raged with unabated fury from the time we left Havana,' Virginia Birch, a dance hall girl from San Francisco, knows that it was quite pleasant when they left. This other fellow may have been in his bunk when they left, but the ladies were out on the hurricane deck, enjoying the scenery. Then Virginia says, 'On Thursday, we passed another fearful day, the vessel rocking and pitching violently,' and yet another passenger reports, 'The storm arose Thursday night, increased until Saturday morning.' Until you get thirty-three of these next to each other, it's really hard to figure out what the approximate truth is."

Bob kept trying to squeeze objective data out of the subjective observations of a largely unscientific group of observers. He found dozens of clues that would help: how high the water rose within the vessel and when it got that high; how hard the wind blew on the second day and the third day and the fourth day out of Havana; when the foremast was cut away, or the drag sail launched, or the spanker torn to shreds off the mizzen. He stuffed his head with thousands of bits of arcane history, and his mind never stopped making connections. As he read the accounts, he would walk up and down, gazing at the matrix, weighing the information, imagining the scenes, pacing a little more, thinking about how each scene was related to another, trying to reconcile that one observer said that at this time a heavy gale blew, while another opined that the gale then seemed only moderate. Or that witnesses reported sighting the *Marine* shortly after noon, or around noon, or about two o'clock, or about three o'clock. He rationalized that with six hundred people on the ship, not all of them were at the same spot at the same time to witness the same event. Some were in cabins, others in steerage, some were on deck, others below; as the storm reached its height, most were bailing. And that was another thing he thought about as he scrutinized the stories.

"A lot of this testimony came from men who first of all had worked for thirty hours bailing water, then they were in the water for eight or ten hours, with no hope of salvation whatsoever, at night, in the middle of nowhere, obviously going to die. And then they're rescued by this strange foreign vessel with a captain who has a heavy German-Swedish-Norwegian accent, who tells them some kind of story about this mysterious bird that led him to the spot."

Coordinates were the most important information and therefore the most scrutinized. When Bob saw a coordinate, he considered the source and imagined the circumstances. Most of the coordinates from the *Central America* progressed from Havana up the East Coast, and those from other captains fit a pattern of rescue in the vicinity or subsequent encounters with survivors. But one coordinate continued to baffle him, the same coordinate Tommy had found early on in the official report by the New York Board of Underwriters. The men who drafted the report had included it because the captain of the *El Dorado* gave that coordinate as the point where he had encountered the *Central America* only ninety minutes before she sank. It was the coordinate that had confused Bob and Tommy earlier, because it was too far away from two coordinates taken aboard other ships. It was the same coordinate Harry John apparently had relied on for X-marks-the-spot in his unsuccessful search for the steamer a year and a half earlier. Bob wouldn't use the coordinate until he could find out more about it, and right now it still made no sense. As he dug deeper into the documents, he pieced together this story:

Upon sailing into Boston, the captain of the *El Dorado* had met with scandal: Ninety minutes before the *Central America* sank, he had been close enough to shout across the water and be heard, and Captain

Herndon had requested him to lay by till morning, that he was sinking, but the *El Dorado* captain had failed to rescue even one of the desperate men in the water. Such behavior during the worst ship disaster in American history raised the public ire when the *El Dorado* arrived in port, and the captain spoke only briefly with reporters, then disappeared. With the captain in hiding, the first mate had to answer reporters' questions, and during this public interrogation, he produced the ship's log, which recorded several coordinates, and right there in the log was the site of the sinking: 31° 25' north, 77° 10' west, the same numbers that appeared in the official report. But to Bob, the coordinate looked suspicious.

"I'm sure that nobody else looking at this thing had ever thought that it doesn't really look right, but it doesn't; 31° 25' latitude, and 77° 10' longitude, very precise coordinates rounded off to the nearest five minutes." And that was the problem: They were too precise.

Bob studied the *El Dorado* log and noticed that with one exception, every other coordinate entered was latitude only, how far north of the equator the ship had progressed. Latitude was the easy number to calculate, the part that a captain could determine by merely spotting familiar landmarks on shore or by "dead reckoning," a sixth sense sailors developed at sea that enabled them to read the surface for wind speed, factor in the performance of their ship in similar seas, consider the distance from port to port and the latitude they had shot the day before at the solar meridian, and determine within five miles where they were.

But longitude, the position of the ship east to west, was so difficult to determine that in 1714 the English Parliament had passed the Longitude Act, offering twenty thousand pounds, the modern equivalent of millions of dollars, as a prize to anyone who could devise a method for determining longitude to within half a degree, or thirty miles. "Discovering the longitude" became synonymous with attempting the impossible. After forty years of continuous effort, John Harrison had finally claimed a portion of the prize with a chronometer, or watch, so complex that a renowned watchmaker required two years to duplicate it. Even in the mid–nineteenth century, most mariners couldn't afford a chronometer, and only the experienced mariner, like Herndon, could calculate longitude using a chronometer and sextant. Captains sailing the East Coast of the United States often reported only latitude. In the *El Dorado* log, Bob saw entry after entry like these: "blowing perfect gale from north northeast, latitude 29° 50' North." Only once, when the *El Dorado* neared the Florida coast and the captain identified a landmark, did he record longitude as well as latitude.

"Until he's sailing past the *Central America,*" thought Bob, "in the middle of the storm, at a time when they could not have gotten a coordinate, and then he renders this extremely precise coordinate." How could the captain of the *El Dorado* take a reading of anything at six-thirty in the evening? Even if the wind had been calm and the water like glass, there was nothing to shoot with his sextant. "This seemed kind of weird to me," said Bob.

He restudied every account concerning the encounter between the two ships, and he found some confusing language in the way the story was reported. Most of the accounts said that Captain Herndon had yelled "our situation," but one account in the *New York Tribune* said that Herndon had shouted to the other captain "our position." Bob surmised that in mid-nineteenth-century English the two could be synonymous; and synony-mous or not, the writers of the other accounts could easily have misconstrued the seaman's language spoken by some of the witnesses; and whether the words were synonymous or misconstrued, they supported Bob's growing suspicion that the captain of the *El Dorado* had gotten that coordinate from someone else.

He imagined the scene: Over five hundred men gathered on the deck of the *Central America,* the steamer sinking beneath them, the storm still high, and out of the storm comes another ship and heaves to around their stern. Captain Herndon shouts through his bullhorn as each captain strains to hear the other, the roar of five hundred desperate men deafening, Captain Herndon wanting the other captain at least to know where he is, and somehow the numbers loft across the water from the *Central America* to the *El Dorado,* and either the other captain or his mate takes them down.

"They must have gotten it from Herndon," concluded Bob. And that answered the question of how such a precise coordinate ended up among all of the half coordinates in the *El Dorado* log. But that raised another question: Did Herndon get the coordinate at six o'clock in the evening from a pitching and rolling deck in the middle of a hurricane with five hundred men facing a watery grave? If not, where did he get it? And in the answer to that question lay the key to finding the ship.

I N HIS RESEARCH, Bob had seen references to the *New York Herald,* the major newspaper of the day, and although he had read none of its articles, he knew the *Herald* had covered the sinking of the *Central America* in great detail. The OSU library did not have the *Herald,* but in the spring of 1985, Bob traveled to New York to visit a brother and to look for sources not available in Columbus. There he found dozens of articles in the *Herald,* and he read them over and over. Some were verbatim accounts he had seen in other papers, others were fresh information, and one of the latter appeared on September 27, 1857, a three-thousand-word interview with Judge Monson. A third of the way through the article appeared this paragraph:

> On Friday afternoon Capt. Herndon came to my stateroom and asked permission to remove his instruments there from his stateroom on the upper deck. There was a possibility, he said, that his stateroom, from its exposed position, might be swept away, and his instruments. I complied with the request of course, and the instruments were removed accordingly.

To Bob that made sense. The captain's quarters on the main deck were built of two-inch lumber, and Friday afternoon the storm had approached its height. Already it had ripped one of the lifeboats from its davits, sent it scudding across the deck, and slammed it against the deckhouse. Soon the deckhouse itself could weaken, buckle, and be swept away, and with it the most important items Captain Herndon kept in his quarters, his navigational instruments. Bob continued reading, slowly, trying to visualize the scene.

> Shortly after daybreak on Saturday morning the clouds cleared away somewhat, and the passengers and crew felt greatly encouraged. The storm, they believed, had passed its culminating point. Capt.

Herndon stated to the men at the pumps that he thought the storm had abated, and that if they would keep at work bailing till noon the steamer might be saved. The captain made the same statement in the cabin in hearing of all the passengers who were there.

Monson's account went on to describe how the captain's words brought cheer to the men at the pumps and sent joy and gladness to the hearts of the lady passengers, and how the men worked with increased vigor and for a few hours had gained on the water rising in the hold. But soon, reported Monson, the storm had returned with an even greater fury. Then followed a curious paragraph.

About 8 o'clock Saturday morning Capt. Herndon came into my stateroom. I had been an old acquaintance of his, occupied a seat on his left at table, and had been his partner at whist on the previous evening of the trip out of Havana. The captain told me then that there was no hope for us unless the storm abated soon or some vessel hove in sight. I presume I was the only person on board to whom he communicated that fact. The captain was perfectly calm, and intimated that it was but to keep up the courage of the passengers and crew until the last moment.

The copy was on microfilm, white on dark gray, the letters slightly puffy, a few of the words illegible. Bob had to read each paragraph several times, deciphering as he went, and as he was reading this last paragraph for the fourth or fifth time, he asked himself, "Now, did the captain go there specifically to act depressed and talk to Judge Monson?" That didn't seem likely. He had a ship full of people who were panicked and exhausted, and he would not be in Judge Monson's stateroom unless he had good reason; baring his soul to the judge was not good enough.

Bob knew that on Friday afternoon the captain had stored his navigational instruments in Monson's room, and that was the only reasonable explanation for his return the following morning, to retrieve his instruments. But this raised another question: Why would Herndon be concerned with navigational instruments in a storm at eight o'clock on Saturday morning? Sailors shot the solar meridian at noon, and that was only on days when they could see the sun. But there was another clue from Monson in the previous paragraph, an observation supported by the accounts of several other passengers.

Early that morning, the wind had abated and the skies had begun to clear; the storm at last appeared to be passing, and for the first time in nearly four days the passengers and crew could see the sun lifting off the eastern horizon. "Everybody got all cheered about it," figured Bob, "and thought, Yes, we are going to conquer this thing, and surely all will be saved," and as he was picturing the skies clearing and the sun rising, he remembered another piece of information he had read a year earlier in the *Philadelphia Ledger*.

Arriving in Norfolk, Captain Johnsen of the *Ellen* had told his story about rescuing forty-nine men, but on the train to Philadelphia, other reporters asked the Norwegian captain again to recount his story. Johnsen then told virtually the same story he had related in Norfolk, with one major difference: Someone this time asked him where he had picked up the survivors, and he replied, "As the morning advanced, and we found ourselves at a considerable distance from the place where the steamer sank, I bore down for it again. It was in longitude 76° 13', latitude 31° 55'. I first took an observation at eight o'clock in the morning and again at noon."

The celestial event at noon was obvious, the solar meridian. But Bob had wondered then, How did he get a coordinate at eight o'clock in the morning? Was it dead reckoned? Did Johnsen take the force of the wind and the height of the seas and the direction of his sail and the number of hours he had spent tacking back and forth and simply distill it all into an approximate location? Was it one of those, as Bob called them, seat-of-your-pants-I-don't-know-where-the-hell-I-am coordinates? Not likely, thought Bob, because longitude had to be determined using a celestial observation. But what celestial event could possibly have occurred at eight o'clock in the morning?

Bob tracked down a nautical ephemeris from the mid–nineteenth century, and reading the charts that logged every celestial event, he realized that Johnsen had not determined the coordinate by the seat of his britches. According to the ephemeris, a significant celestial event had occurred just after eight o'clock on Sunday morning, and that is what Johnsen had shot with his sextant. And now, as Bob read Judge Monson's account over and over, he recalled that story, and suddenly it all made sense: That same celestial event had also occurred a little after seven o'clock on Saturday morning. After shooting the event to take the coordinate, Herndon had probably returned his navigation instruments to Monson's stateroom, and that's when Monson saw him. And that was the same coordinate that Herndon had then shouted across the waves to the captain of the *El Dorado* nearly twelve hours later, *his last known position*.

"It was that moment of excitement," said Bob, "when I figured, THAT'S IT! I remember telling Harv about it. I remember talking in a very animated fashion to him about how I felt I had reached a very key piece of information!"

Bob told Tommy that Herndon was trying to save his ship. He hadn't seen the sun for four days, and he had no idea where he was. He was a great explorer, and for three years he had worked with his brother-in-law Matthew Fontaine Maury at the navy's Depot of Charts and Instruments in Washington, charting astronomical calculations for Maury's work in marine navigation. He was fully versed in wind and current charts and all the latest navigational science.

"And being the intrepid guy that he was," said Bob, "he was not about to waste an opportunity to figure out his position."

He might be close to shore, and if he's close to shore, they might have more hope for being rescued. Monson didn't say that Herndon had his instruments with him, he said that Herndon was depressed, a little upset.

"Well," said Bob, "you know why he was depressed? He's just figured out that they're two hundred miles offshore! I think he had hoped that maybe he was closer to a shipping lane, that maybe the Gulf Stream had carried them farther north, I don't know. But if Captain Herndon were offered an opportunity to determine his latitude and longitude, he would not have missed it, and I said to myself, simply, there was not another opportunity."

The storm that had raged for three days was asymmetrical, and its eye was passing over the *Central America* early that Saturday morning. After this brief clearing the wind again would rise, and the waves would tower above the ship, and the storm would drive her deeper into the sea. But for a short while on that morning, as the clouds parted and the sun shone low in the east, Captain Herndon saw directly above him, cresting at its meridian, the pale white disk of the moon.

The coordinate that had appeared in the underwriters' report, that had come from the *El Dorado,* that Herndon had shouted across the waves, Herndon himself had taken almost twelve hours earlier during the eye of the storm, and in those twelve hours the *Central America* had drifted from that point to where the two ships met just prior to her sinking, and her wreckage had drifted on to the coordinate reported the following day by the *Ellen.* Finally, Bob had reconciled the odd coordinate, and it fit neatly into the pattern created by the others.

EARLY IN WORLD WAR II, U-boats from the German Wolfpack would drift silently in the Gulf Stream along the East Coast of the United States, waiting to torpedo American convoys as they crossed the Atlantic. The German High Command orchestrated the stalking from Germany, transmitting orders across the Atlantic by radio. The U-boat commanders responded by radioing their positions back to Germany. Although the communications were coded, the radio signals from the submarines were powerful and broadcast in all directions. To intercept these signals, the United States and Great Britain developed a net of High Frequency Detection Finding radio outposts encircling the Atlantic Basin, from Cape Hatteras up the eastern seaboard of the United States to Newfoundland, around Greenland and Iceland and over to Ireland and England, down to North Africa, across to South America and up the Caribbean to Florida. Often, when a submarine commander radioed back to Germany, at least one outpost picked up the signal, and the operator then estimated the direction from which the signal was broadcast. The direction was always approximate; on a chart it looked like a V-shaped beam emanating from the outpost and spreading, and somewhere within that beam lay the precise bearing of the submarine. With only one outpost picking up the transmission, the operator couldn't estimate the location of the U-boat, only the direction in which it lay; but sometimes two stations locked on the same signal, and the V-shaped beams intersected. They now had circumscribed a larger, roughly square patch of ocean, and the submarine from which the signal had been transmitted lay somewhere within that patch.

The German U-boats had limited fuel capacity and were powered by suffocating and smelly diesel engines. The commanders dived the vessels and maneuvered with the engines only when preying on enemy ships. The rest of the time, they conserved fuel and vented fumes by drifting on the surface, where antisubmarine aircraft could spot them from the air if the pilots knew where to look. But the ocean was big and the U-boat was small, and the winds and the sea and the currents moved the U-boat, so that from the moment two radio outposts locked on a signal, the likelihood that the U-boat lay in one part of the patch instead of another was constantly changing. Since the aircraft had enough fuel to search only five or six hours, a pilot had to concentrate his search effort in the area most likely to hold the submarine. But how did he determine that area, and how much effort did he put into searching it before he spread his search to other high-probability areas to maximize the probability of finding the submarine in those five hours of search?

To deal with a variety of scientific problems early in the war, the navy recruited an elite corps of physicists, mathematicians, and chemists to form the Operations Evaluations Group. High on the group's list of problems to be solved was how to counteract the German U-boats. One of the first reports the group created in 1942 was a rudimentary concept called "search theory," which applied mathematical optimization techniques to varying sea conditions, such as wind and current, to predict the probability that a submarine would end up in a certain location. From that they produced a formula for how intensively a pilot should search a given area. The navy then brought in a well-known mathematician named Bernard Koopman to refine the research and form a coherent theory, and within a year, the navy was relying on Koopman's theories to sink German U-boats. Koopman's final report, "Search and Screening," remained classified until long after the war had ended.

One of the late-night letters Tommy wrote at Robbie's house in 1984 had gone to Dr. Lawrence Stone at Wagner Associates in Sunnyvale, California. Stone had graduated from Antioch College with a B.S. in mathematics and from Purdue with a master's and a Ph.D. in probability theory. After Koopman died, Stone inherited the mantle as the authority on search analysis. In 1975, he had written *Theory of Optimal Search* for the United States Navy, and the book had remained the premier reference in search theory.

During World War II, Koopman had used a slide rule and sines and cosines to render his formulas. When Stone began refining Koopman's work in 1967, he had a computational tool not available to Koopman: the

computer, which could process information far more numerically intensive than Koopman could ever have conceived. While Stone was advancing search theory, his colleague Tony Richardson was using computers
to reduce the theories to coordinates on a grid. Richardson had worked
with the navy during the recovery of the H-bomb dropped off Palomares,
Spain, in 1966. Two years later, Stone and Richardson directed the successful search for the navy submarine *Scorpion*. Then the two men had
developed a system for the Coast Guard to find persons lost at sea. They
used principles similar to those used to find the *Scorpion;* only now the
targets were moving. If the search was unsuccessful the first day, they had
to reconsider wind and current and project ahead to the next day. Although the persons missing often had been lost for weeks in the Pacific,
the system sometimes still found them, and the Coast Guard attributed
several spectacular successes to Stone and Richardson. From there, they
developed computer-assisted search systems for the navy to track Russian
submarines, which moved according to a patrol pattern. For the past ten
years, the Office of Naval Research had kept Stone under contract as the
principal investigator working on problems in search theory.

Tommy knew of Stone's work, and he had wondered for some time
if the same methods Stone used to monitor enemy subs and rescue people
lost at sea could find a ship sunk long ago and now resting on the bottom. His letter in 1984 requested a summary of Stone's background.
Tommy kept the letter vague, simply planting the seed that he might
need Stone's services. When Stone sent Tommy the company literature,
Tommy called, and they kept in touch by phone over the next year,
Tommy learning more about search theory from Stone and eventually
revealing to Stone that he was looking for a shipwreck. Then in the early
summer of 1985 Stone planned to attend his twentieth college reunion
at Antioch, which was in Yellow Springs, Ohio, only fifty miles west of
Columbus. "I called up Tom," said Stone, "and suggested that it'd be a
good time to get together."

Stone arranged for them to meet in a private classroom with desks
and blackboard. When Tommy and Bob arrived, Stone made an inno-
cent mistake that many others would make, though no one would make
it more than once. Tommy had not discussed the target of his search
other than to describe it as a shipwreck, but Stone knew that no one

would spend the time or the money to look for a shipwreck unless there was something valuable on board, so he assumed the ship had sunk with a substantial treasure. Early in the conversation, he referred to Tommy as a "treasure hunter."

"He bristled," recalled Stone, "and he corrected me rather sharply, so I tried to strike that from my vocabulary."

Tommy was friendly but slow and methodical in his speech, careful to present only the information Stone needed to determine if his system would work for this kind of search. To help them Stone would have to have the best information they could produce, but this was the core of the project, the most secret information they had, and before he would let any of it out, Tommy had to get a sense of how far he could trust Stone.

First, Stone had to sign a nondisclosure agreement, then Tommy explained a little more of the project, that a ship had sunk in a hurricane and about sixty of the survivors had given their stories to the press. Based on those stories, they had created a matrix broken into time segments. He wanted to know if Stone could apply search theory to that information and mold it all into a probability map marking the areas where they should concentrate their sonar search.

As Stone listened, one thing impressed him: Tommy's commitment to the science. "He was going about this in careful scientific fashion. He wasn't in it strictly for the money. That came through strongly, and it's very consistent with the way he's acted ever since."

Still, Stone was skeptical about how well organized they had their information and whether he could turn history into mathematics. In his cursory explanation, Tommy only alluded to the historical data Bob had gathered from the old articles. Bob now carefully unfolded part of the Data Correlation Matrix, the handwritten original, smeared with erasures where information had been plugged into the wrong time slot or Bob had found better data on the same event. Bob told him that that matrix was a working copy and that if Stone found it valuable, he would send him a better draft on a Lotus spreadsheet.

Stone perused the matrix and asked how many accounts they had. Bob told him that fifty-nine survivors and sea captains of other vessels had made statements to reporters onshore. Of those, he had gleaned thirty-three that offered information concerning weather, the condition of the ship, or the ship's location at various times, things that he felt would help them locate the wreck site. "It didn't mean a lot to me until I got a chance to sit down and read it carefully," said Stone, "and then I realized what an incredibly careful job he had done."

Bob told Stone that from the historical accounts, they had pulled the coordinates of the ship shortly before it sank. He asked Stone how he would use that piece of information. Stone sketched on the blackboard, explaining basic search theory. "The trick to doing all these things," he said, "is to take all of the uncertainties and all of your knowledge and quantify them, whether they come from objective or subjective sources." The second step was to combine the numbers in analytically correct fashion to produce a probability distribution of the target location. Stone told Bob he would start with the last known position and factor in a number of variables like wind and current, which he would throw into a computer to create a range of probability.

Bob said, "Okay, I've got another piece of information here. We know that survivors were picked up sometime later by another ship and that ship estimated its position. How would you use that piece of information?"

Back when Stone and Tony Richardson were finding people lost at sea for the Coast Guard, Richardson had pondered different ways to use whatever data they had. "So I pulled that out of my hip pocket," said Stone, "and explained how we could go backward in time." That would be a new scenario and he would take the coordinate of the rescue ship and run everything in reverse, back to the site of the sinking.

By the end of the meeting, Stone had only one reservation about the idea. "Tommy's had trouble getting money together," said Stone, "and I'm asking myself, Suppose I do all this work for him and he doesn't pay the bill. I can't even sue him. He probably doesn't have any assets, right?" But if Tommy would agree to pay for the first set of analyses in advance, Stone concluded he could turn their historical information into a probability map, which would give them the most efficient way to search for the *Central America*.

THE LAWYERS KELLY and Loveland filed the Certificate of Limited Partnership with the Franklin County Recorder, which required them to disclose the names of the partners, and no sooner had they filed than

a reporter for Columbus's newsweekly, *Business First,* routinely checking new filings, saw the roster of elite Columbus citizens and was on
the phone calling them for an interview. She called Tommy, and he
didn't want to talk to her; but worse than that he didn't want to appear evasive. All he could do was steer her away from the romantic
and the sensational toward the scientific. "I'm really a scientist," he
told her. "We're combining multidisciplinary efforts—historic research, ocean engineering, marine geophysics, deep-water biology, and
side-scan technology." On June 24, 1985, the words Tommy dreaded
appeared in the headline on the front page: "TREASURE HUNT
ATTRACTS PROMINENT EXECUTIVES." Although the article
never mentioned the *Central America,* it revealed far more about the
project than Tommy wanted known, and the three investors quoted
used phrases like "take a flier," "crap shoot," and "sunken treasure,"
the swashbuckling tone Tommy had always been careful to avoid. One
compared the project to *Raiders of the Lost Ark* and said it was "top
secret stuff."

This was Tommy's first brush with the media, and he saw how
quickly the public's perception of the project could be skewed if he did
not control the release of information. If the media whipped the public
into a froth with "visions of tall sailing vessels, swashbuckling pirates and
treasure maps," as the article had opened, then everyone would perceive
the project as a treasure hunt, and he would be embroiled in the same
problems he had seen topple the treasure hunters. Tommy needed someone to help him convey to the public the serious and scientific aspects of
the project; he needed someone to help him communicate with investors
and suppliers, and with the science and history and archaeology and technology communities; he needed someone to take all of the deep-water
shipwreck theories he and Bob and Robbie had roughed out in the concept paper and distill them into one clear, comprehensive document for
the average, intelligent investor. He needed a writer, someone good at
communicating ideas, someone who understood the media, someone he
could trust. Someone like his old friend Barry Schatz.

After Tommy left Key West in 1979, Barry had spent six months
in Mexico and Central America. To a journalism colleague he wrote
from San Cristobal, "Here I am holed up in the seven-thousand-foot

level of the Sierra Madre range, going to gory bullfights and generally rolling in the mud and the blood and the beer." He had published
a lengthy magazine piece on Haiti and one on Quebec, written several
short stories, and committed himself to a novel. After six months of
wandering, though, he had returned to Florida, this time Gainesville,
where he finished his degree at the University of Florida, studying
creative writing. He had thought about applying to law school but
instead enrolled in a master's program in Latin American Studies. He
now served as an editor at the University Presses of Florida while
pursuing his master's half-time.

Over the past three years, Tommy had called Barry occasionally to
tell him about working at Battelle and about his interest in recovering
deep-ocean historic shipwrecks, but he had given Barry few details. When
he phoned in late June 1985, he told Barry he had preliminary funding
for the deep-ocean project, and he asked if Barry could come to Columbus. "Harvey needs people to think with," said Barry, "so he called and
asked if I would help out with the press stuff." Tommy emphasized that
there were no guarantees. "It's an exciting project," he said, "a project that
I believe in." And he could pay Barry a small wage. But it could all be
over in six months. Barry didn't care; the idea sounded intriguing, and
after four years in Gainesville he was ready to move on.

When Barry arrived in Columbus, Tommy took him to meet his
new friend and confidant, but even Barry's own leanings toward the
bohemian did not prepare him for Bob Evans. "He was so disorganized
it was amazing," said Barry. "He could keep stuff in his head, but he
could not find his shoes."

Bob had no misgivings about Barry. "It was a very dynamic, very
productive relationship," said Bob, "right from the outset. I immediately
felt like he really could contribute to the project."

Barry spent the next two weeks working twelve- to sixteen-hour
days, trying to unravel and recast the technical language of the original
concept paper into a project game plan. Then together, Tommy, Barry,
and Bob dissected every sentence, searching for nuance. The words had
to inform but not reveal too much, be clear but not condescending,
emphasize technology but not mire in the technical, express confidence
but not certainty, sound adventurous but not swashbuckling, make the
project seem challenging but not impossible.

"We had convinced investors to put up at least enough money to find out if it really was viable," said Bob. "So this was the actual act of creating the go-ahead plan. All those concepts had to be refined and reduced to words."

They stripped it of all flash; this was not a treasure hunt. It was a scientific pursuit, logical and thorough. It would not be used to attract financing; it would be distributed to those who already had invested to assure them that they had bought into a sound operation. More than anything else, Tommy wanted the document to emphasize that finding something of value on the bottom of the deep ocean did not have to be a roll of the dice, as so many had described it. "That was the theme," said Tommy. "If you were going to do one of these, how would you do it so it wouldn't be blue-sky?" Throughout the treatise, the theme surfaced: By methodically identifying and quantifying risk, they could allocate the resources and the time to solving the most difficult problems and thereby achieve the highest probability of success.

Day and night, they brainstormed, wrote and rewrote, made graphs and charts, reduced three paragraphs to one, distilling it all into one simple document. It was their manifesto, with methodology, risk analyses, and flow charts. The concept paper grew into "A Multi-Disciplinary Approach to Historic Shipwreck Recovery," or as Tommy, Bob, and Barry called it, "the Blue Book." For partners and financial dealings, they named their company Recovery Limited; but for the public, they took the name Columbus-America Discovery Group.

WITH SOME OF the seed money, Tommy had rented an old, three-story, redbrick Victorian on Neil Avenue, a street lined for blocks with trees and other old Victorians. The house had been built for entertaining, with twin formal parlors at the entry, leaded windows, and a fireplace in every room. A carved oak staircase led to the second floor and then on to the attic, where the original owner had installed a dance floor among the eaves for Sunday afternoon teas. Tommy used the house as an office, a place to work, to meet with investors; a place for Barry to live. After they finished the Blue Book, every Tuesday night and sometimes several other nights a week, Tommy, Barry, and Bob convened in the formal dining room to discuss the direction of the project. They called the meetings "round tables."

The round tables began at seven and often ran till long past midnight. They were the intellectual heart of the project, where the three men dreamed out loud. To help him ruminate, Bob stuck to his Wild Turkey neat, and Barry joined Tommy in sipping warm tequila. Tommy sometimes had a short agenda of things he wanted to cover, but typically he got no more than halfway down the list before their thoughts had drifted elsewhere. Rather than steer them back to the agenda, Tommy encouraged the divergence. He wanted an atmosphere like Battelle's, but with even less red tape and more personal motivation, "Where you really get into the guts of it," said Tommy, "the power to make something succeed. And the key to it is to find people who like to think like that."

Barry came to the round tables as a journalist and insatiable traveler, with a head full of Latin American writers like Borges and Paz, a gourmand's palate, a writer's eye for detail, childhood experience with Tommy's way of thinking, and ready to embrace a new adventure. Bob now played keyboard in a nighttime blues band organized by an ex-con; packed inside his head was a sizable understanding of fossiliferous strata from the Paleozoic to the Cainozoic, experience with the intricate variations in syncopation from jazz to rhythm and blues, a near photographic memory, a penchant for historical sleuthing, and training in the scientific method. Tommy brought the perspective of an engineer and inventor with knowledge enough to impress the heads of top-secret government efforts, an instinct for quantifying risk, an obsession with experimenting, and a brain still searching through the junkyard of ideas for spare parts to hook together to create something no one had ever seen before.

"We began to realize," said Bob, "that the more minds from divergent areas of thought you can get to agree on a given subject, the sounder the idea."

No one worried that he would be judged by what he said in front of the other two. Ideas were not divided into good and bad, only those that worked and those that didn't work. And although an idea might not work for several reasons, one part of that idea might have merit. But you couldn't pluck the ripe part if the whole idea went unspoken, and Tommy had observed subtle ways that that happened, that words, looks, even demeanor from others could prevent someone from stretching into an area new and creative. "That's where the right chemistry allows people to take stands and not be ostracized for their positions," said Tommy. "It's all part of the process: Be sensitive to each other, yet give each other enough resistance."

Bob could get so revved up he would shake off one point for another and then another, until he was so far removed from the original conversation, he had to stop and think his way back. He always had a reason for the story, but it often took him a half-hour to get there. He started out one night telling Tommy and Barry how the year before the *Central America* had sailed on its final voyage, Captain Herndon's book, *Exploration of the Valley of the Amazon,* had rendered the exotic Amazon so vivid in the imagination of a young man in Keokuk, Iowa, that the young man had quit his job in a print shop and left for Brazil. Tommy and Barry wondered where he was headed with this one, but they had been at this juncture before, so they sat and listened. Bob told them that before the young man left he had written to his brother, "I shall take care that Ma and Orion are plentifully supplied with South American books. They have Herndon's report now." Bob could quote this stuff from memory. He told them that the young man had gone by way of Cincinnati, down the Ohio and the Mississippi. The way the young man figured it, said Bob, when he got to the end of the great river, he would book passage on the next ship out of New Orleans bound for Para, Brazil. Once there, he would work his way up the Amazon and thence into its tributaries to experience the region Captain Herndon had rendered so lively in his report. However, in New Orleans, said Bob, the young man discovered not only that no ship was leaving for Para, but also that no ship had ever left for Para, and that was the point: the guy was stuck at the dock in New Orleans at the mouth of the Mississippi River. The intrigue and romance of Herndon's chronicle had lured him onward on this long journey, but there he was stuck at the dock . . . with little more than the experiences culled from traveling the length of the Mississippi and a recollection of boat men sounding "mark twain." Had Herndon's book not inflamed the imagination of young Samuel Clemens, he might never have traveled the backwater where he later set so many of his stories, might never have changed his name, might never have gone on to write of Tom Sawyer and Huckleberry Finn.

Tommy and Barry let him run because they never knew where it might lead. And that was Tommy's point: Don't categorize at the outset; don't look for answers prior to mixing it up; don't confine yourselves to merely the task at hand. Don't discourage Bob from exploring arcane bits of history just because they have nothing to do with locating the wreck site. "With most people there's a tendency to diverge slightly," said Tommy, "and then come to a convergence, and by not diverging all the way before converging, you've missed out on all kinds of opportunities."

So they let Bob run, and Barry sat in the round tables night after night listening to Bob ramble about history before he began to appreciate the richness of the stories. "One night it dawned on me," said Barry, "how useful these stories were outside of Bob's own personal fascination with them."

By allowing Bob to wander far beyond the details that might help them find the ship, Barry and Tommy realized the richness of the history and its value to the project. Here was our country in one of its finest hours, and few people had ever even heard of the *Central America*. Tommy encouraged Bob to find out more about the California Gold Rush and the burgeoning of San Francisco and the San Francisco Mint and the political climate back east and the personalities on board the ship. The information might not help them find the ship, but it would help them understand the significance of what they were about to do and would enrich the whole experience for everyone.

As they met night after night to talk and dream, Barry began to appreciate a mind-set Tommy had been cultivating for a long time, something he had honed at Battelle, "this wonderful research institute approach of setting out deliberately to do something that hadn't been done before and taking a very methodical approach to discovering how to do it."

Working on the bottom of the deep ocean wasn't impossible, it was only considered impossible, and that was the distinction Tommy had learned: Other people labeled things impossible not because they couldn't be done, but because no one was doing them. He had revisited all of the old assumptions, found many of them no longer valid, and saw ways around the others. Bob had talked to Tommy for years about the project but not till now did he begin to see the whole thing as "a series of incredibly nonexistent barriers."

Realizing that impossibility dwelt only in the imagination was the gateway to a new world of thinking, and this was the world in which Tommy lived. The idea of finding the *Central America* and recovering her "treasure" quickly became a rich metaphor for all that was possible: We find the ship, we recover the gold; what can we learn along the way?

"We were into the idea," said Tommy, "that within this project, all kinds of things can happen. We might be successful at this; we're going to give it our best shot. In the meantime, we're going to learn all these other things."

WHILE BOB AND Barry and Tommy dreamed and theorized, Mel Fisher finally found the hull of the *Atocha,* on July 20, 1985, somewhere between sixteen and twenty-one years after he had begun the search, depending on whose start date you chose, and one decade to the day since his son and two others had drowned. He had spent most of that decade searching the area where his son had found the nine bronze cannons. "We were right there for umpteen zillion years, it seemed like," said Tom Ford. "We airlifted and we dug and we searched and went back and forth by the hunch method. We just beat ourselves senseless." Finally, over nine miles southeast of the cannons, one of Fisher's other sons had anchored on top of the treasure for another day of diving, and when he sent two divers down in fifty-five feet of water, they found thirty-two frozen black masses of silver coins and nine hundred large bars of silver, some small gold bars, and about four hundred emeralds, all of the treasure mounded into a reef now encrusted with coral and over-run by lobsters.

Fisher's vision stopped at the treasure of the *Nuestra Señora de Atocha,* and finally his dream was over. Along the way he had fought incessantly with other treasure hunters and the State of Florida and the IRS and incurred the wrath of creditors and suffered through the deaths of five young people. He still faced protests from archaeologists and environmentalists and lawsuits from investors.

Tommy, Bob, and Barry could have taken the same approach. Nothing prevented them from finding the wreck site of the *Central America,* scooping up the gold with a clambucket, bringing it back, melting it down, selling it for bullion, divvying up the proceeds among the investors, and going back for another wreck. In the beginning, some of the investors proposed they do just that, what Barry called "the thief in the night syndrome." But that wasn't the responsible way, and nobody learned from it.

Barry likened it to the difference between Columbus and Prince Henry the Navigator, an idea that had intrigued him in Daniel Boorstin's *The Discoverers*. Columbus explored blindly; Prince Henry consciously pursued something that had never been done.

"Columbus's approach was fraught with much higher risk," explained Barry, "and don't forget he died not realizing what he'd done." Columbus led the way to the gold, but he never saw the implications. Prince Henry also had heard of gold, off the West Coast of Africa. He made the gold his draw, but his purpose was to build a trade route and shipping network all the way to the wealthy destinations of the Orient. He instructed his explorers to go to the West Coast of Africa and find the gold, but also to learn everything they could along the way about the routes, the winds, the seasons, the people, the other resources. In the long run, that information had greater value. "That's what entrepreneurial vision is all about," said Barry. "We didn't go after the *Central America* as an end in itself, but as a way to learn how to work in the deep ocean and then to discover the resources there. It's just the beginning. Harvey had already thought about it."

The round tables continued through the summer and fall into the winter of 1986, and gradually thousands of ideas had sprouted and received a warm welcome and been explored and stretched, then revised and refined and finally distilled. They had anticipated possible failure modes and formulated plans to negate them; they had considered possible success modes, then devised ways to increase their likelihood. As the project moved forward, they had only to keep alert for more opportunities.

"What was exciting to me happened on paper," said Barry. "It doesn't have anything to do with the gold itself. It was in discovering how to discover."

To raise enough money for the seed phase, Tommy had needed twenty to forty investors to part with $5,000 to $10,000 apiece. Now he needed fifty to one hundred investors to put up $14,000 to $28,000 each to raise $1.4 million for the SeaMARC search. But three things happened that summer that made the *Central America* project seem even more attractive: Fisher found the *Atocha;* a deep-ocean contractor retrieved the flight deck recorders off Air India flight 182, the 747 that exploded in the air and plummeted into sixty-six hundred feet of North Atlantic water; and Bob Ballard and his crew located the *Titanic* on a depth sounder and glimpsed one of her boilers. "It was like all these things that Tommy had been talking about really can happen," said Ashby. "It had a very positive impact on the people that Tommy was talking to." Out of the thirty-eight partners in the seed phase, all but half a dozen anted up for the search phase.

One of the new people Ashby introduced Tommy to in the second round of fund-raising was a former naval officer named Mike Ford. Ford had bounced around on cruisers and destroyers and now managed money for pension funds and trusts. As soon as Tommy summarized the project and his experience at Battelle, Ford realized how much Tommy had at stake. "Man, you do not leave the hallowed halls of Battelle," said Ford. "Battelle is like the papacy. If he came out of there and fell flat on his face, that failure would put him back to his graduation date." Ford liked that; he wanted Tommy's neck well into the noose.

Ford had chased submarines, and he knew you could lose one even when you knew where it was. He understood the vastness of the sea and the problems in trying to find something only three hundred feet long. He also knew Mel Fisher's story and the problems with investing in treasure hunts, but there was something different here. "Fisher blundered around in the same waters and finally found it and couldn't believe he did it," said Ford. "That to me was luck. This wasn't the same thing."

Ford liked to see what American ingenuity could come up with next, and he liked Tommy. First, the young man was extremely methodical. Second, he had a dream; most scientists didn't. Third, he had a way of convincing you without making a great effort. "Occasionally," said Ford, "a smile would race across his face for an instant, and it was that smile that gave you the signal that this was something special. He was living this thing day in and day out." Ford had invested in a lot of companies, but he had never seen this combination in the key man. He persuaded a friend to invest, bought half a Search Phase unit himself for $14,000, and eventually put in over ten times that amount.

By the end of 1985, the Search Phase, nearly one hundred investors, was fully subscribed, and Tommy had $1.4 million to go find the *Central America*. Already, Tommy, Bob, and Barry had found her in thought and on paper; now they had to find her in the deep blue sea.

Two weeks after Larry Stone had met with Tommy and Bob at Antioch College, he signed another nondisclosure agreement, and Tommy sent him a check for ten thousand dollars to begin the first analysis of the data. Bob then mailed four things to Stone: the Data Correlation Matrix on the full Lotus spreadsheet with all data organized in three-hour time slots; a map, complete with coordinates, that traced the final voyage of the *Central America;* a three-page historical narrative, beginning when the *Central America* left Havana and noting key events in the time line, like when the storm hit and when the engines stopped and when the *Marine* arrived; and a wind chart that estimated wind direction and velocity each hour, beginning when the steamer lost power and running till Captain Johnsen of the *Ellen* abandoned the search for survivors the day after the sinking.

Bob had created the wind chart from information culled from survivors' observations. The actual mariners' accounts were the more reliable ones, like the second officer's account and the engineer's account and the accounts of other sea captains. If a mariner used a term, it had specific meaning. When Captain Badger noted, "A perfect hurricane was blowing," he had read the surface of the sea and he knew that the wind whistled upward of sixty-five knots.

"There's a whole spectrum of definitions," said Bob. "And it's quite different if somebody out there who is not a mariner says that a 'light breeze' was blowing. That just means, well, it felt nice. But a 'fresh breeze,' for instance, is different than a 'light breeze.' For a mariner, 'fresh breeze' means a wind of twenty knots." With the wind chart, Bob included the amount of sail set from one hour to the next, until Friday morning, when the wind shredded the storm spencer. Then he estimated the force of the wind exerted on the three bare masts and rigging, then on the two bare masts after they chopped one down, trying to quantify everything he could.

Stone had done search analyses on dozens of persons, ships, and other things lost at sea, but no one like Bob Evans had been there to organize the data for him; Stone had had to sort out all of the inconsistencies himself, and he sometimes had to study intensely for a week just to begin to make sense of the problem. But Bob had done this for him already. "Bob Evans seemed to have a flair for historical investigation," said Stone. "I was taking a much more technical view, and then he came up with this rich collection of interesting detail to go with it. That made the analysis much, much easier. He had done an incredibly detailed and careful job of it, and I mean, I was very impressed."

To understand the time line so he could picture the events and formulate the scenarios, Stone read Bob's narrative summary of the sinking twice and then studied the map, the matrix, and the wind chart. For his initial analysis, he produced rough maps, one of each of the three scenarios they had chosen: one around the *Ellen* coordinate, one around the *Central America* coordinate, and one around the *Marine* coordinate. He took all of the information Bob had given him, converted it to mathematics, and computed each scenario separately by tossing all of the numbers into a computer, which absorbed uncertainties by producing sets of pseudo-random numbers. Then he modeled each scenario ten thousand times. From this he got a scatter of points, each a possible location of the target. Over the scatter he drew a grid of two-mile-square cells, counted up the number of points that fell into each cell, and divided by the same ten thousand; and that gave him the probability that in that scenario the *Central America* was located within that two-mile-square cell. But once he had mapped each scenario, he was not happy with what he found. He sent transparencies of the maps to Tommy so Tommy could lay one over the top of another to see the problem: The three scenarios did not overlap as they should, which meant that the information somehow was inconsistent. A piece to the search map puzzle was still missing.

Off the
Carolina Coast

⚓

Sunday, September 13, 1857

WHEN THEY HEARD beams exploding upward in the hold and water splashing just beneath the saloon, a fireman named Alexander Grant and nine other men abandoned the bailing lines and the pumps and cut away a large section of the hurricane deck and lashed it with ropes. Then Grant hurried below to search for more life vests, but he felt the big steamer lurch suddenly in a heavy sea, and when he made his way back on deck, his companions already had launched the raft off the leeward bow. Grant saw the raft as his only salvation. He leaped, hit the water, surfaced next to the raft, then reached back with his dirk to cut the last line holding the raft to the steamer. The waves lifted the raft and dropped it again, pulling it away, ten men clinging to the ropes. Moments later, the third wave had hit the steamer and the roar of her succumbing to the sea had drowned all other noise, and she had gone down amid cracking and hissing, sucking under men and debris. But the waves had pushed the raft with the ten men just beyond the vortex.

Seven of the men were crew members, mostly coal passers and fire-men, like Grant; the other three were passengers. So heavy were the men upon the narrow timbers that the raft floated nearly two feet beneath the surface. No man could sit upright, but each had to lie on his stom-ach, with his feet dangling in the water, his hands gripping the ropes they had lashed. The waves lifted them and rocked the raft and then washed over them, sending them under, where they swallowed the sea-water. Throughout the night, they prayed that the schooner or the brig would suddenly tack near them and the crew would hear their shouts and strong arms would pluck them from their raft. But both stayed a long way off, and all they heard in the distance were the cries of their comrades.

Grant already had survived three shipwrecks. In one he had tossed in a storm so violent that the masts had snapped and toppled across the deck, crippling the mate, and the waves had swallowed the brig only moments after the crew had leaped to a passing ship one hun-dred miles off Boston. In another, he had floated in the icy North Atlantic with one other shipmate on a fore-hatch, standing with water up to his waist, holding on to a rope. Three times sails had appeared and bore down upon them, first topmasts, then mainsails, then a hull coming into view, and they had blessed the ships and were poised to shout to them, when suddenly the wind had shifted, and just as the ships reached where their crews could spot the fore-hatch, the crews had jibbed the sails and blown forth on another tack. For three days and two nights, they had stood in the cold water with no food and nothing to drink, the bleak Newfoundland wind raking the North Atlantic. And then at noon of the third day, they had seen another sail, again bearing down upon them, and every few minutes she had grown larger, but the sea was running high, and their legs were numb with cold, and half a dozen times the waves had swept Grant off the hatch. Then, when the ship was no more than a mile or two distant, they saw a flag run up the yardarm, as the ship drew closer and closer, until, just before dark, she plucked them from the sea.

"I thought then," said Grant, "that I had suffered as much as mortal man could."

The third time Grant was shipwrecked the sea had slammed his sailing vessel onto a reef in the Bahamas. One of the crew shipwrecked with him was a black man named George Dawson, who was a steerage passenger aboard the *Central America* when it went down. At daybreak on Sunday, a black man floating with a life preserver and small boards under each arm approached Grant's raft and the survivors clinging to it. It was George Dawson. Since the raft already was overloaded and floating beneath the surface, the men said he could not get on, but that he could take hold of one of the ropes and float along with them. So with the pieces of board still under his arms, Dawson grasped a rope and floated beside the raft.

WHEN THE *Central America* sank, Dawson had been sucked straight down headfirst into the vortex, but his life preserver had popped him quickly back to the surface. A man who could not swim had grabbed him by the neck, and only by ducking under the man and prying his arms loose had Dawson escaped his grip. Then he had found three pieces of board and placed the largest under one arm, the two smaller pieces under the other arm, and he had floated on the three boards until late in the night he had seen the lights of a bark tacking back and forth. The bark had sailed within a hundred feet, and Dawson had cried out for help, but no one heard his cries.

Soon after Dawson reached the raft early Sunday morning, the men saw the bark four or five miles distant, too far to try to signal. They watched her until midmorning, when she disappeared, and they never saw her again. The men had been in the water now over twelve hours. The sea had dropped and the clouds had parted and the sun shone brightly, but the men were thirsty and exhausted. Many nodded on the verge of succumbing to sleep and had to be roused, but the others lacked the strength to prop them up or even tie them to the raft. As the morning wore on, delirium began sprouting inside the heads of some of the men. They babbled of cool springs bubbling and rich feasts set just beyond their reach, and lost in their sublime hallucinations they often came near to upsetting the raft. Before noon one fellow dropped his face into the seawater, drank heartily, raved for moments, then rolled from the raft to drown before the eyes of the others. By nightfall three more had sunk down too exhausted to lift their heads, drowned, and floated off into the sea.

Dawson still clung to the ropes lashed to the raft, but with four of the original ten men now gone, he abandoned his boards and pulled himself onto the raft. During the night, despite his pleas and the pleas from Grant to fight sleep and not swallow the seawater, four more men perished. By Monday morning, there remained only Grant, Dawson, and one other man.

The raft now was so light they could kneel, the water being barely a foot deep, and in this kneeling position they braced themselves and tried to sleep. That evening just before dark, the three of them picked up another man floating alone. But the following day, now Tuesday, the new man grew despondent, and his despondency deepened into delirium until he, too, began to hallucinate. He said that by God he was going down to the pantry and the mess and get himself some food and water. The steward himself had told him he could have some water if he'd just come down for it. Then, as though the delirium were contagious, the other man entered the same hallucination, his talk just as deranged as that of the first, and the two ranted with the imaginary steward about the food in the pantry and the water below. Dawson and Grant tried to talk them down, to bring them back, but by evening the two men had whipped themselves into such a frenzy, they suddenly pushed themselves from the raft and swam off together in the dark.

Now alone on the raft, Dawson and Grant had had nothing to eat for five days and nothing to drink for four. Although small fishes often came up to the raft and swam about between the beams, they moved too quickly to capture by hand. But on Wednesday, a fish weighing several pounds jumped onto the raft and before it could flop and slither back over the edge, Dawson grabbed it by its tail and beat its head against the timbers until it quivered and died. With his knife, Grant cut it up and gave small pieces to his comrade, but the flesh was so tough and unpalatable that hungry as they were, neither man could chew it. After a day of lying in the sun, the fish was more tender, and they forced themselves to eat small portions.

On Thursday, a man floating on a plank saw the raft and paddled slowly in their direction. When the plank thumped against the raft, Grant and Dawson helped him on board. He had been floating alone for four and a half days, and within hours of his arrival, he too suddenly became boisterous. Grant and Dawson tried to console and encourage him, but his suffering was so intense he seemed not to hear, for he soon lay down in the water and slipped away.

No one knows if this man was James Birch, or if one of the two men who swam off into the night was James Birch, or if one of the other two passengers originally on the raft was James Birch; but James Birch, the former president of the California Stage Company, the man who had established the first transcontinental stage line, at some point was on the raft with Dawson. An hour before the sinking, Dr. Obed Harvey had talked to Birch, and Birch had said he had little hope of being saved. Wearing a long, heavy overcoat, he seemed saddened but resigned. Ansel Easton had finally persuaded him to exchange the overcoat for a life preserver, and when the ship went down, Birch ended up on either the ten-man raft or a stout plank with a single possession: a small silver baby cup, with which he had boarded the *Sonora* back in San Francisco three weeks earlier, a cup ornately engraved as a gift from a friend to Birch's new son, Frank.

Birch's stage line stopped in Oroville, where Dawson worked in a hotel, and perhaps they had come to know each other from regular stops. Maybe the relationship began and ended that night. But sometime in the night, Birch explained the significance of the sterling silver cup to Dawson, handed him the cup, and asked that if Dawson be saved he deliver it to Birch's son in Massachusetts. Then Birch either lost consciousness and slid into the sea or became delirious and swam away, Dawson and the others too weak to stop him.

Now Dawson himself began swinging through moods of despondency, and in one of these moods he called to Grant: "For God's sake, look out and see if you can see anything." To appease Dawson, Grant lifted up and peered out at the horizon, and in the distance, he saw a lifeboat with an oar raised, and from the oar flapped an overcoat.

JOHN TICE, THE second assistant engineer on the *Central America,* had had no time to grab a life preserver before the steamer sank. When the first of the final waves had hit and the stern had rapidly begun to drop, Tice ripped loose a board about ten feet long and an inch and a half thick and sprang as far out from the ship as he could, then paddled hard to get away. He had cleared the stern by about forty feet when he saw the waves closing over the bow. By the time the steamer had begun its slow spin, he was beyond the vortex. The last thing he saw of the ship was Captain Herndon standing on the hurricane deck. Then the steamer disappeared and the sea began to boil where she had been, and in the midst of the boiling, debris from the wreck shot to the surface.

His chest on the center of the board, Tice paddled with the wind. He saw lights way in the distance, and he continued to paddle toward them for two hours until they disappeared. Two more hours passed when he again saw ship's lights, only this time they were much closer than before. In the lights he could make out the black silhouette of a hull heading directly toward him, but at less than a quarter mile, the vessel altered its course. First the hull, then the lights disappeared, as the others had earlier. During the night, Tice encountered seven other men, all floating on pieces of the wreck, and they spoke to him, but he remembered none of their names.

The following day, though the sea still swelled, the storm had passed, and the sun blazed through parting clouds. Once again he saw the bark, but she was standing off, and as the sun climbed higher in the east the bark moved closer to the horizon in the west. By eleven o'clock, the day was hot and the bark was gone. Tice floated alone on the swells, rising and falling, seeing nothing but water and sky and the front of his plank. Occasionally, a wave would lift fragments of the wreck into his sight, and once he saw a life preserver dance over the summit of a wave, then disappear. In the darkness of Sunday night, his fatigue overcame him several times and his head dropped and he fell into a deep sleep, only to awake suddenly to find his hands frozen at the sides of the plank. The awakenings frightened him, for moments would pass before he realized where he was, and then the deep loneliness would engulf him so completely that after a while he fought sleep. All through Monday and Monday night, he drifted.

He had been in the water over sixty hours, when on Tuesday morning he saw something floating in the distance that looked like a small boat. His body flat against the plank, he scooted himself to the end and began swimming his hands in the water. By now the winds were no more than light breezes and the sea almost calm. Still, he paddled for more than three hours, gradually closing the distance, three miles, two miles, a mile, until he recognized the object. It was one of the lifeboats from the *Central America,* gently floating and apparently empty. Tice paddled up to the side and held on to the gunwale for a long while before he could muster enough energy to lift himself off the plank and into the boat. Water sat halfway up its ribs, but in the water he found three oars, a rusty pan, a corroded pail, three old coats, and an oilcloth jacket. Tice bailed the water, tied one of the coats to an oar, then hoisted the signal at the bow. In the boat he could sleep without fear of being swept away, but still he had no food and no water.

He continued to watch the horizon for bigger ships, but he saw only small parts of the wreck floating in the ocean. Once he spied a wicker flask rising up and over a swell. Hoping the flask contained liquid, he used one oar to scull toward it and discovered that the cork was loose and the flask contained but a few spoonfuls of more salt water. That night he slept an unquiet sleep, and all the next day he saw nothing but his little boat, the sky and the sea, and the hot sun shining down upon all of it.

Midmorning on Thursday, Tice spotted another piece of the wreck, this one larger than most of the pieces he had seen drifting. Again, he took an oar and tried to scull off the stern of the lifeboat in that direction; as he got closer, he could just make out a raft carrying two or three men. He continued sculling toward the raft.

Grant had seen the boat about three miles off. It now approached so slowly that neither he nor Dawson could tell if anyone was inside. Grant stripped to his underwear, Dawson helped him tie a life preserver around his waist, and Grant lowered himself into the water to begin paddling toward the boat, which now was about a mile away. Tice saw someone leave the raft, so he pulled the oar back and forth off the stern, steering toward the swimmer, and within an hour the two met. Tice helped Grant into the boat, then the two rowed to the raft to pull Dawson off. As he struggled with Dawson, Tice noticed that one dead body remained entangled in the ropes.

During the first minutes together, the three men briefly mumbled of their experiences since the steamer sank. Then Dawson used what little strength he had to peel off his clothes so they could dry. Tice advised them both to wet a handkerchief and keep it wet and tie it around their head. After that they again fell silent, weak and dehydrated, no energy to speak except an occasional muttering, a wondering out loud at the chance they would be rescued. Even if they had had the strength to row, they would not have done so, for they did not know where to head. They could see nothing but sea and sky, they had little idea how far they were from shore, and other than following the path of the sun, they had no sense of direction. So they drifted, day and night, scalded by the hot sun as they scanned the empty horizon, sleeping fitfully at night, and borne onward by the wind and the sea.

Two more days passed, one week now since the steamer had gone down. Still no food, no water, no sail. Boils and blisters had bubbled up from their skin and some of those had popped, leaving large raw sores on their backs and arms. They had lost their sense of hunger, but their lips had dried and hardened and cracked, and their tongues were parched and beginning to swell from want of water. Then late Sunday morning, to the northeast, they saw the sail of a schooner, and she appeared to be on a southerly course. They took to the oars and tried pulling for the schooner, watching the wind puff her sails, but when they got within two miles of her, the distance between the ship and their small boat began to increase. The schooner held steady to her course, and in another two hours had sailed over the horizon.

That night, Dawson lay in the bottom of the boat wishing he could die. Then the next day, for the first time since the hurricane, they drifted into a rain shower. They opened their mouths and caught the drops and sucked the cool moisture from their clothes. They filled the small silver cup, and between the pail and the pan collected another quart; but they had gone for so long without water that when the rain shower passed, it had provided little to ease their suffering.

None of the three could move his limbs. They all sat with their heads on their knees, drifting, and waiting silently to die. But behind the rain shower, one of them spied a brig a few miles distant standing before a light breeze. At the risk she would turn out an apparition conjured only by their exhaustion, or real enough yet sail away, they again felt hope, and the ship gradually drew closer. Then they saw her topsails unfurl and her bowsprit swing directly at them.

Grant and Dawson sat side by side and wrapped their cracked hands around two oars and with nothing but hope to power their efforts, they tried to row toward the brig. But Dawson gave up. Then he tried again to row, and again he quit. Finally, he began rowing again, and soon thereafter, the captain of the brig hailed them across the water and in minutes they were alongside. She was the brig *Mary* out of Greenock, Scotland, recently departed from Cárdenas, Cuba, bound for Cork, loaded with molasses and sugar. Dawson's white shirt had caught the eye of one sailor.

The sailors slowly hoisted the three men on board, being careful not to strike their emaciated bodies against the timbers. Then they carried the three men across the deck to the cabin, where the captain refreshed them first with a glass of warm claret sweetened with sugar. When they had swallowed the sweet wine, they begged for water, which the captain wisely refused. After they had rested a while, he gave them small portions of thin gruel, which they followed with another fit of begging, which the captain again refused. As days went by, he fed them a little more and allowed them to have sips of water. Slowly he nursed them back with small amounts of water and gruel, and slowly they gained strength.

After the three men had been aboard the *Mary* for a week, the captain encountered the bark *Laura,* bound for New York, and transferred them for safe passage to their original destination. When Grant, Dawson, and Tice arrived in the city on October 5, a relentless press hounded them for interviews and described the condition of their bodies to a readership agape at the details. *Frank Leslie's Illustrated* said their suffering was "unparalleled in the history of shipwrecks." Even after two weeks of care and convalescence, their cheeks still were sunken, their limbs emaciated and covered with large sea boils. Flesh still peeled from their hands. One periodical described Grant: "His large, manly face was white and almost fleshless, showing the bony outlines with ghastly distinctness, and his black, scarred lips looked as though in his agony he had frequently bitten them through. But the most shocking traces of suffering were in his eyes. Naturally large, they were now preternaturally distended and wore a fixed, straining, sleepless expression as though still looking from the raft along the dreary horizon for a friendly sail."

"I do not like to speak about our sufferings," Grant told the reporter in a voice hoarse and hollow. "They were all but death."

A reporter for the *New York Times* wrote that the three men were "almost suffocated" by the crowd and could not answer even half of the questions put to them. "The colored man, Dawson," observed the reporter, "evidently impatient of the distinguished attention shown him, soon found an opening through the crowd, and limped away."

The *Mary* had rescued the three men at four o'clock Monday afternoon, eight days and twenty hours since they had been cast adrift as the *Central America* went to the bottom. The captain of the *Mary* logged the coordinate at the time of their rescue, 36° 40' latitude, 76° 00' longitude, which meant that since the sinking, Grant, Dawson, and Tice had drifted to the northeast almost five hundred miles.

When he was able to walk again, George Dawson, either in New York, or having traveled to Swansea, Massachusetts, kept his promise to a dying man: He presented the silver cup to James Birch's wife and to Birch's young son, Frank. Birch's wife added another inscription: "Saved from the Steamer Central America, Lost September 12, 1857." The cup is now on display at the Hearst Mining Building, University of California, Berkeley.

Bob had discovered this extraordinary story early in his research, and despite his curiosity at how the men could have mustered the will to survive, he saw immediately the mathematical value of the tale. Already he had consulted an authority on the Gulf Stream at the University of Miami. Presenting the story hypothetically and using different coordinates, he reconstructed an equivalent scenario and posed this question to the expert: If something adrift at this point on the ocean ended up at this point, would it be in the Gulf Stream? "And the answer," Bob said later, "was definitely it would be."

Although the course and width of the Gulf Stream alters from time to time, this information greatly narrowed the site of the sinking, and Tommy guarded it closely; he had not revealed it even to Larry Stone. But when Stone's probability maps failed to overlap, Tommy told Stone to reestimate everything, assuming a three-knot current always headed northeast. He wanted to see if that made a difference. "And it did," said Stone. "Once we put that in, then the maps started to overlap rather nicely. So that was encouraging. That means that the information we were using from these three independent sources was now consistent."

THE DEEP BLUE SEA

200 Miles off the
Carolina Coast

———————— ⚓ ————————

June 1986

T HE STEEL BOW of the *Pine River* shot upward eight feet, then slammed down, shuddering all the way to the fantail. Then the bow leaped up and slammed down again. A forty-knot wind whipped the sea white beneath lightning tearing at the night sky. When the stern dropped, water exploded over the fantail and rolled a hundred feet up the flat deck to crash against the control room, where the sonar techs sat trying to figure out what was wrong with the SeaMARC. They could look out a small window cut into the steel and see the wall of water headed up the deck.

"At times," said sonar technician John Lettow, "it seemed as if there was ten foot of ocean and you were under it."

The *Pine River* was a flat-bottomed mudboat from the Louisiana oil patch built to ferry drilling mud and supplies out to the oil rigs in the gulf. Tommy had found her in a shipyard in Orange, Texas, while scouting with a former navy commander named Don Craft. In his late fifties, Craft had retired after thirty years in the navy with an Unlimited Master's ticket: He could skipper any vessel in any ocean. In late 1984, Tommy had called Craft because the commander now consulted for offshore operations and he knew which vessel and what equipment Tommy needed to run a SeaMARC search in deep water two hundred miles off the coast.

At first, Craft was leery. But Tommy sent him a check for his fee, and the check cleared, so Craft met Tommy in Houston. For four days they drove, talked, ate seafood, and stopped at every bayou shipyard from Orange to Jennings to Lafayette, from Cameron to Patterson to Houma, showcase spots along the gulf where the offshore support industry displayed its rustbucket mudboats for charter. Tommy and Craft ventured out on scaffold piers in search of one that could be sucked from the muck, sandblasted, overhauled, and refitted for a deep-water survey of the Atlantic Ocean.

For the first day and a half, Craft wondered if Tommy would ever stop talking. "He asked me every goddamned thing you can think of, on every subject you can think of," said Craft, "vessels, ROVs, operational techniques, equipment, shipping companies, methods used in the gulf, how seafarers used to do things and what problems they had. We covered it all during that four days."

In Orange, Texas, they found the *Pine River,* which Tommy liked because it had a helo-deck. Under the helo-deck was good control space, a small shop, and some storage, which appealed to Craft. At 165 feet, it was smaller than what Craft had envisioned, but he was satisfied.

Craft then had ripped out a lot of old equipment from a previous charter, got the vessel cleaned up for transit, measured the fuel on board, and topped it off at sixty thousand gallons. When he left Orange, he had the *Pine River* on charter, beginning the 14th of May and running through the end of July. Eight days later he arrived in Jacksonville, ready to place the tow point and the winch, weld a modified log boom to the deck for launching and recovering the twelve-hundred-pound SeaMARC, fire up the galley, fill the ship with groceries, and await Mike Williamson and the sonar techs for mobilization.

Twenty-two men lived on the ship. Six men from the Louisiana bayou kept it clean, running, and pointed in the right direction. A cook ran the galley. Don Craft oversaw the operation. Twenty-four hours a day, Williamson and his sonar crew of eleven manned the electronics in the control room. Bob had remained in Columbus with the handle "Info Bob," a source for additional information. Barry documented the search on film and video, helped Tommy write letters to the partners, and was responsible for ship-to-shore communications to assure a steady flow of supplies, spare parts, and information. Tommy was Williamson's client, and as the client he was supposed to watch and listen. But he had $1.4 million in his pocket from partners who counted on him, and he would no more leave Williamson alone to run the SeaMARC survey at sea than he would have left Larry Stone alone to produce the probability map.

That spring, Tommy had called Stone frequently to question the assumptions they had relied upon to create the map. He would ask Stone, "How critical is that assumption? Can we get better data on that? How would we go about improving it? If the hurricane that hit the *Central America* was actually 110 knots instead of 78, how would that affect the eventual position of the sinking?"

"He just was so detailed and careful about what he did," said Stone. "He would keep revisiting parts of the analysis, trying to poke holes in it, seeing where the soft spots were, and 'Let's see if we can fix 'em before we go to sea.'"

Tommy wanted to explore the chopped-down mast, the dumped anchor, the drag sail, anything that might affect the ship's drift. He wanted to tighten their assumptions about wind and current. He wanted Stone to investigate the accuracy of nineteenth-century sextants and chronometers. He told Stone to call experts. Just get them on the phone, he said, they'll talk to you, and Stone was surprised at how well it worked. "It only takes about three or four phone calls to find an expert about any subject in this country," said Stone, "and that was something he was always instructing me to do."

A former Coast Guard officer told them that celestial readings taken with sextants and chronometers of the mid–nineteenth century would be off by no more than four nautical miles. Another expert at the University of Miami advised them on ocean currents. A meteorologist at the Naval Postgraduate School created a computer model to estimate how hurricane force winds might affect those currents. A professor at Florida Atlantic University calculated the leeway factor, how fast those same hurricane winds would blow a hull filled with water across the surface of the sea.

But Tommy wanted even better estimates of the parameters they were using; he wanted sensitivity studies to see how using different estimates might change the probability distributions. "He was like a bulldog," said Stone. "He just kept worrying this problem. He kept looking it over and thinking about the assumptions and picking them apart and trying to deal with the problems and making me deal with them."

The final assumption Tommy wanted to examine was that the captains' coordinates for each of the three scenarios Stone had used were equally reliable. After he, Bob, and Stone had discussed this, they assigned the *Ellen* scenario a weight of 72 percent, the *Central America* scenario 23 percent, and the *Marine* scenario 5 percent. They weighted the *Ellen* coordinate heavier because Captain Johnsen had shot his celestial fix after the storm, and he had recorded it in the ship's log; Herndon's coordinate from the *Central America* was passed along orally in the middle of the storm, and they still weren't certain of its origin; and the *Marine* coordinate from Captain Burt was dead-reckoned. With each scenario weighted, they finally combined the three into one probability map, and Stone was pleased with what he saw before him.

"Everything looked consistent. And it gave me a very warm feeling, and I think it gave Tommy additional confidence in these probability maps, because we seemed to be able to resolve most of the inconsistencies and somehow make these maps all come together and overlap. Now, they had uncertainties in them, but they overlapped very nicely."

The probability map Stone presented to Tommy was neat and precise; specific little numbers in perfect little cells, each representing a two-mile square of ocean. At the end of each line forming the grid, top and side, Stone had included the latitude and longitude down to the nearest minute. Within most of the cells appeared a number from 0 to 73, indicating the probability out of a thousand that the ship would be located in that two-mile square. The highest cell had a 7.3 percent chance of containing the shipwreck site; the many cells marked zero had some probability, but less than one chance in a thousand.

Mike Williamson and a colleague had then taken Stone's probability map, factored in topography, weather, and the velocity and direction of currents, and designed a search map, the most efficient way to run track lines back and forth through the highest-probability cells on Stone's map. They laid out each track on a grid with finely tuned coordinates and direction of sail. But, as they were experiencing in these first few days of the search, the currents never ran true and the equipment never operated without flaw and the ocean never lay still.

THEY HAD LEFT Jacksonville just after midnight on June 3, the *Pine River* bucking ever higher waves and rising wind. On the morning of June 4, they were in hilly water near the start of track line 1, two hundred miles at sea, ready to begin the search for the *Central America*. But two hours into the search, they had to abandon the track line and recover the SeaMARC: The navigation didn't work, so they couldn't tell where it was or how high it flew, and if they didn't know where the SeaMARC was and how high it flew, they would have no idea where the shipwrecks were when the SeaMARC saw them, which was the reason for being out there. By that evening, they had the guts of the SeaMARC spread across tables in the control room. Outside, the weather continued to climb.

"It was a hurricane," said Lettow, "and then it was downgraded as it moved offshore out of the Gulf Stream. But there were storm seas and lightning strikes virtually everywhere around us." The eye of Tropical Storm Andrew blessed them late one night with fifteen minutes of calm water and clear skies, before the wind and the sea built again, buffeting the *Pine River,* her aft deck, according to the log, "frequently awash."

The techs took advantage of the bad weather to test one system after another, but they couldn't even agree on the problems.

One said, "There's a missing logic gate; we got to put this gate in."

Another thought, "The system worked before without the gate. It's the whole logic process that's screwed."

A third wanted to throw the first overboard for distracting the second, who might have a solution.

"Hours are going by," said technician Will Watson, "days are going by, and you're just watching this."

They stared at card after computer card and gradually eliminated a few more problems only to discover that on a previous sea trial another crew had rewired the system and shifted all of the cards. When they reconnected it and aligned everything according to the schematics, the tow fish still would not process signals from the topside controls. Finally, they discovered that the schematics themselves were incorrect.

Andrew passed and the wind slackened, but the seas remained confused and rough. On the afternoon of June 7, they reassembled the SeaMARC and launched again, but no sooner had the tow fish hit the water than they discovered more problems. They talked about "blown buffer chips" on the "tone burst generator" and grounding problems in a "24-pin Burton," an almost incomprehensible argot, but the gist was clear: One problem after another after another after another, each time something different. They recovered the tow fish and again tore it apart.

Tommy was always there, watching the techs, trying to work with them, seeing hours and then days spin by on the clock. "Mike was trying to let the techs work on it," said Tommy, "and I'd talk to the techs and try to help them think through the logic and diagnostics, and so I would be included in the circle. Mike's telling me, 'They'll get it, they'll get it,' but when I'm working with the techs they're saying, 'We don't know this part of the circuit. If we go in there, we could damage it and not ever get it fixed.'"

One night as Tommy looked at the SeaMARC electronics spread across the control room and listened to the techs trying to figure out the problems and felt the pitch and roll of the ship in storm seas and looked back over an entire week of great frustration with nothing accomplished since they had set sail, he said just loud enough for Will Watson to hear, "I was afraid of this."

Tommy was dealing with men who knew more about electronics and sonar than he did, but they had never used electronics or sonar to find a wooden-hulled shipwreck in the deep ocean. No one had. And they hadn't spent most of their lives studying ways to solve problems. "Typically," said Tommy, "they just took one path at solving a problem, and then they'd get to the end of that and go, 'Okay, that didn't work, now what do we do?' and they'd go down another path."

Tommy would ask the techs, "How long before you think you have it apart and figure out if that's the problem?" Someone would give him an estimate, and Tommy would say, "What if you can't fix it?"

They'd say, "Oh, then we'll do this."

And Tommy would ask, "Can't we be doing that right now? We've got eight hours here and not everybody can be working on the fish. Can we be making phone calls? Can we be getting information? Can we be ordering integrated circuits? What integrated circuits do you think? Are there five chips that are suspect? Why don't we order those now?"

If the techs working on the problem were wrong after eight hours, at least the others would be far along a parallel path. And if they were right, they could always cancel the parts they had ordered or send them back. If a supplier charged them a few bucks, so what? Tommy was spending twenty thousand dollars a day on the best sonar experts in the world, and until their sonar worked they were surveying nothing.

The afternoon of the 8th, they redeployed the SeaMARC and began towing again along track line 1. The weather that had moderated slightly since the eye of Andrew passed over two days earlier began building back up, the seas to six feet and the wind to twenty knots. Trying again to tow along those neat little track lines on the search map, the *Pine River* fought strong currents and heavy seas, sometimes crabbing almost sideways. Once, a blast of lightning cracked down to the water so close to the ship it blew out a piece of cable that carried the signal from the submerged SeaMARC to the ship topside.

Tommy asked Lettow, "What do you think?"

Lettow said, "I think we should have gone back to the beach and drank beer three days ago instead of mucking around in this."

ON THE FIRST track line, they would follow a course through probability cells equaling just over 25 percent of the distribution: one chance in four they would image the *Central America* in that first thirty-mile run.

They would then turn around and run the next track line back, over-lapping half of the first track and imaging a mile-and-a-half strip of new bottom, turn around, and head back again on track line 3. At the end of that line, they would have covered nearly 50 percent of the probability.

But hours into their second try at line 1, the wind blew the bow of the *Pine River* off course. Trying to force the bow back, the captain cranked the rudder over farther, until it was almost sideways, pushing water. This slowed the ship and the tow fish headed for the seafloor. Seeing that the fish was dropping, the winch operator in the control room wound in slack so the SeaMARC wouldn't hit bottom, and that extra weight pulling back on the ship slowed it even more, which made it more difficult to keep the bow on track, so the captain tried more rudder.

The fish was all over, up and down, its speed constantly changing, in and out of the track, leaving holidays in that first run through the highest-probability cells. Tommy was back and forth between the control room and the bridge, trying to get the two crews to realize they needed to communicate with each other. "That created a little bit of friction," said Tommy, "but we had to figure out a way to make the ship work."

Tommy worried that the *Central America* was in one of those high-probability cells in that first track line where the SeaMARC was not working; or that everyone was so tired and nauseated from the work and the weather no one would know where they were when the SeaMARC did image a big target; that the *Central America* with those gigantic paddle wheels and all that gold stuffed in her belly would glide right on by, and no one would notice.

"I'd sometimes go in there in the middle of the night," he said, "and everybody'd be asleep on the EPC recorders."

When clients called Mike Williamson, they wanted him to find something on the bottom of the ocean. They told him where they thought it was, then turned Williamson loose to find it. That was their role as clients. In Tommy's mind, his role as client was "to help direct when I thought they weren't thinking clearly about the problem," and he pushed Williamson to rethink everything. If Tommy was up in the middle of the night and Williamson was napping in his bunk, Tommy got him up. When Williamson's crew started scratching their heads over what was wrong with the SeaMARC, Tommy got on the satellite phone at ten dollars a minute and called the guy who created the SeaMARC.

Don Craft had skippered several operations where Williamson was the sonar expert. He knew that when Williamson and his crew came on board, the sonar responsibilities would be handled as efficiently and professionally as possible. "Mike Williamson," said Craft, "is a world-class sonar operator." And Williamson had handpicked his crew after years of working with the best. "We brought to the party," he said, "what we thought was the best team in the world."

But Tommy was thinking, I have a hundred partners who have bet a lot of money that I know how to find and recover a shipwreck at depths no one has ever worked in before. My first test is to image the shipwreck on the SeaMARC IA. I had forty days at sea to get that image. Eleven of those days are now gone, lost to weather, towing foul-ups, navigation problems, and other malfunctions in the SeaMARC, and all we have is one track line from a lurching tow fish constantly being dragged off course.

Ted Brockett, who had designed the sled for the tow fish, understood Tommy's concern, but he sympathized with Williamson. "Harvey was always underfoot," said Brockett. "If something wasn't working right, he'd get in there with a calculator and a pencil and he'd be designing, calculating the stresses. He was down to the nuts and bolts on *everything,* so I spent a fair amount of time soothing egos and trying to keep these two guys calmed down. Nobody was throwing punches, but the discussions got fairly heated, and on a small boat when you can't get away from each other, those things tend to build."

OFFSHORE NO ONE functions at 100 percent. Williamson figured you could take anyone who performs well on the beach and put them in a small ship at sea and their productivity would drop by 90 percent. Ships the size of the *Pine River* pitched and rolled. They were powered by diesel engines, which were noisy and belched fumes that filled the head when the head already felt light from the rocking of the ship. Anyone who said they never got seasick was lying; seasickness incapacitated some people, greatly reduced the abilities of others, and dropped the productivity in all. Every crewman had a wave train out there somewhere with his name on it. A storm like Andrew produced enough wave trains with enough frequencies to nauseate the entire crew. When you haven't slept well for days, and the engines are groaning and causing the hull itself to vibrate, and the very place you plant your feet is slick steel constantly in motion, and the entire space you have is much less than half the size of a football field, and crowded into that space are twenty other men, at least half of whom you've never seen before, and you're trying to get your work done in a sailor's three point—two feet on the deck, one hand holding on—you get to where you just can't tolerate certain things.

"Little insignificant things," noted Williamson. "Like you can't stand the way somebody ties their shoes."

Food takes on a special significance at sea; it becomes almost sacred, certainly symbolic. Because after the weather has gone to hell, after the job has bored everyone senseless, after all of the knees and foreheads have been banged against pitching steel and all the little niggling personal habits of everyone on board have wound everyone else's internal strings as tight as a banjo's, food is the only thing left that makes life aboard ship tolerable. In his hands the cook holds the morale of the ship.

On the *Pine River* that summer, first the food was bad, then, as the storm ended and the sea lay down, and they finally could tow straight track lines, they ran out of food altogether. And the cook's hands shook so bad he couldn't hold a spatula, let alone the ship's morale.

Charlie was his name, and he wasn't a terrible cook; he was an okay cook, but he was an old army cook, a former prisoner of war in Korea who still cooked like he was mess captain at a concentration camp. One of the techs was a strict vegetarian; most of the others tried to stay in shape, and they watched what they ate. They requested a lot of fish, fruit, and vegetables. Charlie was a little guy, high strung and irritable, and he deep-fried everything. One of the techs said he couldn't fix a chocolate sundae without bacon grease. Three times a day they came into the galley and had to face Charlie the cook. One moment he was friendly, "Yo, how's it goin'?" and the next he seemed about that far from relieving your neck of your head. Craft knew that Charlie had stashed several pints of his favorite beverage around the ship, but he couldn't find the stash and he couldn't catch him nipping.

"He was a low-level drunk throughout the day," said Craft, "getting just enough to keep himself percolating along, and occasionally he would get a little over that, and then he would be totally helpless."

Ten days into the sonar search, Charlie ran out of steaks to deep-fry for dinner and started deep-frying chicken. The crew ate deep-fried chicken every night for three nights. "People were getting zits and everything," said Tommy. "It was horrible." Finally, the techs took a stand: They would not eat deep-fried chicken one more night. When Charlie heard this, he started serving deep-fried chicken every meal. Even breakfast. Then someone looked in the freezer and realized there was no food left, except for twenty-eight cases of chicken.

"There was no flour to make bread," said Craft. "There was no cereal left, no juices, no Cokes, no meat, nada. Chicken was it."

Using a bogus list as the inventory for the ship's food stores, Charlie had slipped by Craft, the skipper of the ship, and the company's port captain, and loaded only as much food as he needed to feed the crew for as long as *he* intended to stay at sea, which was about fourteen days.

Charlie locked himself in his bunk and was afraid to come out. Rumors circulated among the crew that they were heading in to reprovision, but Tommy calculated that if they returned to port now, they would lose another six days. He couldn't go to port, and Williamson agreed. Williamson told the crew he would eat saltines, but they had to finish the job. Tommy called his shore support in Wilmington for a boat to haul groceries and a new cook out to them, but the only boat shore support could find to sail two hundred miles offshore was a shrimper called the *Joe Christmas*.

When Tommy and Bob and Barry had gathered at the round tables and dreamed up ways to accomplish the impossible, they foresaw many difficulties. They studied how others had failed, and they calculated all the ways their own project could fail, and then they set about minimizing their risks. They knew weather could get them, they knew suppliers would be a problem, they knew they might have competition. But in their wildest failure-mode nightmares, they had never come face to face with the *Joe Christmas*.

When they found out their food was on a shrimp boat with a top end of about five knots, Ted Brockett and Will Watson caught some bluegreen dolphin on a red bass plug, but that was not enough to feed everybody. The crew still had no food, except chicken. Then someone had the idea of using the chicken to catch one of the sharks that constantly circled the ship. John Lettow fashioned a hook out of stainless-steel rod, and they fastened it to lightweight aircraft cable, and they wired on a whole chicken and threw it overboard.

"There's like a stream of fat coming off the back of the chicken," remembered Tommy, "and a shark came up and put his mouth on the chicken, and then he just kind of rolled over and opened his mouth and let the chicken out and swam away. They eat garbage cans and everybody knew that, only they wouldn't eat this chicken."

As the shark rolled over and swam away, somewhere in the Atlantic Bight between Charleston and the *Pine River* wandered the *Joe Christmas,* now one day overdue, both her captain and his mate drunk, trying to chart a course but off by a full degree, which had them meandering toward a coordinate sixty miles to the east of the *Pine River.* After they were a day overdue, Tommy had to hire a pilot to fly out and find them.

When the tech crew saw her, no one could believe she had ever set sail. The *Joe Christmas* was forty feet stem to stern and belonged in bays and harbors, fishing for scallops. Brittle shrimp and lobster shells formed scales across a deck that hadn't been sprayed and cleaned of fish parts for months. Tommy got a whiff when she was still half a mile away. "That boat just stunk to high heaven," said Tommy. "It made everybody sick."

Charlie the cook didn't care what the boat smelled like. They hardly had a line on the shrimp boat before he was aboard and hunkered in down below, and no one on the *Pine River* ever saw him again.

The tech crew was in the middle of a track line and couldn't stop. Tommy and Craft figured they could transfer the groceries with the two boats moving slowly in tandem, but the shrimp boat skipper, into the sauce and happy to be alive, couldn't get the hang of it. The techs had to stop in the middle of the track line, record where they had ended, winch in nine thousand feet of cable, and let the tow fish dangle, until they could resume after the *Joe Christmas* departed.

In an hour they had swung the cargo slings and pallets of food over with the crane, and with everyone expecting the skipper to cast off, his demeanor changed from happy to be alive to terrified of beginning the long trek back to the beach in the dark. He got on the radio and screamed, Oh my God, that he was sinking, that all of a sudden water was rising in the bilge and his pump wasn't pumping fast enough to keep it out. An engineer from the *Pine River* replaced the skipper's jerry-rigged pump with a regulation pump, and the water disappeared.

Hours had now passed since the new cook and the groceries had arrived safely on board, but the skipper wouldn't leave. He yelled that he wasn't too sure about that new pump, water still seemed to be coming in from somewhere, and now his steering felt loose.

"It's like a huge curse came on us," said Tommy. They couldn't leave the skipper out on the ocean with vision and judgment impaired. But they had track lines to run. Who was going to make up for the time if they didn't cut the ropes?

It was now almost midnight on June 17. They took the *Joe Christmas* in tow, headed back to the last track line, and the following afternoon resumed where they had left off. For the next two days they ran their track lines and monitored the recorders now that the system seemed to be working, all the while hauling behind them a stinking shrimp boat, her soused skipper, his soused mate, and their soused new friend. "They were all over there getting shit-faced," said Watson. He could hear them screaming at each other, and then they'd get on the radio and yell obscenities.

Then the skipper notified Tommy he was diabetic. He hadn't figured on getting lost, so his supply of insulin had run out. Without his insulin he could lapse into a diabetic coma and die. Tommy radioed the Coast Guard, and as he was wondering how much this exercise was going to cost his partners, a Coast Guard Lear jet flew out and dropped a spare bilge pump, syringes, and a refrigerated bottle of insulin.

T HE OCEAN AROUND the intersection of latitude 33 and longitude 77 is a lonely patch of water: not much to see, not many visitors, an occasional freighter passing through bound for somewhere else. Days went by without a glimpse of another hull. No birds, a few sharks, some dolphin flashing silver in schools. Not much else. Even the bottom is monotonous, mile after nautical mile of nothing but sediment.

The afternoon after the Coast Guard dropped the pump and the insulin, John Lettow was on the day tower watching four EPC recorders while reading a book. Lettow looked at the recorders, read a couple of sentences, looked at the recorders, read a couple of sentences, his eye rhythmically bouncing from one to the other in a kind of half beat. "You're clicking in a way that you become syncopated with the machinery," he said.

That afternoon he was deep into a new book, Tom Clancy's *The Hunt for Red October*. "This is pretty ironic," he said. "Clancy's describing this elaborate system for identifying submarine sonar signatures that the navy uses to track enemy submarines, and he's discussing the increase in Russian sub activity off of our coastline as the *Red October* was making its break to come to the U.S."

About that point, Lettow glanced up at the recorders, and right where the stylus belt sweeps the paper he saw puffs of black smoke. So much energy had suddenly shot back into the recorders from somewhere that it overloaded the system and burned the paper black.

Lettow turned down the gain on the recorders to stop them from smoking and pressed the button on the intercom. "Bridge, we got any rainstorms on the horizon?"

The bridge said, "No."

"Any ships passing close by?"

"No."

"Any marine life out there splashing around?"

"No."

"Well, then," figured Lettow, "we're having some problems with the sonar." Whatever the source, it was destroying their data along this track line.

Snapping shrimp clicking backwards in big schools sometimes smoked the paper. Lettow had seen that before. Rain falling faster than about a half inch every hour would also do it. But neither seemed to be the problem.

Five minutes later, two destroyers steamed over the horizon and a fast-attack nuclear submarine surfaced nearby, a hunter-killer group engaged in antisubmarine warfare, and they surrounded the *Pine River*. The crew went out on deck to look at them, and Lettow figured out what had happened. The navy had picked up their sonar signal coming off the SeaMARC, and they couldn't recognize the signature—it's a prototype—so they crept closer. Whatever the reason, they had wiped out the sonar records coming up from the SeaMARC, creating big gaps in the track line. The *Pine River* was still towing the *Joe Christmas,* and if the navy continued its maneuvers in the vicinity, Tommy would lose even more days of search time.

Don Craft got on the radio and tried to call the task force commander; he had been a task force commander himself, so he knew the procedure. But even though he got through to one officer and then another, they wouldn't put him through to the commander. Sometimes they wouldn't even answer. Meanwhile, one of the destroyers cruised up abreast of them no more than a quarter mile away, and every time its radar came around, the computers in the control room on the *Pine River* crackled and went dark.

Craft got on the SAT COM and called the CINCLANTFLT Watch Officer at Mayport Naval Base outside Jacksonville. Craft told him they were trying to run a sonar survey and the navy ships on maneuvers were destroying their sonar returns. He wondered if the exercises could move somewhere out of range. The officer said he would see what he could do, but he wasn't too optimistic. When Craft did not hear back, he called again, and the officer gave him a negative; he said that the task force commander would continue his operations, and that the navy really wasn't concerned about Craft's survey; the officer doubted that what they were doing was truly a survey anyhow.

Craft reminded the officer of a provision under federal law that held the navy liable if a ship ran over lobster pots not in the way of safe navigation. The same applied to fishing nets and other commercial operations. Craft told the officer they had a day rate of over twenty thousand dollars and a scientific need for the information that was so big Craft couldn't even estimate the dollar cost. He said he would certainly appreciate it if the officer would try to get their situation reconsidered. The officer said he would pass it on, but if he were Craft, he wouldn't get too optimistic.

By now, Craft was feeling cantankerous. He thanked the young officer, and then he said, "By the way, this same situation came up on the West Coast not long ago, and the navy spent $250,000 to pay the claims. What is the name of the OTC out here, so we know how to properly address this?"

The officer informed Craft that the names of navy personnel, especially the officer of tactical command, could not be released. He said he would be in touch. Three hours later, the hunter-killer group disappeared over the horizon.

But they still had the *Joe Christmas* in tow. They finally understood that the skipper had no intention of leaving. He had a new pump, a fresh supply of insulin, a little hootch, nowhere else to go, and a couple of friends who saw things just about the way he did. His only beef was that the Coast Guard had forgotten the cigarettes he had ordered with the insulin.

Tommy radioed the Coast Guard again, and the Coast Guard agreed to send out a cutter to escort the skipper back to the beach. A day and a half later, the cutter arrived at 3:00 A.M., and the last thing the crew remembered hearing from the skipper was, "Oh wow! Look at those blue lights, man! Where'd you guys come from?" Then the cutter put the *Joe Christmas* in tow, and together, like a mother duck and her ugly duckling, they putted back to Charleston.

Tommy had twenty days left to image the *Central America*.

AFTER THE WEATHER cleared and the swells and the wind subsided, they finally had the SeaMARC calibrated and tuned and the two crews coordinated on the towing. For the next two weeks, even with the *Joe Christmas* in tow and with interference from the navy, they had run thirty-mile track lines.

Mowing the lawn, they called it: moving through the water at one to two knots, back and forth, overlapping each track so as not to miss anything, thirty miles down, five hours to turn around, crab over twenty-five hundred meters, then thirty miles back. The tow fish flew a few hundred meters above the ocean floor, its ears listening for a sonar return, a bounce of its sound waves off a solid object anywhere within its range of five thousand meters.

Enclosed at the head of the back deck, the control room was filled with racks of electronics, recorders, and computers. Williamson ran two towers a day, the day tower from noon to midnight, the night tower from midnight to noon. On each tower was a pilot who controlled the altitude of the tow fish by winding in or paying out on the winch; a navigator who monitored the ship's navigation computers; a sonar operator who watched the charts rolling out of the recorders; a SeaMARC technician who recorded each change in the settings; and a watch leader, who had to ensure that no one got so involved in his own little project that the fish crashed into a mountain or the seafloor. It had happened before.

As the SeaMARC passed slowly and silently over the ocean floor a mile and a half beneath the surface of the sea, five graphic recorders, looking like large IBM Selectric typewriters, ran full time, the sonar operator continuously tuning them and changing the paper and the stylus belts. Like the fan belt on a car, the stylus belts whirred around two small spools, the tiny metal styli hypnotically zipping a faint gray line across the paper chart every four seconds. The room smelled of ozone and graphite from the styli burning the special coating off the paper, and the recorders sat covered in black dust.

If a target appeared on the bottom, the styli would spark several times and leave dark marks on the paper. The tech monitoring the recorders would alert navigation he had contact and note the time of contact in the log book and in the margin of the strip chart. Immediately, someone would bring a ruler and try to gauge its length. The paper moved slowly on, the stylus belt still etching a line every four seconds, more sparks coming off the styli, the figure taking shape as they watched, and then the figure would begin to taper off and the styli would go back to laying down faint gray lines as the darkened image scrolled by.

When the SeaMARC had passed a target, they radioed the bridge for the ship's speed, plotted its position, checked the altitude of the tow fish, checked the gain on the signal, read the cable gauge for how many feet of tow line they had out, and recorded all of it in the logs. Later they could return to that spot for a closer look.

"No target acquisition data has any validity unless it's repeatable," noted Craft. "You have to be able to go back to the same spot; otherwise, you haven't learned anything when you see a target."

THE SeaMARC sent back to the surface far more information than the EPC recorders could reproduce in shades of gray with a stylus belt and paper. Tommy wanted to recapture some of that lost information, for it might reveal enough to help him distinguish the *Central America* from other shipwrecks.

As the SeaMARC flew above the ocean floor and shot signals back to the EPC recorders, simultaneously the computer started to paint a color version of the same target. Then they could off-load the information to an optical disk and later recall the target, manipulate it, blow it up, subdivide the screen into four different images, paint one black and white like conventional side scan, tweak the intensity of the colors and the scaling ratios on the other three, and determine that *this* target is not what they're looking for. Or that maybe this one *is* what they're looking for.

On the optical disk, a gray sketch no bigger than a thumb on the strip chart could be blown up to a picture the size of a man's head. And each tiny pixel, only a pinprick in the picture, could be assigned a color according to a narrow range of density, and that color could be further subdivided into 256 intensity levels, so they could tell what the target was made of. By the time they had blown up, colored, thresholded, filtered, highlighted, and otherwise stretched the information on that target, what appeared to be a thumb print on the strip chart might look like a steamship with paddle wheels on the computer screen.

Will Watson was monitoring the EPC recorders late one night when they acquired their first large sonar target, an anomaly on the seafloor that could be the *Central America*. Tommy was in the control room, and Watson motioned him to come over. "Here's our first target."

"Harvey said, 'Wow,'" remembered Watson, "and you just saw the sparkle in his eye. He knew at that moment that this system could find old wooden-hulled shipwrecks on the seafloor."

For twelve hours one tech team watched the strip charts roll by; then they switched, and the other tech team watched them roll by for another twelve hours. Then they switched again. Ten to twenty times a day the styli drew the black-and-white image of an anomaly, something that stood out against the sediment on the ocean floor. But most anomalies they quickly discarded: cargo containers ripped loose in a storm, a German U-boat sunk in World War II, sailboats and fishing boats gone down, clumps of fifty-five-gallon drums lined in concrete and filled with radioactive waste. "Lots of junk down there," said Williamson. Every two or three days, one of the two tech teams would see a single target about the right size, about the right density, that could be the wreck of the *Central America.*

Williamson, who was watch leader of the night tower, likened the operation to flying a helicopter, the pilot trying to fly at a uniform height above the bottom, continuously jockeying with the winch, cable in, cable out, compensating for currents or subtle variations in the ship's speed, trying to find that sweet spot where the water flowing by the long arc of cable did not push it up too high or allow it to sag too low, but kept the SeaMARC at a constant altitude, flying straight and level. Williamson had been doing this for a long time, and he never got bored.

Craft could be in the control room three minutes before tears from the pain of boredom came to his eyes. "If the gear's in the water and everything is operating properly," said Craft, "the people on watch are sitting in there watching rolls of paper go by. When they paint a target, it's a big event."

The lucky shift was the one on duty when they reached the end of a track line and had to turn the ship around. This maneuver took five or six hours, and the techs had nothing to do in the control room until the ship was back on track and headed down another line. If the turn came during daylight, they got their sunglasses and headed up to catch a few rays on Steel Beach, the helo-pad that stretched out over the back deck.

The techs got along pretty well, but Will Watson noticed a subtle transition at sea. After a few weeks, every guy out there had his mainspring torqued good and tight. Watson figured it all had to do with loss of control. Onshore, you had control; offshore, everything unfolded any way it damn well pleased and you went along—with the weather, with equipment breaks, with lack of sleep, bad food, tedium, worrying over wives and girlfriends—until your mind entered an altered state, where a normal, easygoing person became suspicious, impatient, and sometimes paranoid. When you got back to the beach and gradually resumed control over your life, you couldn't believe you ever felt that way about something so insignificant.

One afternoon three or four weeks into the survey, a few of the techs were watching paper roll by in the control room, while several others were sunning themselves on Steel Beach or fishing for mahimahi, when suddenly they heard gunshots.

One of the techs had been standing on the back deck yelling for help, and a couple of guys in the control room had raced out and found him hooked into a seven-foot white-tip shark that was angry and thrashing blue water white. He didn't know what to do. Lettow pulled the line in enough to take a turn around the mooring bitts. Then the tech picked up an aluminum gaff with a five-foot handle, and he gaffed the shark, landing the point in the shark's eye. The shark popped its tail, rose up out of the water, and bit off three feet of the gaff. The tech leaped back several feet, the aluminum stub in his hand, and then with three other techs watching wide eyed and slack jawed, he dived his hand into his vest and drew out an opera-size .22 revolver and started blasting the shark.

"None of us would have felt very good the whole trip," said Lettow, "had we known he had a gun."

With a meat hook stuck in its jaw, one eye gaffed, and a couple of hard thumps from the bullets, the shark glared up at them with its one good eye, jerked its head, snapped the cable, and swam away.

When Craft heard about the incident, he was furious. He had three rules when he sailed: No drugs, no alcohol, no guns. No excuses. No conditions. Ever. He told everyone an hour before departure, and he warned them he would *not* proceed as a court of law, that he would search lockers, even if he only *thought* someone may have violated these rules. The drugs, the alcohol, the gun would then be confiscated, and the owner would be dumped on the dock at the next port o' call. Craft was already mad that Charlie the cook had managed to keep his liquor supply hidden.

"As soon as I found out about it, I crawled aboard him," said Craft. "I've always operated on the theory that to be effective, discipline must be swift." Craft threatened to kill him if he ever pulled a stunt like that again. "He was a real kook," said Craft. If they had been headed into port during the next few days, Craft would have bounced the guy and his expertise off the ship. He didn't care if the guy was God's gift to sonar. But they were so far behind already, they couldn't afford a trip in, or they would lose another two or three days.

They never saw the gun again, but they caught the same guy later hoarding Popsicles, eating five or six of them a shift, claiming he had had only two, then secreting little stashes of them in the backs of freezers around the ship. When confronted, he claimed they belonged to him. Will Watson took them away and passed them out to the other techs. It was ninety degrees out on the water, they didn't have a whole lot to look forward to anyhow, and those Popsicles made life a little more tolerable.

W HEN W ILLIAMSON CAME on the night tower, he analyzed the data from the previous twelve-hour shift. No one questioned Williamson's eye. He could see things in a sonar return no one else could see. As he reviewed the strip charts, he would call out targets for the techs to bring up on the computer screen, and they would play with the image: blow it up, filter it, alter the colors, measure it. The strip charts still were the first stage of elimination, where they could approximate length and width; where they could draw inferences about wood and steel. But with the computer they could see things in a target no one had ever seen.

At last the weather had moderated and the sea was calm, only gentle swells rocking the *Pine River* as they mowed the lawn up and down the search map. They had now completed half of the track lines on the search map; yet, because those lines ran through the cells of much higher probability, they had already covered eighty-six percentile on the probability map, and they had imaged hundreds of anomalies. Williamson dismissed most of them because they obviously were too short. He eliminated most of even the larger targets for being too round, too hard, or apparently some form of geology. The few left became key targets. Williamson called this short list the "hit parade" and every target on it had a reasonable possibility of being the *Central America*. He next arranged the hit parade in order, according to each target's resemblance to the models Tommy developed of what the wreck of the *Central America* would look like after 130 years on the ocean floor. And from the moment he saw it, one stood out, even on the strip chart: It appeared to be a sidewheel steamer, resting upright on the bottom, a dark humped shadow amidships indicating paddle wheels. Williamson designated the target "Sidewheel."

They had already imaged so many targets in the high-probability cells that Williamson now wanted to shut down the broad-swath survey, drop the SeaMARC to a lower altitude, tighten the swatch from five thousand to one thousand meters, and go back for a much closer look at the good targets, especially Sidewheel. With the swath at five thousand meters, all they could tell was that something large sat on the ocean floor and stood out in contrast to its surroundings; at one thousand meters, even images produced by sound would look almost like a picture.

Williamson himself realized the irony in what he wanted to do: He was always the one to insist they complete the broad-swath search before they looked closer at any of the targets, because the target that appeared promising on sonar might not be the target they were looking for.

"We had made a big issue of this with Harvey," said Williamson, "that the search plan would be followed. This is the bible for this cruise, and we will not deviate from it."

Now Williamson was the one suggesting they deviate from the plan, and he had good reason. The weather had finally died down, and the one-thousand-meter search required these smoother conditions. They could run the five-thousand-meter search in much rougher weather. "It was my judgment," said Williamson, "that we should take advantage of the good weather and do the stuff that we could do only in good weather, and then when the weather got bad again, we could go back to doing the stuff we could do in bad weather." All they had to do was flip switches topside to convert the SeaMARC to the high-resolution mode.

Tommy saw another irony: The world's best sonar technician, who had once accused him of being a treasure hunter, now seemed himself to be succumbing to what Tommy called "treasure hunter syndrome." The SeaMARC had flown through every high-probability cell on the map, and right at the juncture of a quad of high-probability cells they had imaged Sidewheel, a dark, pencil-shaped target with little humps in the middle. As tempting as it was to think that the *Central America* was already safely imaged in their computer and now lay waiting to be explored, Tommy wanted to avoid the mind-set of the treasure hunter, that every promising clue was the thing itself.

Williamson argued that they had searched 80-something percent of the probabilities, and that if they had hot targets already, ship shapes about three hundred feet long, one with a hump in the middle, why not shoot all of them at much higher resolution, maybe even drop a camera on Sidewheel? He wanted to give Tommy not just scrolls of strip charts and piles of diskettes, but pictures. If it was the *Central America,* they were through; he could move on to other clients, and Tommy could come back next year with a robot and recover it.

But Tommy reasoned, If it *isn't* the *Central America,* and we waste all this time shooting high-resolution sonar of it, and the season ends, and we haven't completed the probability map, and we come back next year with a robot ready to recover that site, and Williamson is off with the SeaMARC looking for black boxes and bombs when we discover it is *not* the *Central America,* what do we do? He told Williamson that he had paid to have the sonar crew search the entire probability map, and regardless of the likelihood that one of the targets already imaged was the *Central America,* he wanted the rest of that map searched.

Tommy wanted to complete the map, then analyze each of the targets, compare them with one another, and pick carefully the order of verification with the high-resolution passes. Williamson and his crew were the best sonar technicians available, and Tommy knew they were the best. But not even Williamson and his crew knew what the *Central America* was supposed to look like on sonar. No one had ever imaged a deep-water wooden-hulled shipwreck and then gone to the bottom with a camera to see what it really looked like. Just talking about how much Sidewheel looked like a sidewheel steamer concerned Tommy. "They could talk themselves into thinking that's the *Central America.* I wanted to make sure we didn't get into delusions of grandeur. We were extremely excited about the results so far, but you don't want to be like everything you see is the main pile."

Tommy had structured the search phase based on sixty days at sea, and he contracted for the SeaMARC from the first of June to the end of July. In May, Williamson had called him: The National Oceanographic and Atmospheric Administration also wanted the SeaMARC in July; could Tommy cut his search down to forty days? The NOAA job was a good contract for Williamson's young business, so Tommy agreed to a new arrangement. "We'll just use that as analytical time," he told Williamson, "and we'll start up again August sixth and do the last twenty days."

But about halfway into the cruise, Barry started seeing telexes coming to the ship from Amoco, who wanted the system in August. Williamson was negotiating with them via the fax machine. "It became apparent through the faxes," said Tommy, "that they were going to do the Amoco job and just blow off our option to come back in August."

Williamson told Barry, "In this business, there's the oil companies, the oil companies, and the oil companies. You can't deliver for one of them, you think they're going to call you again?" Besides, and Williamson kept coming back to this, he had already imaged the *Central America,* and Tommy was just being stubborn tying up the SeaMARC with an option so he could *maybe* spend that last twenty days looking out in all of those less-than-one-chance-in-a-thousand probability cells. "But the odds were high," Tommy said later, "that if I needed that twenty days to finish the map and couldn't get it, that would be the end of the project."

For a couple of hours each day, Tommy told the captain to turn off the radar sweep, and he and Williamson would sit on lawn chairs up on the fly bridge and try to resolve their differences. The techs remembered Williamson coming down from the meetings flexing his jaw, so frustrated he could not repeat what he and Tommy had just discussed. Williamson was an easygoing, likable guy and a professional, "an officer and a gentleman," one of his men described him, yet Tommy would leave him in the middle of the control room, or standing on the back deck, or up on the bridge, with his jaw dropped, shaking with rage, blue veins popping out in his neck. "Beet red with anger," said navigator Alan Scott. Things got so bad one day that with Tommy sitting on a table in the galley four feet away, Williamson turned to Barry and barked, "Barry, you tell Harvey . . ."

Tommy just had this way of blinking slowly, long eyelashes accentuating the pace, and he wouldn't be adamant, he wouldn't raise his voice, he wouldn't gesticulate and gyrate, he would say, "We're not going to do it that way." And something in the way his eyes locked on yours behind those slowly blinking eyelashes, something about the cadence of the words as they came out let you know that what he said was final. He didn't care how much conventional wisdom dictated otherwise, he had already confronted conventional wisdom, he was painting on a bigger canvas, and he knew other truths about the big picture.

Williamson had brought a rowing machine on the ship, and after a session with Tommy, he would descend to the control room, fuming and mumbling to himself and running up a six-thousand-dollar bill on the satellite phone negotiating with Amoco, then hit that rowing machine like a man trying to exorcise his demons. Brockett figured he was forty miles downriver before he cooled off.

Ten of the twelve technicians on board thought about the same as Williamson. Lettow said it made no sense if you were looking for the *Titanic,* and you found a target that resembled the *Titanic* in every way, but you decided to go on searching anyhow. Alan Scott and Will Watson figured that since Tommy was paying the bill, and he was not being completely unreasonable, they should first cover the entire grid in the broad swaths. "The thing is," Watson admitted, "we'd found the thing early on, so it was like being in Eden trying not to eat the apple. It took some discipline."

IN THE CONTROL room, the techs had Sidewheel frozen on the computer screen, taking Polaroids, measuring parts of the ship, and pulling more information from the image. "We were pretty happy," said Watson, "because things were looking pretty good. It looked like a wood vessel, and we could see the sidewheels."

While feelings over the August option simmered, Tommy agreed to let Williamson run high-resolution passes on each of the promising targets, not because Williamson insisted or the other techs pressured him, but because he had calculated how much time they needed to complete the broad-swath search, and they still could get it done in what remained of the forty days, if the SeaMARC held up and the weather did not kick them in the teeth any harder than it already had.

On June 24, they narrowed the swath on the SeaMARC, retuned the recorders, and began reacquiring the targets on the hit parade for the high-resolution flyovers. The techs now had something to play with and ponder: bigger, tighter, finer images of shipwrecks on the ocean floor eight to ten thousand feet below.

Each tower calculated from its previous navigation where a target would be, then aimed for that spot at a ninety-degree angle with the SeaMARC set for one thousand meters. On the night tower, the navigator called it play by play, as the *Pine River,* with the SeaMARC behind, approached a target.

"It focused you," said Watson. "You're coming in, you're looking, and he'd give this entry into the line, like, 'Okay, gentlemen, you're on line da-da, at a heading of da-da, coming in with a fish altitude of da-da.' And then he goes, 'And we're going to get this one LIKE A BIG DAWG!!' We'd sit back and laugh. He was great. By that point we were really having a ball. Everything was so exciting."

By the end of the second day, averaging six to eight hours per target, they had reacquired and shot several, including Sidewheel, at close range. Tommy unrolled blueprints none of the techs had known existed, and for the first time they saw a sharp outline of the sidewheel steamer they had been looking for. They measured different parts of the ship and compared the dimensions to what they were finding on the higher-resolution images. More than one target looked good in this second round.

"They had maintained their integrity," said Williamson. "You could see individual ribs, they were standing proud off the bottom, they were shaped like a ship, looked like a ship, had a shadow behind them." Every time they ran a tighter swath over a promising target, Tommy reviewed it with Williamson and the other techs, but no matter how good it looked, Tommy wanted to continue searching.

When Williamson saw the computer enhancements of these closer, high-resolution passes, he wanted to skip even the close-up work on the other good targets and immediately drop an underwater camera on Sidewheel. Tommy insisted they continue to take advantage of the good weather to shoot one high-resolution pass on the other major targets already imaged at five thousand meters, even though some of those targets lay in low-probability cells in the southwest corner of the search map, forty miles away. Williamson argued that that was a long trip that would eat up at least a day when time already was getting precious, and he wasn't the only one pushing Tommy to drop a camera on Sidewheel.

"We were real sure we had a sidewheel steamer," said Watson. "Scotty had done some pretty good image processing, and you could actually see this wheel on the vessel." Alan Scott favored Tommy's argument that they first shoot close-ups of all the promising targets, but even he couldn't deny the irresistibility of Sidewheel. "I swear," he said, "look-ing at it on the image-processing system, it appeared that the wheel had spokes to it." But Tommy insisted they leave Sidewheel and the others behind, head forty miles downstream to the southwest corner of the search map, and shoot high-resolution images of those targets.

At midnight on the 25th of June the *Pine River* was en route to the new location, and by midmorning on the 26th they had a high-resolution image of another target. By noon, they were in the middle of a new track line, headed almost due west. On the first pass, they saw nothing but more ocean bottom. "No joy," Williamson wrote in the watch leader's log. Into the second run no more than fifteen min-utes, they began to see what appeared to be rock outcroppings or drifts of sand. A little farther, the recorders sketched a target two hundred meters to port, but the return was weak and surrounded by a scatter-ing of little pinpoints of reflection, similar to the geology they already had seen. It looked nothing like the high-resolution returns from Sidewheel and the other targets where they could see the outline of a ship's hull. The navigator wrote on the strip chart, "Contact 200 meters port—geology?"

This was the problem Bob Evans had predicted would arise if the *Central America* had sunk closer to shore in shallower water: an abun-dance of outcropping material that would confuse the sonar returns of a shipwreck. But they couldn't figure out what had caused the bright return on this target when they swept the area on the broader swath a few days earlier. Then as they neared the end of that track, the EPC recorders began to sketch another anomaly, one at the far northern edge of the sonar range. They turned the ship, shifted eight hundred meters to the north, and began their third run at the target.

Early in the line they saw more of what appeared to be geology. Then suddenly the recorders began to paint what someone later de-scribed as "a real banger." The navigator wrote on the chart, "Contact off port 150 meters." But when the recorders stopped painting, the techs studied the image, and the target, for all its brightness, seemed no longer than maybe thirty meters. Someone suggested it might be a shipping container, but it was too long. Whatever the anomaly, apparently it was what they had seen earlier on the original broad-swath run. It was big and it was bright, but not big enough and maybe too bright for a historic wooden ship.

Craft liked the bigger anomaly the sonar techs had imaged on the first run, the one they designated geology. The morning after they shot it, he pulled Tommy out of his bunk and told him, "If I were you, I'd take a closer look at Geo." Still in his pajamas, Tommy followed Craft into the control room to look at the strip chart, and he liked the target even more than Craft. To him it looked like a ship.

Tommy now had at least one high-res look at each of the important targets, and the techs wanted to get back to Sidewheel. "We didn't want to be searching the entire Gulf Stream area," said Lettow, "just to map the sea floor and find every shipwreck out there." But to square off the searched area along the southern boundary, Tommy insisted they raise the swath to 5,000 meters, loop to the east and run a final east-west track. Then he had them run another track line along the western boundary, completing the line in seven long hours of the tech crew staring once again at the ocean bottom.

Finally, late the afternoon of June 28, they steamed again toward Sidewheel. They shot it a second time at high resolution, and on the third run they narrowed the swath width even more and imaged it at five hundred meters. Then they turned and shot it a fourth time, turned and shot it a fifth time, turned and shot it a sixth. The closer they got and the tighter they focused, the more the image resembled a photograph of the *Central America*. Williamson recorded in his log, "This target appears to have the characteristics of our objective." If Tommy would just let him prove it was the *Central America,* Tommy would be happy, the techs would be happy, and he could pursue the job with Amoco. But Tommy still would not let Williamson drop the camera.

The camera sled had no subsea navigation, and the *Pine River* could not hold position in the Gulf Stream current. The only way they could image the site with a camera was to dangle the sled on the cable and drag it back and forth in short tracks, much as they ran the sonar sweeps, except they had to get the camera within twenty feet. The method required patience and too much luck. Tommy had no time for patience or luck. He agreed to finish a suite of high-resolution sonar images of

Sidewheel, but then he insisted they sweep the parts of the search map they had not covered.

To keep the tech crew as objective as possible in the beginning, Tommy had purposely told them little about the *Central America*. But over the days of evaluating targets with Tommy and Williamson, they slowly had gleaned a set of criteria that any target must possess to be a candidate: It needed to be about three hundred feet long, about fifty feet wide, have three masts, two sidewheels, and one boiler stack; it had to be made of wood, although 750 tons of degrading iron and about 200 tons of coal might also litter the site.

As they dropped the SeaMARC closer to Sidewheel, as they shot it from more and more angles, now seven times, eight times, nine times, its rounded features sharpened, and the techs saw emerge from the collective images a ship. They shot it perfectly abeam, they shot it at forty degrees to see up inside a large cavity, they shortened the wave length on the SeaMARC signal so they could resolve even finer detail.

"We were taking Polaroids," said Watson, "and then lining them up and saying, 'This is this with this color palette, and this is this with this color palette.' We were so ripped! 'Oh, you didn't eat today?' 'No, I'll eat tomorrow, let's try this, and, okay, let's try that.' We looked at it a hundred different ways in a hundred different colors from a hundred different aspects. We needed like eight more sets of hands to try all the things we wanted to try."

In the shadows of the images they saw a huge hump amidships that would be the housing for one of the paddle wheels, and they could see that the ship had two prominent masts. This bothered Williamson, because in the blueprints he had seen three masts on the *Central America*. Tommy explained that during the storm the second mate and two other men had chopped down the foremast to ease her listing to starboard.

Williamson had only one nagging problem with Sidewheel: No matter how they estimated, it seemed smaller than the *Central America* was when it entered the storm. They had to account for ways that the image of a shipwreck of a 280-foot ship could appear to be shorter, and the images themselves gave them a clue. They could see damage to the bow, as if the ship had struck the seafloor head-on at high speed. Then the techs learned that some of the five hundred passengers and crew left behind had chopped up large portions of the ship to make rafts, and the techs surmised that upon impact with the seafloor, the weakened ship had accordioned.

JUST AFTER MIDNIGHT on June 30, they attempted to run a second high-res shot of a good target near Sidewheel, but weather forced them to abort the line not a half hour into the run. They swung around and re-aligned, and on the third run they imaged the target only seventy-five meters to port. Williamson studied the strip chart and the computer returns, and before they had even ended the line, he recorded in his log, "S/wheel still best bet."

They returned to Sidewheel and ran another line, and Williamson continued to study that target. He recorded in his log, "Search team recommends deployment of transponders for camera runs." He was even preparing for the camera work, when Tommy interceded, and thirty minutes later they had raised the swath width of the SeaMARC back to the full five thousand meters and were proceeding toward a track line along the eastern edge of the search map, "against recommendations of MEW," wrote Williamson. Tommy wanted to look into every probability cell that contained a number higher than 0; to the west were several with numbers from 2 through 8, and more to the east had numbers from 1 to 7, a few chances out of a thousand that the *Central America* lay beneath that two-mile square of ocean.

Brockett remembered Williamson being furious. "Look," Williamson told Tommy, "this is what I do for a living. I know what I'm doing, I'm an expert. We are wasting our time out here. We need to get some detailed information on these targets so that you will be able to do a good job next year!" And Brockett remembered Tommy insisting, "No, we need to go out here and look in these low-probability areas and complete the search."

About ten hours into that next eastern track line, they imaged another target large enough and bright enough for Williamson to add to the hit parade, and it was labeled "Galaxy."

But Tommy still wanted to run the final eastern track, even though the total of all of the chances out of a thousand that the *Central America* lay along that track was 3. Tommy wanted to run that line and even extend it by nineteen kilometers. "Extending that line would go through a bunch of cells with zero probability or infinitesimal probability," remembered Williamson. "There was quite a heated discussion about that. Harvey just said, 'You will extend this line!'"

For ten hours, they towed the SeaMARC along that far eastern track line, while a following sea and a strong northeast current constantly tried to push the *Pine River* off course. Williamson wrote in the log, "Hope we have not wasted the good weather/sea conditions needed for camera/video runs on S/wheel, which we believe to be our target (the *C.A.*)." Then the wind began to kick up, too, and before they had reached the end of the line, they had to abort the run. "Again," wrote Williamson, "it is the recommendation of the survey team that we return to contact 'Sidewheel', deploy LBS transponders, calibrate the acoustic grid, and proceed with video and camera verification of the target as a 280' paddlewheel steamer we have been contracted to locate."

But now Tommy wanted a high-resolution look at Galaxy, the anomaly they had imaged the day before. Williamson had already studied the broad-swath image of Galaxy and determined it was not a mid-nineteenth-century sidewheel steamer. He recorded in the log, "Harvey Thompson insists that we next run a verification pass at 1km on target 'Galaxy,' which appears to be a steel-hulled vessel with associated debris (bulk cargo)."

On the first attempt, they hit the target in less than four minutes and ended the run in ten, but the image painted only on the strip charts, not on the computer. They tightened the swath to five hundred meters, turned and swept the target a second time, then turned and swept it again for the third and final time. Now Tommy had at least one and often two or three high-resolution images of every major target they had located. He calculated they had now covered enough of the map that with only three or four days remaining, the time would be more wisely spent either running closer looks at the promising targets or trying to drop a camera on Sidewheel. He directed them to quit and sail for Sidewheel to begin laying the navigation grid for the camera runs.

"We ultimately prevailed," said Lettow, "and we put the video camera down on that shipwreck."

* * *

At about six hundred feet, the deep ocean squeezes out the last particle of natural light, and below that everything moves in blackness. The SeaMARC flew through that blackness, shooting sound waves at the deep-ocean floor, and the system recorded the behavior of those sound waves as they encountered hard objects. Then it converted that behavior to information the eye could see: squiggles on the chart recorders and pixels in a color mosaic on the computer display. Even when they dropped the SeaMARC closer to a target and tightened the swath, they still could see only converted sound waves, not the object that bounced them back. That's what was so frustrating about sonar: You couldn't see the thing itself.

On July 4, they retrieved the SeaMARC and left it on deck. For the next two days, they set up a subsea navigation grid, repaired the cable, and worked on the video system. They launched the SeaMARC again to run it over Sidewheel another six times, until they thought they could find it with a camera. Then they deployed the camera sled early on the morning of July 7. By 5:00 A.M., the camera trailed in the bottom darkness nearly nine thousand feet below.

They had clamped the lights, a video camera, and a still camera to a hydrodynamic sled, but the cameras were stationary, aimed straight down, with no thrusters to position the sled and no pan and tilt to control the cameras. All they could do was hope that their navigation readouts were accurate, that they could relocate the site and drag the camera sled across the center of it, and that the remains of a ship suddenly would glide into view.

Williamson called the searches "Brownian motion," a phrase borrowed from physicists to describe the random movement of ricocheting gas particles. Tommy referred to them as "spaghetti searches."

They directed the captain of the *Pine River* to proceed at a specified heading, while they sat in the control room and watched the television monitor. They had shut down the sonar recorders and the rest of the SeaMARC equipment—no gains to adjust, no nav to record, no measurements to take, just a film log to keep, a television monitor to watch, and one guy, the pilot, ready to raise or lower the camera sled dragging behind the *Pine River*.

To see any detail, the camera had to be within twenty feet of the object—ten feet was better—and if the camera was farther away than thirty feet, they could see nothing at all. As the pilot watched the monitor, he paid out on the winch when the boat heaved up and hauled in on the winch when the boat heaved down, trying to keep the sled flying at a steady altitude. If he let the camera rise higher than thirty feet off the floor, they saw nothing on the monitor but backscatter—"Just like looking at a fuzzy white TV," said Lettow. To get the camera back in range so they could see the bottom, he had to pay out on the winch just a tap without going too far and crashing the sled into the floor. Lettow likened it to playing an intense video game.

On the first pass, they towed the camera sled and watched the monitor for two hours and saw nothing but snow. On camera run two they saw the bottom briefly, and they got some footage of the crisscrossing trails of sea cucumbers, but at the end of the run, they discovered that the VCR was not operating properly, and they had lost whatever tape they had on runs one and two. They tried a third camera run, and again they glimpsed the bottom only for moments. They seemed to have good horizontal control, but it was difficult for the pilots to keep the sled in that narrow thirty-foot window without risking the equipment. Already, one of the pilots had crashed the sled into the floor, bending the frame. With the sea beginning to build again, they tried a fourth camera run, and when they still saw nothing, they turned to reposition the ship and decided to change their approach. This time they would drift across the target with the camera almost directly beneath the ship.

They recovered the camera sled, hooked it to a chunk of pig iron, and lowered the whole rig over the side. Since the *Pine River* could not remain still above the target, they tried to align her with the wind and the currents so she would drift slowly across the spot. "Just kind of wander about in the area," said Williamson. "Very frustrating because obviously you're just looking at whatever providence brings into your field of view."

Within an hour, Lettow piloted the camera over anchor chain, huge iron links in the sand trailing off into the darkness. Then a ghostly image of what appeared to be timbers crossed the monitor. They realigned the ship for another drift, and two hours later they again saw a piece of something. And again an hour after that. Then they saw what they thought was debris and a few minutes later something that looked like a wood beam. But in all of these passes they could never tell what they were seeing or where they were on the site.

Each time they thought they saw the ship, the pilot began firing the still camera to shoot as many frames as he could, but every time he fired, the video monitor would go blank or suddenly fuzz with static. When they recovered the camera sled to load fresh film and tape and replace the batteries, they discovered that out of the four hundred still frames they had shot, only one hundred had been exposed, and most of those were either underexposed or double and triple exposed because the frames had not advanced.

They lowered the sled back to the bottom, and immediately they could see the ocean floor, but they couldn't see the shipwreck. They drifted until they were certain they had passed the site, then they set up the drift again and went back across the area and again saw nothing. On the sonagrams, the shipwreck appeared to be about the size of a 400-meter oval of track with a football field and both end zones tucked inside, and they were drifting across it roughly on the perpendicular. How could they be missing it?

The second night into the close-in camera work, Lettow was asleep in his bunk when Watson woke him up. "You got to get up here," said Watson. "These guys just aren't keeping it together." The atmosphere in the control room had gotten so tense for the night tower that Watson and the other pilot had started picking at each other, arguing about why they couldn't find the ship and how each was doing his job, and finally, the pilot had said to hell with it and walked out.

Lettow dressed and returned to the control room and started running the system again, as Tommy and the techs talked and watched sea cucumber trails scroll by on the monitor. The trails reminded Lettow of a dry desert lake where motorcycles had crisscrossed in every direction. In the middle of a long silence, he said, "What are the chances of us picking up that anchor chain again in the middle of all these sea cucumber trails and then finding a shipwreck attached to it?" About five heartbeats later, big and dark and right in the middle of the monitor appeared the anchor chain, and Lettow had the camera running parallel.

"JEEEZUS!" yelled Lettow. "Look at that!"

Watson yelled, "Check it OUT!"

Everybody was watching the monitor now, including Tommy. The camera followed the chain for about a hundred feet, and the chain ran right up to the bow. Lettow already had started raising the sled so he wouldn't crash it into the ship, and it flew up and over the side, and then he eased it back down.

Half the techs in the room started yelling at once. "Get it out of there! Get it out of there! You're going to get it hung up!"

Lettow was already hauling in on the wire, and the sled lifted, but they still could see across part of a deck to what looked like cable shrouds from a mast and some sort of superstructure. As the camera glided forward, the decking appeared to be collapsed, and Lettow thought he saw cargo down in the hold. Then they saw the other side approaching, much of the gunwale eaten away, only jagged posts sticking up, and the techs yelled again to get the sled up or they were going to lose their camera system.

"I flew it up over that," said Lettow, "and you could see the seafloor again. Then we drifted away." The whole scene had lasted less than a minute.

Lettow started singing. "Chain, chain, chain . . ."

And the other techs raised their hands, and started dancing and singing with Lettow, all of them rummy from six weeks at sea.

". . . chain of foo-ools."

Beat, beat.

"Chain, chain, chain . . ."

Nobody knew the rest of the words, but it didn't matter. Then, as if they had just discovered how magicians pulled rabbits out of hats, one tech said, "What are the chances of us seeing," and he looked quickly at the monitor, "some engine works?"

Then another, "Wouldn't it be nice to see a . . . paddle wheel."

And another, "How about a . . . treasure chest."

"We were getting out there on the reality mode," said Lettow. "We were all kind of delirious at that point."

But the magic never worked again. The weather had worsened since they began the camera runs two days earlier, the sea chopping into bigger and bigger hills, the wind rising. Williamson had to get back; he had commitments with oil companies. Besides, the camera was not right, the lighting was not right, the *Pine River* was not right. Everyone could see it. They had reached the limit of their capability for that summer.

"We did not have an ROV available to go to that depth," explained Craft. "There *were* no ROVs anywhere that could make that depth." That was Tommy's next challenge: to create an ROV, a robot, that could work that deep. If he was successful, he could return the following summer with a vessel capable of holding position on the surface and a stable camera platform and accurate navigation so they could drop the camera next to the site every time and reposition it a few meters away with small thrusters on the frame, not drift through the area and hope. They would have better electronics and better lighting and clearer pictures, and the capability of staying for long periods of reconnaissance and evaluation.

Tommy decided he would not need to exercise the option on the SeaMARC in August, that Williamson could pursue his contract with Amoco. He would like to have had more high-resolution work on other promising targets, but he had at least one high-res look at each main target, and he had now dragged the SeaMARC across more than 90 percent of a large and carefully calculated search map, and two teams of the finest sonar technicians in the world were confident they had imaged the *Central America.* He would further analyze the strip charts and the computer images, the stills and the video footage, and he would report to his partners, and he would begin his next round of funding for verification and recovery, and he would modify or build a vehicle that could do all of the work he needed to study that wreck, to verify it was the *Central America,* to preserve the history and initiate the science, and to recover the gold for his partners. For now, even if it was a mid-nineteenth-century sidewheel steamer laden with gold, they had no way to study the site properly, no way to document and preserve it, and no way to retrieve the treasure. And there was no better place to secure it anyhow than under eight or nine thousand feet of water somewhere out in the Atlantic.

Tommy and Barry asked the techs to cast a confidence vote on Sidewheel. Each would put his vote on a piece of paper, fold it, and toss it into a hat. The question was this: What chance does Sidewheel have of being the *Central America*? Tommy and Barry did not vote, although Barry admitted that even he thought it looked good. "It had a beautiful outline to it. It was very thin, very long, pencil shape like the *C.A.,* and it had this curious half-moon rise right under the gunwales and dead in the middle of the ship."

When they tallied the results, the techs collectively had voted that Sidewheel had a 90 percent chance of being the *Central America.*

"The percentages were pretty damn high," said Tommy. "But it still wasn't good to get cranked up."

At seven o'clock on the morning of July 9, they recovered the depressor and the camera sled, secured them both and unloaded the cameras, then pointed the bow of the *Pine River* toward the beach.

Columbus, Ohio

⚓

Fall, 1986

THE PUBLIC'S AWARENESS and excitement over all that was possible in the world beneath the sea heightened in the summer of 1986 while Tommy was quietly searching for the *Central America*. George Bass from Texas A&M's Institute of Nautical Archaeology had uncovered ornate gold pendants, weapons, and stores of bronze, tin, and glass while exploring a wooden vessel that had sunk in 150 feet of water off Turkey thirty-four centuries ago. Mel Fisher had recovered more treasure from the *Atocha:* an additional thirty-two hundred emeralds, more silver coins and bars, ornate reliquaries, and finely cast gold artifacts. Less than a quarter mile off Cape Cod, Barry Clifford had begun recovering the spoils from the pirate ship *Wydah*.

That same summer, Christie's had auctioned the Nanking Cargo, 150,000 simple porcelain pieces Michael Hatcher had found at the bottom of the South China Sea, where it had lain since 1747 preserved in wooden chests of tea. For a presale viewing of the porcelain, over twenty thousand of the curious from around the world had gone to Amsterdam, and for five days, five hundred to a thousand bidders had crammed the Christie's auction room. Every lot had sold, and the excitement over these treasures reclaimed from the ocean had caused some lots to bring more than ten times their appraised value.

But what truly inflamed the public's collective imagination that summer was Bob Ballard's return to the queen of the deep, the *R.M.S. Titanic*. For twelve days in 12,500 feet of water, Ballard and his crew had dived to film her for sixty hours and photograph her thousands of times. Memories of a lifetime ago awakened, and though everyone thought she had been lost forever among the mountains of the deep North Atlantic, in Ballard's photographs and film the *Titanic* dramatically lived again in all of her decaying splendor.

BY LATE FALL of 1986, Tommy had evaluated the SeaMARC records and studied the sonagrams of the key targets. He called Larry Stone to thank him for his work on the probability map and told him the site wasn't in the highest-probability cell but was in one of the highest-probability cells. "Not only that," remembered Stone, "he said, 'We think we detected the *Central America* on the second leg.'"

Tommy now had to raise $3.6 million for the verification and recovery phase. In November, he sent to his partners a letter announcing a meeting on December 13 at the Great Southern Hotel in downtown Columbus. He piqued their curiosity and ensured a good turnout when he promised, "Our presentation will be highlighted by color sonar images of the target shipwreck as she appears 129 years after she came to rest on the ocean floor."

At the meeting, Tommy showed color slides of some of the more interesting sonar targets. He finished his presentation with the suite of high-resolution images they had of Sidewheel, explaining how the sonar techs had interpreted the colors in the sonagrams. Buck Patton was trying to be skeptical, wondering what could go wrong now. "But here's this target," he remembered, "and you can see the sidewheel on it. Tommy'd measured the thing in centimeters and correlated it to the pictures, and this is the *Central America,* a 90 percent probability. It was a great presentation, and we all walked out of there feeling like, 'We've got it!'"

In less than a month, Tommy had sold 41½ of the 50 Recovery Phase units at seventy-two thousand dollars a unit, all to partners who had invested in the Search Phase. The partners were confident now, even excited. But as always, Tommy was advising caution. In a postscript to one letter, he admonished the partners again to keep quiet; their security policy had served them well, had allowed him to consider all options without pressure from potential competition or unwanted publicity. "Such freedom," he wrote, "will become more important than ever leading up to recovery operations."

Harry John already had ventured into deeper water in search of the *Central America,* and Tommy could name half a dozen other men who had led expeditions to find the *Central America* and still had designs on the ship. "If people like that get wind of what we're doing," said Tommy, "we could easily end up with one of these guys coming out there trying to interfere any way they could." In the offering circular for the Recovery Phase, he mentioned the possibility of being challenged by competition and he outlined a backup or emergency plan if they were. "I was taking a chance even bringing it up with the partners. 'What do you mean, competition?' But way before I even got money for that phase, we were already thinking: With that much gold on it, the *Central America* is not going to remain undiscovered forever."

TOMMY ASSIGNED TED BROCKETT, who had sailed as part of Williamson's crew the previous summer, to snoop around in the ocean community and learn about deep-water recovery systems for sale and for lease: He wanted to know their availability, their capability, their limitations, and their cost. He was looking for "a next-generation system."

Brockett was a good choice; he knew the community. In 1978, he had helped design the collectors that went to twenty thousand feet and brought back two thousand tons of manganese nodules for Inco USA, the first successful mining of the deep ocean. For Bob Ballard, he had designed the camera sled *Argo,* which Ballard was using when his cameras sent back the first pictures of the *Titanic* in 1985. He had designed the sled for Williamson's SeaMARC IA, and he had designed and built other deep-water recovery systems working with several of the ocean engineers he would need to consult for Tommy. So his connections were an advantage, but they were also a disadvantage, for he had to proceed surreptitiously. "I was insistent on that," said Tommy, "because I knew the dynamics that could occur if it got out of hand early on."

The deep ocean was such a hostile environment and it cost so many millions of dollars to go there that no one ever went unless there was good reason. Half the time that reason was a top-secret, national security interest of the government, and the other half it was a highly proprietary big business venture. Everybody in the deep-ocean community ran around with little secrets ricocheting off the insides of their skulls like billiard balls at the break and talked like good ol' boys, sizing each other up. And since everybody always wanted to know what everybody else was doing, they listened real close to what you said, so you had to be careful how you phrased your questions: The community was so small and incestuous, the equipment so rare and specialized, that one word too specific and the listener could quickly calculate what you were about to do and where you were about to do it. It all had to do with depth. The deeper you went, the greater the pressure; the greater the pressure, the stronger the seals had to be or salt water seeped into the electronic housings. If you ordered electronic housings good to a certain depth, that provided a big clue. And the cable required to run an ROV was so complex and expensive no one ever ordered more than they had to; if you worked in about four thousand feet of water, you used about five thousand feet of cable. You couldn't use less; you couldn't afford more.

Lights and canisters also had to be ordered to depth. Even the foam used on ROVs contained little glass microspheres that were rated to depth. When you ordered the foam, that telegraphed the maximum operating depth of your vehicle; immediately, in the minds of people who knew the deep ocean, that narrowed the possibilities by 90 percent.

Only defined places in the ocean measure four thousand feet or ten thousand feet, and if you didn't mention 220-volt electrical current, or ask about the availability of spare parts in Asian or European countries, that left two coasts in the United States. If nothing of significance was known to exist at that depth off one coast, that narrowed the focus to the other, and the continental shelf sloped into the Atlantic at such a degree that depth alone revealed the distance from the coast to the work site. Then the talk started. If Brockett said he was looking for a deep-water work package, and he let slip that he was concerned about the watertight seals holding at eight thousand to ten thousand feet, and anything else in his speech or his mannerisms seemed to eliminate foreign countries, the talk in the community would go out as, "Hey, there's a deep-ocean deal at about eight thousand feet off the East Coast."

"That's all you'd need to say to the people who know what's going on," said Tommy, "to warn anybody else who was trying to find the *Central America* to get out there sooner."

To disguise the venture, Tommy and Brockett developed "fuzzy brackets," parameters specific enough for engineers to solve the right problems, but not so specific they could figure out the project. For the deep-ocean people, Brockett framed the problem like this: We're investigating doing work in a wide range of depths, and we've got to be able to cover them with one vehicle. We don't want to have a different vehicle for four thousand feet than we would use at ten thousand feet. And for your cost estimation, it has to be capable of operating off either coast of the United States. If you see any problems going overseas, please quantify those.

Brockett contacted the five biggest deep-ocean companies and met with engineers from Florida to Southern California to Vancouver, B.C., trying to decide which company could use Tommy's ideas to build a ten-thousand-foot system, or which one already had a system Tommy could adapt. Brockett never mentioned the name of his client or the kind of project or the target of interest or the location or the depth. He just jawed with everybody about deep-water recovery and what they saw on the horizon.

The meetings would go smoothly until someone asked the first sensitive question about the project and Brockett had to say, well, he really couldn't talk about it. As soon as he said that, his counterpart grinned and got cagey.

"Now he's more interested in picking your brain about the project," said Brockett, "because his curiosity is driving him nuts."

Two companies were so used to working with the military that the whole relationship was straightforward and formal and conducted with no grins whatsoever. At two other companies, the engineers were especially experienced and sophisticated in their understanding of the deep ocean, and Brockett knew them well. He wanted to bang heads with them, but these engineers also knew the lore and the names of the famous historic shipwrecks, had even tried to recover them, and they yearned to be in the same business Tommy was in. When Brockett framed the problem in fuzzy brackets, they grinned and poked at him with, "Okay, Brockett, we know something's up," and Brockett grinned too and tried to pick their brains while telling them nothing. But that wasn't easy.

"When somebody's sitting across the table that close and grinning ear to ear," he said, "it's difficult to sit there with a straight face answering questions and not let something out."

Then Brockett discovered the perfect cover: Bob Ballard and the *Titanic*. He learned that if he phrased part of his requirements like, "and must be capable of recovering an array of objects in deep water," the engineers always shifted to questions about the luxury liner. Then it was easy for Brockett to say, aw, shucks, he really was not at liberty to divulge the name of his client or the project, which convinced everyone it really was the *Titanic*.

WHILE HE HAD Brockett out schmoozing with the deep-ocean crowd, Tommy continued to experiment with designs for his own vehicle. He had contacted an ROV operator named John Moore, who lived in the little town of Bellingham, a hundred miles north of Seattle.

If you asked anybody in the worldwide offshore community from the North Sea to the Gulf of Mexico, Who is the best ROV operator in the business? they would say, John Moore. If you asked them, Who is the most temperamental, cantankerous sonofabitch they've ever been around? they would say, John Moore.

"I have a foul temper," Moore says of himself, "always have and probably always will. I am firmly convinced that the world is mostly populated with idiots."

John Moore looked like the sheriff in a movie about a few bad outlaws trying to terrorize an otherwise peaceful and law-abiding citizenry.

He stood a lanky six-foot-two, his dark hair hanging near his shoulders in the back, his blue eyes intense above a thick, dark, droopy mustache. He never finished college. He putzed around for a couple of quarters at a small state college in eastern Washington, tried a few more quarters at Seattle University, and finally quit. "I was going to school majoring in drugs and girls, and that didn't work too well," said Moore. Then there were two years in the Coast Guard, which he found frustrating. "I didn't like having to take orders from people that were stupid."

But Moore's stint in the Coast Guard whetted his desire to see more of the world. Not long after he got out of the Coast Guard, he bought a one-way ticket to England and looked up an American deep-sea diving company called Subsea International, where he was told to return the next day in boots and overalls. He stayed with Subsea for ten years, and during his second year, his boss sent him to London to learn about a new field in robotics, underwater Remote Operated Vehicles, ROVs. Some of the clearer heads in the deep-ocean community could see the waning of manned submersibles and the waxing of the deep-water robot. Moore got so good at operating these new ROVs—"big toys," he called them—he began to design and build them himself and started doing things with them that dropped knowledgeable jaws.

Don Craft had once watched him for two hours pass a very small object through a very small hole at an extremely awkward angle, while he stared at a two-dimensional TV monitor and imagined the three-dimensional spatial differences that existed on the ocean floor where the robot sat a few thousand feet below. "Damn!" thought Craft.

Moore did so many wild things with robots that deep-ocean people liked to tell stories about him. Mike Williamson had worked with him on a project in about five thousand feet of water. One day Moore was in the control room at the monitor and something was wrong with the robot and no one could figure out what, so Moore stomped out of the control room onto the deck, leaned over the side, and started strumming the ROV cable with his fingertips.

"He's out there with his eyes closed," said Williamson, "holding this cable, feeling how that vehicle a mile below is responding. Then he's giving course changes and orders to the pilot in the shack without even looking at the sonar screen or the video. I thought that was pretty remarkable."

After years of designing and installing ROVs, Moore decided he should receive more than operator's pay, so he quit Subsea and told his bosses he was now a consultant. "You could hear the screams halfway around the world," remembered Craft. "When it comes to technical matters, he is a tough sonofabitch to beat, I'll guarantee you that. You would have to go to the world's biggest ROV operators to find anybody that might possibly come close to him. Nowhere can you find anybody that has the range he has at that high level."

In his research, Tommy had heard of John Moore, and judging from the man's reputation, he thought Moore might be the one to help him build and then operate the vehicle he envisioned. "I chose mavericks," Tommy once said, "because we had to do 'the impossible.'"

Tommy called Moore in the fall of 1986, and they talked for over two hours. He told Moore he wanted to recover a shipwreck in deep water. He said the ship was probably substantially intact, and they discussed how to dismantle portions of it to gain access. By December, Tommy had Moore under contract, and for the next three months the two men talked on the phone several times a week, sometimes for hours.

Most deep-ocean clients who came to Moore said, This is the problem we want you to solve. Tommy said, Let's talk about all of the problems that might arise and then we'll evaluate a couple dozen solutions to each. "We worked through the methodologies and operating problems at length," said Moore. "We did a lot of what-iffing, and, 'Do you think we could do this? And what if we tried that?'"

Moore told Tommy that the big capital costs were in the cable, the propulsion, and the housings. He confirmed that a battery-operated ROV on a simple co-ax cable was the cheapest way to get to the bottom. Although the co-ax cable had limited capability, they could buy components off the shelf that would push more signals down the cable; they didn't need expensive housings to protect the instruments on the bottom, they could use glass spheres; they didn't have to send power down the cable, they could put battery packs on the vehicle itself. Most people would not design it this way, but it was possible. And it was the way

Tommy liked to think: How can we keep it simple and inexpensive by rethinking the whole approach?

IN THE LATE 1960s, the U.S. Navy created a manned underwater environment called *Sea Lab,* and they hired Don Hackman at Battelle to design a machine shop for it, something that despite the water, extreme pressure, and corrosion would function like a machine shop on land. When Hackman finished, he could do anything underwater that anyone else could do topside, every tool impervious, sturdy, corrosion proof, and easy to use: drill presses, milling machines, grinders, sanders, kipping hammers, impact wrenches, welding explosives. Hackman could do it all underwater.

Hackman had been an engineer at Battelle for nearly twenty-five years. In 1981 he had coauthored for scientists and engineers the definitive treatise *Underwater Tools.* Although he designed medical tools and manufacturing plant automation devices and was lead designer in dozens of other specialized projects at Battelle, his reputation was as the foremost designer of underwater tooling in the world.

In his late forties, Hackman still had dark brown hair. His eyes were also brown and looked small encircled by the large, brown plastic frames of his glasses. He liked things efficient: Every few years he went down to the local Sears in Columbus and bought one dozen pair of black loafers, which he stored in his attic. He kept three pair going: one for the yard, one for church, and one for everything else; as the pair for the yard wore out, he threw them away, then he went up to the attic and broke out a brand-new pair for church, wore the old church pair to work the next day, and worked out in the yard in the pair he had worn to work the day before. The rest of his wardrobe consisted of three things: navy pants, navy shorts for hot weather at sea, and light blue shirts. Another engineer once came to a costume party dressed as "Don Hackman," and no one missed the joke. Besides his wardrobe, Hackman was famous around Battelle for quotes like, "You can't always be young, but you can always be immature," which often led to a dirty joke or a sexist joke or an ethnic joke. He cared not one whit about correctness. "If I knew what I was," he once said, "I'd tell jokes about that, too."

Hackman was one of the engineers who had grilled Tommy on his experiences making things work in the ocean and then insisted that Don Frink hire the young man. For the three and a half years Tommy had worked at Battelle, his office had been around the corner from Hackman's, and they saw each other frequently. Sometimes they worked together on projects, like redesigning exhaust manifolds on marine engines or creating a simulator for testing torpedoes.

Tommy had watched Hackman work, and he liked Hackman's thinking. In his book, Hackman emphasized that the design of underwater tools had been "evolutionary" not "revolutionary," and that was a problem: If divers wearing thick gloves can't turn the little chuck key on an underwater drill, don't make the key bigger, rethink the system. Tommy often talked to Hackman about whole systems that could work in the deep ocean, and gradually, noticed Hackman, Tommy's questions about working in the deep ocean had drifted toward finding and recovering deep-water shipwrecks.

When Tommy talked about working in the deep ocean, Hackman mostly listened. He had worked for the oil companies and for the navy on projects similar to what Tommy talked about, and he considered the work "a real challenge." He held twenty patents at Battelle because he tried things no one had thought of before; yet, when Tommy came to him with some of his ideas, even Hackman shook his head.

"As long as I've known him," said Hackman, "he's had these far-out ideas, and when somebody comes and tells me, 'I'm gonna go down two miles and get some treasure, would you help me on it?' you know? 'Uh, I've got more important things to do. My car needs polishing and my lawn needs to be mowed.'"

Tommy had taken a leave of absence from Battelle, and Hackman hadn't seen him for several months. In the fall of 1986, after Tommy had been to sea with the SeaMARC, he called Hackman and asked if they could get together. Hackman said, "Let's make it after work."

When Tommy arrived at Hackman's office, he told Hackman, "I've found a ship, and I need some serious work done now, and I'm ready for a contract." He showed Hackman some of the sonagrams, and he told him two things about the wreck: It was a wooden ship resting in less than ten thousand feet of water. "I was more interested now that

he said he'd found something," recalled Hackman. "Still very skeptical, but listening. 'Okay, you haven't convinced me, but you might.'"

Battelle policy allowed the engineers to take on outside projects as long
as they filled out the proper papers and explained how they and Battelle
would benefit from the experience. Tommy had analyzed that provision.
After he talked to Hackman, he met with his old boss at Battelle, Don
Frink, and persuaded Frink that developing a deep-sea robot would be a
good experience for some of the Battelle engineers: They would learn more
about working in the deep ocean, and it would cost Battelle nothing. When
Tommy had finished his speech, Frink was nodding in agreement. "My
people would be learning by what they were doing with Tommy," said
Frink. "My clients would be gaining from Tommy's contributions." He
approved the papers for his engineers who wanted to work directly with
Tommy, and for some of the work Tommy agreed to hire Frink's department at Battelle, an arm's-length contract.

Tommy went back to Hackman with a contract for a feasibility
study on deep-water tools to cut through decks. "As soon as somebody
starts paying me," said Hackman, "then I must take it seriously."

"All right," he told Tommy, "here's what I need to know, and here's
what I don't want to know."

The first thing, the most important thing, was depth. Depth is pressure, depth is handling. "I absolutely have to know the depth as close as
you can get it," said Hackman. "Next, I have to know roughly how far
from a harbor. Next, I have to know what the seas are like. Is it like the
North Sea or is it calm? Next, I have to know what the target is made of.
What do I have to cut holes into or get inside of? And how big are the
pieces I have to bring back?"

Hackman compared working in the deep ocean to working in space,
except in many ways, working in the deep ocean was more difficult. Salt
water ate tools. The other major problem was pressure: The *Thresher*
had gone to the bottom in 1963 because a water line broke, frying the
electronics so they couldn't shut off the valves to stop a quick descent.
The submarine had gone into an uncontrolled dive, and when it hit two
thousand feet, the pressure crushed its two-inch-thick steel like a beer
can in the hand of a college boy. Despite the pressure, joints had to bend,
bearings had to roll, impellers had to spin, hydraulics had to flow.

Hackman began sketching tools Tommy could use to enter and recover cargo from a sidewheel steamer with oak beams and pine decks somewhere between eight and nine thousand feet deep in the Atlantic Bight. All Tommy needed was a vehicle to get the tools to the bottom.

In January, while John Moore was out in Bellingham working on components for a deep-water vehicle, and Don Hackman was in Columbus conjuring tools to work from it, Tommy heard the first vague rumor through the deep-ocean grapevine that someone else was preparing an expedition to search for the *Central America*.

AFTER THREE MONTHS of traveling coast to coast, sounding out engineers in the deep-ocean community, Brockett had five proposals from the deep-water ROV companies, and Tommy evaluated each carefully. Not one offered fresh thinking. Some proposals incorporated equipment Tommy thought of as toys. "It's amazing what people said they could do versus what I knew they could do."

Brockett saw this, too; not that the companies made false claims, but they made it sound too easy, "like it was a piece of cake," said Brockett, "and maybe that scared Harvey off."

Some engineers told Brockett he needed a manned submersible. "Look at what Ballard did with the *Alvin* last summer on the *Titanic*," they said. Continuously updated and retrofitted for even deeper water, the *Alvin* was now twenty-two years old. The navy had spent over $50 million to develop it, and the cost of operating the submersible and its support ship was almost $30,000 a day. Despite Ballard's success, the *Alvin* was still just as slow, dangerous, restricted, and expensive as Tommy had considered it a few years earlier. Ballard could only film and photograph small sections of the *Titanic* from the outside and send the little robot Jason Jr., in fifteen or twenty feet to film the inside.

Ballard himself had told an interviewer only months earlier, "Telepresence is still back in the days of a guy yelling into a tin can on a string. I mean, we're still using black-and-white TV cameras for crying out loud. . . . In ten years . . . when I put you in Alvin you'll be disappointed because you're not going to have the freedom of vision that [a robot] can give you. You'll be looking out a little porthole. Compared to the robot's eye view, an Alvin dive will be somewhat like crawling around on the bottom of a cave with a flashlight . . . and blinders around your eyes."

The navy's most sophisticated manned submersible, *Sea Cliff,* was bigger and faster than the *Alvin* and had manipulator arms, but they were so stiff and clumsy an operator couldn't select an artifact in a debris field; if he could, the jaws likely would crush it. *Alvin* belonged to Woods Hole Oceanographic and *Sea Cliff* belonged to the navy anyhow, and Tommy could not have used them if he wanted to, but they were the best manned vehicles technology had to offer, and they still could not sit on the bottom and work with the intricacy that Tommy envisioned. Much earlier, he had considered them poor choices, and in the intervening years he had grown more adamant. "We knew they would not have the capability to do serious recovery work," he said, "let alone the cost and risk to life."

The previous summer, 1985, two robots called Scarab 1 and Scarab 2 had gone down sixty-six hundred feet to film the wreckage of Air India flight 182, recover the cockpit voice recorder and the data recorder, and retrieve twenty-three pieces of the wreck. It was the deepest an ROV had ever performed, and even there, the Scarabs had only imaged the wreckage and either clutched a small piece as a winch pulled it to the surface, or dropped a bridle around a piece, which a crane had raised. Tommy needed to perform work much more delicate and complex and do it almost two thousand feet deeper.

Tommy had talked to Hackman; he had talked to Moore; he himself had investigated systems for working in the deep ocean; and he doubted the companies could do what they told Brockett they could do. He started thinking in his failure mode: What if they can't do what they say they can do? What if we get out there, ship and handpicked crew on hire for the summer, investors with another $3.6 million in the project, the clock ticking, and a vehicle we can't use? "The risk factors would have just gone through the roof," he decided.

Then he worried about security: How does he control information if someone else builds the core of his project? And last, he thought, "It would be real easy for them to try to make money on doing the operation, but not really put their heart and soul into making it work." And that's what Tommy wanted, their heart and soul. He didn't want old machines and old ways of thinking; he wanted commitment and loyalty and excitement and energy and vision. What he really wanted was an organization like the one he already had, a bunch of mavericks ready to look at things in a new way. "Actually," he admitted, "I'd been thinking about another way of doing it for quite some time."

Brockett was still talking seriously to the deep-ocean people, trying to get the dollar figures down to something reasonable, and he had found what might be a workable approach, a combination of two existing vehicles modified with thickened housings and new flotation. He had talked to everyone in the business who knew about these things, and he was convinced the approach would work. "Then one day Harvey called me out of the blue," recalled Brockett, "and he said, 'Nope, forget all that, we're gonna build our own.' We are starting from absolute scratch, right? 'Here's a blank piece of paper, guys. We've got to be in the water the first of July.'" Brockett was incredulous. "That was like February or March," he said. "There's no way in the world you can start from ground zero and design and build an ROV in a few months. You can't get a cable in less than six months."

But Tommy had concluded that no one else out there knew any more about working on the floor of the deep ocean than he did. And that was all he needed to know.

For years, in his head and on sketch pads, Tommy had been designing an underwater robot. "I'd already figured out a lot of what needed to be done to get the kind of capability we needed on the bottom, and I had in my mind a concept design of where I wanted to go with it." Many people had told Tommy that he couldn't do what he wanted to do; they had tried, or they had known others who had tried. But Tommy liked to retreat to the point where technology branched and all thought on the matter had shuffled off down the path that led to Conventional Wisdom. He liked to travel back to the fork and take another look at the landscape. Maybe somebody missed something.

Science and engineering had reached the fork in the late 1940s, between the deep adventures of an American scientist named William Beebe, and the even deeper exploits of the Swiss physicist Auguste Piccard. Diving off Bermuda in 1934, Beebe had descended to half a mile in a hollow steel ball. The bathysphere, as Beebe called it, hung from a thin cable, and every bounce and roll of the ship as it rocked topside jerked the sphere, but it was the first time anyone had ventured into the pure blackness of the deep ocean.

Beebe had dived in the bathysphere because he was frustrated with the limitations of studying life from the deep sea brought to the surface in nets or washed up on the beach. He wanted to study it in its own environment, and in the darkness he saw the lights of intricate deep-sea creatures twinkling and darting. One creature was bigger than the five-foot bathysphere, a fish that glowed an iridescent blue and had rows of illuminated fangs. But Beebe could not interact with the sea life he saw or explore the bottom. All he could do was hang and watch. If he tried to land, the sphere, attached to the ship by the cable, would slam repeatedly against the ocean floor.

Piccard realized that if anyone was going to explore the deep ocean, the vehicle would have to be free of a cable connected to a rocking ship. He called his vehicle a bathyscaphe and named it *Trieste*. He engineered the *Trieste* by counterbalancing the weight of gasoline, seawater, and iron, so he could control descent and ascent without a cable. Above the observation sphere, he put nearly twenty thousand gallons of lighter-than-water gasoline in a fifty-foot tank, which countered the weight of the steel chamber and made the vehicle neutrally buoyant: In the ocean it would neither sink nor rise. To dive, he filled two ballast tanks with seawater; to rise again, he jettisoned nine tons of magnetized iron pellets and left them on the ocean floor. In 1953, on their first dive off Naples, Italy, Piccard and his son descended almost ten thousand feet to the ooze of a lifeless, featureless bottom.

Intrigued by the Piccards and their success, U.S. Navy scientists in the summer of 1957 persuaded the Office of Naval Research to lease the *Trieste* for fifteen deep dives in the Mediterranean, and eventually the navy bought the *Trieste*. Wanting to go still deeper, navy scientists designed and built a newer version with thicker walls and reduced the size of the viewing ports to barely two inches. In January 1960, Piccard's son and a navy submariner dived in the new *Trieste* to the deepest point on earth, the Challenger Deep in the Marianas Trench off the Philippines, 35,840 feet below. But the following year, despite protests from scientists, the navy abruptly stopped further deep-ocean exploration and re-tired the *Trieste,* until 1963, when the *Thresher* sank and the military had no way to reach it.

Over the decades since Piccard had built the *Trieste,* engineers had found simpler ways than using gasoline and iron pellets to achieve neutral buoyancy, and they had worn a path in that direction, which led to the *Alvin* and the *Sea Cliff* and similar vehicles. Yet nearly forty years had passed, and those vehicles could do little more than the *Trieste.* The path had led to what Tommy perceived as a dead end, with marine engineers piling more gadgets on the front, enhancing individual subsystems in small increments, but never addressing the system as a whole.

Tommy started back at the fork, reexamining the assumptions of Piccard and those who followed. Except for a small group of prepro-grammed submersibles, all robots, or ROVs, such as Tommy wanted to build were controlled by a cable connected to a ship. ROV engineers overcame the cable's tugging effect with a combination of buoyancy and propulsion, but the vehicles remained susceptible to all of the other problems: no reach, no power, no real capability. Those were the problems Tommy separated and analyzed, and after years, he thought he could solve them. He would never reveal to anyone the totality of his thinking, but he would discuss portions so that other creative minds, like Hackman's and Moore's, could help him solve the narrow problems, now that he had clarified and isolated them.

Rule Number One in designing underwater ROV work systems was Do As Much As You Can on Deck; keep it all topside—your brains, your power—and send it down to the vehicle on a multiconductor co-axial cable. But a co-axial cable was two inches in diameter, and you still needed several smaller cables to run everything on the vehicle. That was the problem: The central cable alone cost about twenty-five dollars a foot, and each of the smaller co-ax cables ran another five dollars a foot; with design costs included, you could put almost a million dollars into just a few wires and some armoring. And this big wad of cables meant a bigger winch, a bigger crane, and a stronger tow point; therefore, a bigger ship, and a bigger crew, and a more complex handling system was needed to get the whole thing into the water.

One piece of Tommy's concept shared by many engineers trying to work in the deep ocean was keep it simple. But Tommy pushed for simplicity where others either hadn't thought to look or had looked and tried but given up. His first move was to operate everything down below with one thin co-axial cable. Because the cable was the piece of equipment most likely to be damaged from twisting, snapping, or rubbing constantly over the shivs, it had to be cheap and easily replaced. Co-ax had been around for twenty years. "We'd take the component that was exposed the most," said Tommy, "and make it the lowest-tech. Then we would put the complicated things on each end."

But one small co-ax cable couldn't transfer power downside and still have room for the signals needed to run everything else. That's why no one else did it. All those signals chasing up and down at the same time on that tiny copper wire banged into each other and the cable overheated and the electronics fritzed at both ends. "Years ago many people attempted to do this," said Don Craft, "and they discarded it because they couldn't make it work."

For hours on the phone that winter, Tommy ran what he called "thought experiments" with John Moore and other engineers, looking for ways to make it all work on a single cable. "The thought experiment is the conceptual structure for the problem you need to solve," explained Tommy. "'If I change this variable, this'll happen; if I change that variable, that'll happen.'" A big company would build a prototype and test it in a laboratory, but Tommy had neither the time nor the money, so he built prototypes in his head and tested them by banging his head against someone else's. "The whole notion," said Tommy, "is that it simplifies the problem."

Tommy wanted to try putting the power down below on the vehicle, so they wouldn't have to send it down from topside. That had been done before, but the battery packs put out too little power. Tommy wondered out loud if they could somehow stretch that power. They ran more thought experiments. Moore found engineers who had designed a unit to control the flow of signals through a cable. He explained how many signals he needed to send to the bottom, and the engineers said, "We've never done that many signals before, but let's try it and see what happens." They talked about ways to send signals at different frequencies and how to fire off packets of these signals fractions of a second apart, one packet behind another.

After weeks of brainstorming with Tommy and consulting with the engineers, Moore thought he could get enough power to the vehicle and still shoot enough signals down that single co-ax to operate everything Tommy wanted. Then Tommy phoned Hackman and asked him to call around to his suppliers to find out how soon and for how much he could get a one-inch co-axial cable about twelve thousand feet long.

Hackman started calling and on his second call, a manufacturer said: "I've got some co-ax already sitting here that's three-quarters of an inch, thirteen thousand feet of it." Somebody had ordered the cable and then couldn't pay, so it had been sitting around for two or three years and the man wanted to get rid of it. "I'll sell it for a dollar a foot," he told Hackman. That was an 80 percent discount. Hackman called Tommy and told him what he had found; Tommy called Moore and asked him if he could do everything they had talked about on a three-quarter-inch cable; Moore said he probably could; Tommy got back to Hackman, Hackman bought the cable for thirteen thousand dollars, and they designed the vehicle with that cable.

Tommy and Hackman were so busy during the week that through the winter and into spring they often would get together on Sunday mornings at a nature interpretation center called Blacklick Woods Metropolitan Park. There were no telephones at the park, and no one knew how to find them. In the middle of jogging trails and nature hikes sat a small observation room with displays that explained the wildlife living in the surrounding woods and with large smoked-glass windows that looked out over the pond, the ducks, the deer, dove, and squirrels. Often, especially on cold days with patches of snow among the barren trees, the observation room lay deserted, and they would spread Hackman's drawings out on the floor and talk. If a few people wandered in, some of the smaller children might stop to look at their drawings of elbows and robotic arms, but mostly the visitors ignored the two men on the floor talking about torque and flex and tensile strength.

Tommy had given Hackman only two guidelines for design: One, he had to be able to put whatever Hackman designed on any boat available for hire, no dedicated motherships; two, he wanted off-the-shelf components, simple things like the cable itself, nothing specially made that would be difficult to replace.

"You don't want to build a whole brand-new rake," Tommy gave as an example. "You can use a garden rake." On a government project, the engineers would have a rake welded in the machine shop out of stainless steel, instead of going to a hardware store. "The government would spend twenty thousand dollars on it," said Tommy. "We don't mind being practical."

Tommy knew the *Central America* had been built with thick oak beams and heavy pine planking, and he gave Hackman blueprints of the ship's architecture for dimensions. But whether the ship would be virtually intact and standing tall, or collapsed, decayed, and eaten through, Tommy couldn't be sure. Some of the sonagrams, especially the suite taken of Sidewheel, showed long shadows, indicating that masts and hull stood almost proud. Other anomalies cast short shadows, suggesting that they stood no more than a few feet above the sand.

Some biologists told Tommy he might find "bugs," or shipworms, on the bottom. Others said that the site was so deep, the water so cold, and the pressure so great, that no shipworms could live down there; the ship, 40 feet high, 50 feet wide, 280 feet long, should be virtually intact. So they would need eyes to see into the saloon where the women had huddled with the children, to venture down the gangway where the men had lined up to bail, to peek into the purser's office where the passengers had kept their gold. Hackman's original designs were of tiny ROVs, flying eyeballs, like Ballard's Jason Jr., that could swim inside and tele-explore.

Once they had explored and documented the site, Hackman figured they would have to pick things up and slowly eat their way into the center of the ship. They would have to cut carefully through decks, and since the *Central America* went down with at least two hundred tons of coal in her hold, they would have to move coal. Hackman sometimes had an idea for a chain saw or a gripper or a digger, but he would tell Tommy, "You can't do something like this on any vehicle we have today." Tommy would say, "That's all right, let's keep looking at it." Hackman had learned to try it Tommy's way and to keep trying it

Tommy's way, even if it didn't work at first. "He's asked for an awful lot of crazy things," said Hackman, "but an awful lot of those crazy things have helped a lot."

TOWARD THE END of February 1987, the March issue of *Life* magazine appeared on newsstands in Columbus. On the cover were two hands, the wrists wrapped in gold chains, the palms and fingers uplifted and covered with gold and silver coins and chunks of emerald as large as a thumbnail. The hands belonged to Mel Fisher, and the treasure came from the *Atocha*. The cover read, "The Search for Lost Treasures: Eight Great Mysteries of the Americas." Inside was a one-and-a-half-page spread of Fisher in a bathing suit, reclining in a hammock, a tropical drink in his hand, a smile on his face, gold chains around his tan neck, the hammock stretched over a pile of huge silver ingots, more gold chains, gold finger bars, and other gold artifacts. The caption read, "This Lucky Man Found His, But Others Lie Unclaimed."

The article, by Linda Gomez, then told of the *San Jose,* with its billion dollar treasure lying in turbulent waters off the coast of Columbia since 1708. Gomez also wrote of the Lost Dutchman Mine in Arizona, the Money Pit on Oak Island off Nova Scotia, the funerary cache in the pre-Incan city of Chan Chan in Peru, and the treasure troves on Cocos Island, off Costa Rica. Number two on the list of eight was titled "The Gold Rush's Saddest Claim," and in one short paragraph Gomez told the story of the sinking of the *Central America.* In a large inset was a photograph of a Charleston man named Lee Spence, a bearded fellow with a round, pleasant face. The article said that Spence had located several Civil War ships on the bottom, and that for the past fifteen years he had been researching the story of the *Central America.*

"He believes he has discovered where the *Central America* lies," wrote Gomez, "and hopes to beat competitors to the prize within two years. 'All the pieces of the puzzle are there,'" she quoted Spence. "'It's a rush to the finish now.'"

Virtually every piece of mail Gomez received regarding the story was about Spence and the *Central America.* Fifteen readers wrote for Spence's phone number or address. They wanted to invest. A week after the article appeared, Spence flew to New York, where Jane Pauley interviewed him on the "Today Show" about his plans to recover the gold shipment from the fabled sidewheel steamer.

Tommy's partners, many with $50,000 to $100,000 in the project, got nervous. "We didn't think anyone else knew about the ship!" said Jim Turner. The article prompted Tommy to write a letter to the partners on March 7. He had monitored Spence's efforts for months, and Spence was still trying to raise money. "We will continue to monitor Spence," he wrote, "as well as several other groups that have not yet gone public but have the potential to become serious competitors."

No one wanted to believe that another group could interfere with what they were doing, but in his failure-mode thinking, Tommy had to wonder and to plan for the possibility. From the beginning, he had warned they might have competition. He had put the caveat in his original offerings for the partnership and ever since had reminded the partners of the importance of keeping quiet. Much earlier, he had raised his own antenna, listening for hints in the deep-ocean community that someone else might be launching a major expedition to find the *Central America*. And to ensure that he would be there first, he had even conceived the Emergency Plan, a fall-back strategy he could activate immediately to get out to the site quickly if he had to.

After Tommy had heard the first rumor of competition back in January, he had considered going to the Emergency Plan, but that was a difficult decision. The E-Plan was an all-out sprint to get to sea the moment the weather window opened around June 1. It meant cutting out so many links in the time chain that they would have to alter the vehicle's design drastically, make it far simpler, far less capable, and leave no opportunity to test it. Then they would have to step up production schedules on everything else and be at sea longer than planned, so the whole thing would cost the partnership another million dollars. That was a lot of money, and Tommy didn't know if the partnership would be willing to add that to the financial burden they already shouldered. Although the partnership consisted of over a hundred silent partners, about twenty were close to Tommy, kept in touch, and watched his progress. They were businessmen, mostly self-made millionaires, who understood that problems arise, and that any business venture worthwhile had risk. They were toughened to the ideas of competition and secrecy, and Tommy could talk to them. When the *Life* article appeared, he scheduled a meeting with this group.

They held the meeting at Buck Patton's office, and Patton later described the discussion as "very emotional." Tommy had the E-Plan ready to activate if he needed it, but he didn't know if the partnership was ready for it. He estimated there was a 5 percent probability that someone else could mobilize a search expedition, find the site, and claim it before August, when he planned to be out there. That was the dilemma he presented to the partners at the meeting: Do we spend a million dollars to accelerate the project, eat up our cushion, and have to go back to the partners for more money, just because there is one chance in twenty that competition will beat us to the *Central America,* image it, confuse our legal claim, and undermine our financing, which could create myriad headaches for us?

Jim Turner wasn't sure if Tommy was asking permission to overstep his budget, or if he was trying to gauge the backlash if he did.

"Tommy always did a wonderful job of laying out positions," said Turner, "without letting anyone know where he stood. I think he was asking, 'Should I go ahead?' My response was, 'Why are you asking the question? Let's go do it!'" All along, Turner had figured Tommy would have to come back for more money, anyhow.

"But it's a pride issue with him," said Turner. "He not only wants to bring this thing up, he wants to bring it up in budget."

A year earlier, serious competition would have ended the project. But now they probably had an image of the ship, and as Tommy had told them from the beginning, the search was the biggest risk. The biggest risk now was that somebody else would also image the ship on sonar and file an admiralty claim, leaving the partnership with a handful of pretty sonagrams. When Tommy distilled the problem to a million dollars versus a 5 percent probability, Patton said, "Tom, the point is, if that probability comes true, we've got 100 percent of nothing."

"But a million dollars," said Tommy, "is a million dollars."

"In a project this size," said Patton, "that's just an insurance policy to make sure we're clean with the board of health. If we don't spend that million dollars, eight years of your life and this great big dream, and a lot of investors' dreams at this point, are just going to go down the tubes."

The meeting lasted for half the day, a heated debate, with Patton waxing passionate for the million dollar insurance policy and Jim Turner and Bill Arthur waxing only slightly less so. The lawyers Kelly and Loveland agreed: Spend a million bucks, get a piece of the ship, take it to court, file your claim, and then you're safe; you've eliminated your biggest risk. Not to spend the money now was foolish.

Tommy had heard what he wanted to hear. The partners were still with him, even if he had to dip into their pockets again to protect the site. Now he could tuck that ace away and pull it out only if he needed it, and he wouldn't need it to ward off Spence. The rumors Tommy had heard in the deep-ocean community concerned someone far bigger than the self-taught underwater archaeologist from Charleston.

Shortly after the meeting, Tommy heard a second rumor that someone was mounting an expedition to find and salvage the *Central America* that summer. But the rumor was so vague, so many times removed, Tommy couldn't be sure, and he didn't know who it might be. His sources were friends in the business or new suppliers with whom he was just forming a relationship, so he had to be cautious: Information can flow two ways. The somebody who told Tommy that another party was trying to lease a big winch might be the same somebody who told the party renting the big winch something about Tommy. "It's kind of tricky," said Tommy, "because our main product is information."

One more thing Tommy wondered: Was he hearing about himself? Was he the competition? Was he about to commit a million dollars in a mad race to beat himself to sea? He had to consider that, but even before the first rumor, he had been careful to expose nothing that could be narrowed to him, the Atlantic Bight, or the *Central America*. He had meticulously crafted the fuzzy brackets with Brockett, and because he knew what Brockett had put out, he was almost certain that what he was now hearing was not his own echo bouncing back through the industry.

THE FIRST TIME Ted Brockett went to sea with tools he had designed at his office—cleverly designed, he thought—one came back smashed, two came back with the aft sections broken off, and one never came back. At sea, he realized, you didn't delicately pick something up off the deck and quietly place it in the water. "It gets swung around and bashed on the ship," he said, "and beat up against the crane and dropped on the bottom."

The experience had taught him a lesson in deep-ocean design: Make it dumb. Skip the frills, the high-tech niftiness, and make it as dumb as you possibly can, because only dumb will survive the sea.

"That was the approach we took with the vehicle frame for Harvey," said Brockett. "It had to be dumb, and it had to be simple."

Brockett had to design the framework for the design engineer to hang all of their ingenuity on, and the more Brockett listened to Tommy and thought about this vehicle, the more he realized that the key to its design was flexibility. One day it had to be a passively towed camera sled, long and skinny, to reacquire the target; and the next day it had to be squat and strong enough to sit on the bottom and work in the debris field, pick up ceramic cups and small gold coins, or thousand-pound anchors, and put them into a basket or tie ropes around them and bring them back to the surface. And the day after that it had to be tall and thin enough to drop through a narrow hatch and look around for gold.

For weeks, Brockett had thought and sketched, and each thought got simpler, each sketch more refined until his thinking and his sketching evolved into something that looked like a child's Erector set: discrete module frames changed at will by removing a few bolts. "It was very crude," said Brockett, "and it looked ugly as sin, I don't deny any of that, but it allowed us to reconfigure it and have thrusters on one day and not have thrusters on the next day, and to be tall and skinny or short and fat."

And that's what Brockett was working on, refining his sketches into drawings, arranging and rearranging on paper all of the pieces to this simple but homely underwater marvel, when Tommy called him on March 20. Tommy had just heard the third rumor, and he was shifting to the E-Plan.

THE RUMORS WERE still no more than that, and Tommy didn't even have a name, but now he had heard the same story three ways from three independent sources, people in the industry calling with bits of information like, "Somebody's been looking at equipment for a job off the East Coast," and the equipment was a winch that held seventeen thou-

GARY KINDER

sand feet of cable, which Tommy knew could only be used in a deep-water sonar search. At one time it even seemed like there might be two other groups planning expeditions, people looking for equipment or specific capability, hiring expertise to do things for that type of operation at those depths.

"I was looking for three independent sources," said Tommy, "and I finally got the third right around March 20. But it wasn't like somebody I knew well had talked to the guy who was going out. So I did a lot of fretting because I couldn't verify it. Everybody around me was saying it couldn't be done, we couldn't leave that early, but I decided we couldn't take the chance, we had to get to sea."

Tommy worried little that someone else would find and recover the *Central America* that summer. But if a competitor was poised to mount an expedition earlier than August, he saw two problems: First, if the competitor somehow imaged the Sidewheel site, it complicated the legal issues so that Columbus-America could lose control of the site or have to share the recovery with the competitor; and second, if the competitor was a typical treasure hunting group, it would try to raise money through the media by claiming to have found the *Central America,* which might scare off Tommy's investors.

Tommy had wanted to go to sea with good operating capability toward the end of July and then have August and September to verify the ship as the *Central America,* thoroughly photograph and document the site, and retrieve some of the gold. Now he had to announce to the partnership that he had decided to halt plans to build the full-up vehicle he and the engineers had designed, to forget about recovering significant amounts of gold that summer and just rush to sea in two months. The E-Plan. They had to be in the water by June 1.

The lawyers had explained to Tommy that before he could claim the Sidewheel site he had to have an artifact to take into court, and that single requirement became the design goal for what Tommy now called the Emergency Vehicle: maybe not find gold, maybe not even explore the ship, just retrieve an artifact and bring it to the surface and take it to court.

Brockett already had told Tommy that trying to build any kind of vehicle before the end of the summer was a futile exercise; now Tommy said he needed the vehicle in May. Brockett called in every I.O.U. of good will he had with his suppliers. "I know I'm being ridiculous," he told them, "but here's the drawing, have it to us day after tomorrow." John Moore flew to an ROV trade show in San Diego, looking for ways to build the components and assemble them more quickly, and every commitment he could get from every manufacturer was "no faster than ninety days."

Hackman later remembered Tommy coming to him about this time and asking how long he thought it would take to build a simple vehicle that could go down eight thousand feet, relocate the site, and bring back one artifact. That's all Tommy wanted. Hackman said, "Two years."

IN APRIL, AFTER weeks of phone calls and faxing, Don Craft found Tommy another mudboat out of Louisiana, the *Nicor Navigator*. One of the crew described her as, "Like a big, giant, long tugboat." She was sixty feet longer than the *Pine River*. Her deckhouse bunched up on the bow, and from there back ran a flat deck open all the way to the stern. On that wide-open deck, Craft had to create a control room for the techs, a laboratory for Bob Evans, a communications shack for Tommy and Barry, sleeping quarters for the tech crew, storage, large and small repair shops, and install a winch, a crane, a deployment arm, a handling ramp, Star Fix satellite navigation electronics, and a dynamic-positioning system.

Part of what made the deep ocean an "impossible barrier" was the inability to anchor the surface ship. Anchor chain had to be five times longer than depth, and you needed four or five of those anchors for a two-hundred-foot ship. In only two thousand feet of water, that would require about fifty thousand feet of anchor chain to keep your ship from drifting, and how did you set all five of those anchors so that the ship lay positioned precisely over the spot you wanted to work?

Dynamic positioning had become the key to all work in the deep ocean: Despite winds, currents, and seas, a combination of navigation systems, interfacing computers, and two or three large propellers would keep a ship constantly nudging its way back to a prescribed spot. Ship captains used it to berth supertankers. The deep-water folks used it to keep a ship stationary at the surface while their vehicle worked below.

Craft got quotes from the four major DP manufacturers ranging from $750,000 to $2.5 million to convert the *Nicor Navigator* into a dynamic-positioning vessel. Craft figured he could do it for a lot less, but every expert in the business told him no. Then rubbernecking at a workboat show in New Orleans, he discovered the Robertson DP system and decided it would work on the *Navigator*. He had Hackman draft the specs, sent them to the Robertson facility in Norway, and the Robertson engineers put a system together, flew it into the United States, and installed it in less than a month, all for $130,000.

In New Orleans, Craft also found a "crackerjack yard foreman" to ramrod the ship conversion. He told the foreman he wanted it all done right, but he wanted it all done fast. Get it on, bolt it down, wire it up, test it out. Five days after the *Navigator* went on charter May 5, they took her into the gulf at the mouth of the Mississippi and ran sea trials for a day and a half on the new DP system. Everything checked out, and the *Navigator* held station. Three days later, the captain eased the *Navigator* back into the gulf and aimed her bow at the Florida Keys. Three and a half days after that, Craft had the *Navigator* docked at the Atlantic Marine shipyard on the St. Johns River about ten miles out of Jacksonville, the navigation and dynamic-positioning systems wired in, the crane pedestal beefed up, the crane mounted, a new winch in place, the deployment arm and blocks installed, and five huge shipping containers, the Elder vans, bolted to the deck starboard and port.

"The primary objective throughout the mobilization period," said Craft, "was to get the stuff on the ship, and get the ship to sea, on site, to be there before somebody else got there."

Moore had shipped a tractor-trailer load of aluminum and electronics from Seattle to Jacksonville, and other shipping containers were arriving from Columbus. He and the rest of the techs were in the yard, beginning to assemble the vehicle, when Tommy, Bob, and Barry arrived on the 18th of May. Two other members of the tech team had already joined them in Columbus and were now in Jacksonville.

One was an underwater acoustics expert and software wizard Tommy had worked closely with on the search cruise the previous summer. Alan Scott, thirty-seven, had a master's degree in underwater acoustics from Catholic University in Washington, D.C., and for ten years had worked for the navy as a civilian. He would be the navigator, the one responsible for getting all of the navigation and DP computers talking to each other, so he could direct the ship's captain to the proper point on the ocean and guide Moore along the bottom to the shipwreck site.

Scotty stood out on a ship because he always looked crisp. When everyone else at sea was rumpled and wrinkled and unshaved, with tiny holes fraying along the elastic neck of their T-shirts, Scotty looked fresh and starched right down to a part in his hair and a crease in his jeans. His mind was as orderly. Craft knew Scotty from other deep-water jobs, and he marveled at the younger man's intellect. "I once sat at a table where Scotty and another tech had a conversation," he recalled, "and I listened to them very, very intently, very carefully for ten minutes during this interchange, and I never once understood a single thing they said to each other." Craft had seen Scotty and another tech take apart two PCs, then put them back together and do things with them at sea that Craft knew had never been done before. But Craft didn't care how good a guy was at the computer. "He has to be able to take care of himself and help take care of the ship," said Craft, "or he's more a pain in the ass than all his expertise is worth." Which is why Craft not only respected Scotty for his skill, but also liked having him on deck. Scotty was a dedicated runner, thin, strong, and he performed on deck like a seaman, "one of the happy guys," Craft called him. "You would've thought he'd been a deckhand all his life."

The other member new to the tech team was an old acquaintance of Tommy's from the *James Bay* days, John Doering. Tommy, Barry, Bob, and the other techs thought of the search for the *Central America* as an adventure of ideas and technology; to Doering it was a treasure hunt. "I don't make any bones about being a treasure hunter," said Doering. "That's what I am and what I do." Tommy had brought Doering in for two reasons: One, he knew how to run the crane for launch and recovery; two, he had an eye for underwater artifacts coupled with a drafting background—he could look at worn, bent, broken, corroded, collapsed, dilapidated, silt-covered, storm-tossed ship wreckage on the bottom of the ocean, somehow reconfigure it in his head, and understand what he was seeing; then he could draw pictures of what the original looked like.

The last member of the tech team to arrive was Don Hackman, who showed up a few days later wearing navy shorts and a light blue shirt.

For the next week, the techs stored navigation grid equipment, sorted out parts of the vehicle, wound thirteen thousand feet of cable onto the winch, equipped the laboratory and the shop and the communications shack, and installed the computers and the monitors in the control room. Bob set up his laboratory with equipment and reference works to study samples of sediment and wood fragments at the site, and to collect and catalog the artifacts. Doering and Hackman created a darkroom in one of the shower stalls, so they could immediately develop and analyze the film. Scotty worked to get the Star Fix navigation system talking to the Robertson DP. Barry helped Tommy draft letters to the partners and consulted with a public relations firm out of New York. Tommy seemed to be always on the phone.

Moore and the rest of the techs sat in the yard surrounded by unopened boxes of electronics and stacks of four-foot cubes framed in aluminum. They would have no quiet warehouse to assemble the vehicle, no controlled atmosphere to test it. They would have to mount all of the parts and route all of the cables and hoses and connect all of the electronics and hydraulics in the yard, at the dock, and on the back deck of the *Navigator* as it headed out. The vehicle's first test would have to be at sea. As the techs worked, welders and machinists and carpenters and electricians hustled around them from the dock to the ship and back to the dock again.

"It was absolutely nuts," said Brockett. "It was a madhouse."

Every morning, in the middle of the chaos, a nice man, a squeezy-huggy, grandfatherly type, showed up at the ship asking for a list. This was Bob Hodgdon, the logistics expert who could find anything. A long time ago, Hodgdon had decided to become a logistics expert, because he hated to wear long pants. He and Craft knew each other from earlier top-secret operations for the government.

Hodgdon had arrived in town a week earlier than everybody else, and he did the same thing he had done back when he managed logistics for the CIA's covert attempt to raise the Russian submarine: got out the yellow pages of the local phone book to see what goods and services were available, then visited the shops to find out what they could do and what they had on hand—machine shops, welding shops, rigging shops, electronics shops, electrical shops, even grocery stores. Then each morning, he walked around the yard and the shop, collecting everyone's parts list, and he returned each evening in a van loaded with welders and wrenches and hammers, transistors and resistors and capacitors, computer chips and screws and wire in a selection of gauges, nails, bolts, fuses, tape, saws, vises, circuit breakers, filters, floats, potting material, radiator hoses, heater hoses, hydraulic fittings, pipe cutters, filling the ship with everything they might need at sea.

WHILE THE E-PLAN mobilization continued into late May, the lawyers flew to Jacksonville to meet with Tommy and decide one thing: Should they file an admiralty claim with the court? If they did, they had to reveal their coordinates. If they didn't, they risked someone else's finding the ship and filing a claim first.

As soon as Tommy had returned from the sonar search the previous summer, he had met with Kelly and Loveland to begin brainstorming on legal matters. The two lawyers advised Tommy on how to file a claim and where to find outside counsel experienced in the ancient laws of admiralty. They kept coming back to this problem: Before a court could award the *Central America* to Tommy, the court had to have jurisdiction over the shipwreck; before the court could have jurisdiction over the shipwreck, Tommy had to bring the *Central America* into the courtroom. This, of course, was impractical, so the law allowed for a bit of fiction: Tommy could bring an artifact from the *Central America,* a stanchion perhaps, or an anchor, or a bell, or maybe some part of the rigging, into the courtroom and have it represent the ship. A federal marshal would then "arrest" the artifact, the court would assume jurisdiction over the site, and the court would award the *Central America* to Tommy. But there was a catch: The site had to lie within three miles of the beach; that was the physical boundary of the court's district. The *Central America* lay at the far reaches of the Economic Zone, almost two hundred miles offshore. No one had ever tried to recover an historic shipwreck so deep it lay beyond the three-mile boundary. Tommy could bring a piece of the *Central America* into the courtroom, but no one knew what would happen next.

GARY KINDER

After studying the problem for some time, Kelly and Loveland recommended filing the claim in the Eastern District of Virginia, which
sat in Norfolk. Besides having case law that seemed to favor their claim,
Norfolk had a good harbor with good facilities for ship repair. There
was a historical connection, too: Most of those who survived the sinking of the *Central America* had sailed into Norfolk.

Kelly and Loveland then searched *Martindale Hubbell,* the state-by-
state reference on lawyers, reading the biographies of lawyers who prac-
ticed admiralty law in Virginia. They needed someone with experience,
but they didn't want an old hard-nosed, ham-fisted trial lawyer who
would try to pound precedent into the judge's ear. There was no precedent. The case was unique. They needed someone who could foresee at
once the future of five areas of law—international, maritime, conflicts,
procedure, property—and understand how their natural progression
would affect jurisdiction over shipwrecks far at sea. Someone who could
deftly remove from the judge's thoughts that single sticking point about
the three-mile limit, examine it for the judge logically and equitably
while at the same time ever so delicately rotating it for a fresh perspective. They needed someone with a creative legal mind, someone articu-
late, quick on the feet, someone who projected honesty and fairness and
sensibility. They needed someone who could persuade a federal judge
to do what no federal judge had ever done: accept jurisdiction over a
wreck site on the far side of the Gulf Stream, two hundred miles at sea.

If they had created the perfect lawyer for what they needed as Norfolk counsel, Kelly and Loveland likely would never have thought to
include everything that came in the package of Rick Robol, a short, wiry
man with a straight nose and a puckish grin.

Robol was thirty-four, an admiralty litigator practicing in Norfolk for
eight years. He had graduated first in his class at the University of Virginia,
received a Fulbright fellowship to study international law and comparative government in Italy, and had gone on to Harvard for his law degree.
He was Phi Beta Kappa at Virginia, a Moot Court National Oralist at
Harvard, and managing editor of the *Harvard International Law Journal.*
After law school, he had clerked for the chief judge of the Eastern District of Virginia in Norfolk, the same court in which Kelly and Loveland
hoped to press Tommy's claim to the *Central America*. At Harvard, Robol

had won the Addison Brown Prize for a treatise on admiralty jurisdiction. In that treatise, written almost ten years earlier, he had examined the need for federal courts to extend their admiralty jurisdiction beyond the three-mile territorial limits and into international waters.

In the fall of 1986, Robol flew to Columbus to meet with Loveland, Kelly, and Tommy. Wayne Ashby was also at the meeting, and he seemed to have the only misgivings about the admiralty lawyer. After he met Robol, Ashby leaned over to Loveland and whispered, "This guy's too young."

Loveland leaned over and whispered back, "He's older than Tommy."

Robol confirmed that the Eastern District of Virginia would be their best forum. Around the country, it was known as "the rocket docket"— no waiting for five years to be heard. The judges knew admiralty law well, and the decisions that had come out of that district generally supported what Columbus-America was trying to do. Robol saw Tommy's case as the next step in the evolution of jurisdictional law: If exploration in the deep ocean were to go forward, then the law had to go with it; someone had to assume jurisdiction over what was happening, or chaos would ensue. All Robol had to do was make an enduring leap seem but a small and natural step.

One more thought they discussed: Could Robol use the dozen sonar images of Sidewheel to persuade the court to skip the artifact requirement, assume jurisdiction over the ship right now, and grant it to Tommy? Robol thought that someday the law might recognize telepossession, but it wasn't there yet. The leap was too big, and he didn't want to argue to the court that having a sonar image of a ship was the same as having a piece of the ship, especially of a ship sunk on the other side of that three-mile boundary.

He had told Tommy, "If you want the legal right to recover that site, you have to have an artifact." Even then, Robol was concerned that the court might assume jurisdiction over just that artifact and not the rest of the shipwreck. "We didn't know what was going to happen," said Robol. "It was not apparent that the court would take jurisdiction over the shipwreck, that they would apply that fiction to something beyond the three miles. That was the major conceptual leap forward that was so important to us."

Although filing a claim on the site meant they also had to file their coordinates, they decided there was greater advantage in having an early audience with the judge and in filing with a court of their choosing. To lessen the exposure of including their coordinates, they would file two sets: one general set for the public, and one precise set under seal for the judge.

When they met again in Jacksonville just before the *Navigator* set to sea, Robol talked with Tommy. He knew the threat of competition had forced Tommy to go to his Emergency Plan and that his vehicle was no more than a bunch of parts still in boxes on the back deck of the *Navigator.* But he reiterated that retrieving an artifact was so important that even if interlopers did appear at the site, and even if they did try to interfere, "even if it is on the brink of crisis," said Robol, he would not ask the court to assume jurisdiction over the site until Tommy had raised an artifact.

As is customary under admiralty law, Robol sued the shipwreck itself: *Columbus-America Discovery Group v. The Unidentified, Wrecked and Abandoned Sailing Vessel, her engines, tackle, apparel, appurtenances, cargo, etc., located within a circle having a radius of 10 miles, whose center point is at coordinates 31 degrees 52 minutes North latitude and 76 degrees 21 minutes West longitude.*

Robol filed the complaint in the Eastern District of Virginia in Norfolk on May 26. That night down in Jacksonville, the *Nicor Navigator* departed Atlantic Marine, the ship secured for rough seas and the bridge alerted to watch and record all traffic.

Aboard the
Nicor Navigator

⚓

Early Summer, 1987

B<small>ILL</small> B<small>URLINGHAM STOOD</small> five-foot-four, but he seemed six-foot-two. You could be talking to him, looking down, and somehow feel you were looking up: He had a seventeen-inch neck, with matching forearms and chest. Only thirty-three, yet a sea captain for twelve years, he also had a commanding presence, bright blue eyes, tussled blond hair, and a thick reddish blond mustache. He read constantly and had so many arcane bits of information rattling around in his skull that he was the only one on the ship who could beat "Info" Bob Evans at Trivial Pursuit. He was a "hands-on" captain, said one of his crew, a captain who worked alongside the men, and Tommy liked that. Tommy also liked that Burlingham was strong willed but not closed minded.

Burlingham was the captain of the ship, a graduate of New York Maritime Academy, and another of the men from the deep-ocean community who had worked with Williamson and Brockett on government projects, one of the names Craft kept in his little book of good people to call for the next operation. When the owner of the *Nicor Navigator* insisted Craft keep a Nicor captain on board, Craft told Tommy and Tommy said fine, but Bill Burlingham will be in charge of the ship.

The *Navigator* ran through the night at ten knots, the night breezes bringing in the cool salt smell of the Atlantic. Under bright lights, the tech team moved back and forth between the shop van and the two aluminum cubes Brockett had bolted together for their initial close-in sonar and camera runs. Far into the night, music blared from the deck speakers, accompanied by the wet hiss of spray arcing from the bow.

By sunrise the following day, they had entered the deep blue ribbon of the Gulf Stream. Though occasional squalls of moderate wind and rain blew through, all that day and throughout the next night the weather remained fair, the seas to four feet, the breezes light. Large swells came in from the southeast as the bow of the *Navigator* plowed through on course to a deep-water site 120 miles out, where they would test the crane and the winch.

But at the test site late the following morning, the crane swung too far and they couldn't stop it, and the winch took six hours to wind in three thousand feet of cable, five and a half hours longer than it should have. By eight o'clock their first night of operations, Burlingham had set his course back to Jacksonville.

Three days later, on the evening of May 31, they departed Atlantic Marine for the second time, destined for the same deep-water test site. This time, the crane held, and the winch wound the cable back in at a good pace. But Hackman was watching the cable come in when suddenly it began crossing over itself and heaping up on one end, like a bird's nest in a casting reel, and he realized that now the weight of the cable had crushed the steel drum. In all of his years as an engineer working in the ocean, Hackman had never seen that happen. With no winch to wrap their electronics cable smooth and tight, they couldn't operate.

Burlingham headed back to Jacksonville for the second time. Replacing the drum would be expensive and require three or four months for delivery, but Hackman thought he could solve the problem in a couple of days for pennies: Roll two steel half cylinders and weld them over the old drum. It was Battelle's creed at work: Give me a farm kid who can fix a tractor with a coat hanger tonight and have it back in the field at sunrise tomorrow. They arrived at Atlantic Marine around seven on the evening of June 3, and Hackman and the shipyard crew started immediately on the winch, working throughout the night and all the following day.

While the *Navigator* sat dockside, Tommy chartered a plane to run a surveillance flight over the Atlantic. If anyone slipped into their search area while the *Navigator* was in port, he wanted to know about it. Bob Evans flew with the two pilots, calling out the coordinates, until they had completed an air search of that part of the Atlantic. They returned after dark, and they had seen nothing.

By midnight of the second day, they had started winding the cable back onto the repaired winch. At 3:30 A.M. on June 5 the cable was on and the drum seemed sturdy. Burlingham shifted from shore power to ship power and secured the vessel to sea, and for the third time in ten days, a river pilot guided the *Navigator* out of the Atlantic Marine headed for blue water.

"We were exhausted," said Tommy, "just totally exhausted going to sea."

At 6:15, with the sea buoy abeam and the sun rising ahead, the *Navigator* entered a calm sea, and Tommy told Burlingham to direct his course for a second test site. From the moment they had departed Jacksonville the first time ten days earlier, Tommy had expected Burlingham to call him at any moment and say that the bridge had just spotted another vessel in the vicinity, not clipping by at ten to fifteen knots, obviously headed somewhere, but creeping along at one to two knots and displaying the colors and shapes of a vessel with an object in tow, obviously searching for something.

On the back deck sat a stack of welded aluminum cubes and myriad unopened cartons and crates filled with electronics and cameras and cables and thruster parts and manipulators and sonars and transponders and batteries and glass spheres and junction boxes and barrels of hydraulic oil and piles of brackets and various computer items, the makings of an ROV they hoped could dive to the bottom deeper than any ROV had ever been and retrieve an artifact.

Moore and Brockett were up early, continuing their work on the back deck. After thirty straight hours of rebuilding the winch drum, Hackman had turned in and slept till midmorning, when he joined them again. He had arranged his Battelle schedule to spend three weeks at sea, but now as he watched Moore and Brockett working among the boxes of parts, he wondered why he had come at all.

"Here I am, doing what I vowed I would never in my life do: go to sea with something that hadn't been tested. It not only hadn't been tested, it hadn't even been bolted together, and that never works. It's just too hostile an environment; there's just too many things that can go wrong; it's too dangerous. You can't build complex systems and assemble little electronic things on a rough ship, and you can't test things at that depth that have never even been wet before. You never, ever do that."

TOMMY HAD SWITCHED their destination from the first test site to the second test site partly because the second site was close enough to monitor the Sidewheel area on radar. They could build their equipment, test it, and keep an ear to the target area at the same time. If another ship came into the area, they could move quickly to set up over their primary target and try to protect it. Halfway out, Tommy switched the destination again, to a third test site, cutting the distance to Sidewheel in half, so now they could see with their own eyes anything entering the patch of water above their primary target.

As Tommy had learned more about his competition, he kept simplifying the vehicle design. "If we run out of time," he reasoned, "we've got to move from the E vehicle to the E-square vehicle, and then if we run out of time on that, we're going to end up with the E-cube vehicle, which is what we did. That's all we could get going."

The E-cube vehicle now sat on deck, a drastically stripped-down version of a drastically stripped-down version of the vehicle parked inside Tommy's head: a simple sled with just video and close-in sonar. Moore couldn't even run both at the same time.

At the third practice site, from the morning of June 6 through the morning of June 10, they tested the dynamic-positioning systems and the navigation systems and the systems on the camera sled. They launched and recovered the sled seven times. The crane now rotated smoothly under control and the winch recovered evenly and Hackman's rebuilt drum held strong against the pressure of thirteen thousand feet of steel cable. They now shifted their operation a few miles to the Sidewheel site to begin their search for the *Central America*.

O̲N̲ ̲T̲H̲E̲ ̲A̲F̲T̲E̲R̲N̲O̲O̲N̲ of June 12, Scotty gave Burlingham the range and bearing for the first track line, and Burlingham made way to the start-ing point, shifted the *Navigator* into DP mode, and about one o'clock they deployed the vehicle. An hour and a half later, the vehicle had reached the bottom and Burlingham commenced the first track line.

The control room was cool, air conditioned for the computers, and dark except for digital readouts and the light blue glow of the Mesotech sonar screen. At Scotty's direction, Burlingham tapped in new DP co-ordinates every several minutes to keep the ship creeping forward along a prescribed bearing at half a knot. Nearly nine thousand feet below, the sled glided through the darkness with the Mesotech sonar sweep-ing ahead a hundred meters.

When they reached the end of the first track line an hour later, they had seen nothing. Scotty called Burlingham on the intercom and adjusted the range and bearing for the next track line, and in forty-five minutes, Burlingham had turned the ship 180 degrees, shifted his course slightly, and headed back for the second run.

After another hour of sweeping the darkness, nothing had ap-peared on the screen. Burlingham turned for a third run, and Tommy and Barry left for the COM shack. As Scotty monitored the naviga-tion readouts, Bob, Doering, Brockett, and Hackman stood or sat in the dark and watched the Mesotech screen. Moore had his hand on the controls, keeping the sled flying over the bottom a few meters up. When the *Navigator* reached the final stretch of the third run they still had seen nothing.

Afternoon turned to evening. Moore raised the sled a few hundred meters from the bottom, and the techs repaired to the galley for supper. When they returned to the control room, Scotty continued to call up new ranges and bearings for Burlingham, and Burlingham ran the track lines, until they had crossed the area several times, and still nothing had appeared on the Mesotech. They decided to extend each track and over-run the target by half a mile.

Nearing midnight, Tommy and Barry were in the communications shack, when they heard Scotty's voice on the intercom. Using Tommy's term for historic shipwreck, he said, "We have a possible cultural deposit here."

When Tommy and Barry left the control van hours earlier, the atmosphere had been subdued; when they returned, Hackman, Brockett, and Moore acted like they'd been bungee jumping off a railroad bridge, about as giddy as grown men will allow themselves to be, and Bur-lingham had turned the ship for a repeat of the last track line. They had seen the ship, and up close on the Mesotech, it looked even more like a sidewheel steamer.

"It showed up right about eighty meters or so out," Brockett told Tommy.

"Just before you came in," said Hackman, "it was a perfect boat shape, just absolutely pointy at the ends and everything, just absolutely perfect."

Bob was telling Tommy how they were going to try to get the ve-hicle to swing back near the target for another look, but before he could finish, the sonar began to paint again. Hackman watched the screen over Moore's head.

"That's the way it came on first," he said, "that very first bleep."

"That's exactly how it came in," agreed Moore. "Right at eighty meters it started."

Everyone watched the blue monitor as the sonar swept the dark-ness and thick, irregular blotches of white appeared on the screen to the right, then moved to the left, immediately splitting into two bowed but roughly parallel lines, which continued sweeping back.

"That's where the boat shape comes," said Hackman.

Moore was laughing. "This is where there's no doubt in anybody's mind what you're looking at!"

Brockett pointed a pen on the sonar screen at a slight hump right about midships. "I didn't notice that," he said. "You can see a little bit of a wheel right there."

"That's a wheel there," said Hackman. "There's no doubt about that. In this next sweep, I think we can see the mast."

Tommy was still analyzing the little hump Brockett called a wheel. "Where's the wheel?" He touched his finger to the screen. "Right here?" He was skeptical. He asked Hackman, "What mast do you see? You mean that thing?" He pointed at the screen again.

"Yeah," said Hackman. "Last time through—"

"That's standing up too straight for a mast," interrupted Tommy.

The spots continued painting, now narrowing to form the stern of the ship. Brockett noted in a sly voice how much this image resembled the SeaMARC image from the previous year. He accused Moore of having created the Mesotech readout over the winter to match the SeaMARC sonar images. "He's been waiting for the right moment to put it on the screen."

"Good one!" laughed Moore.

It was now after midnight. Tomorrow they could find the site again with the Mesotech, get a tighter fix on its position, then go in for a closer look with the camera. At 2:45 the morning of the 13th, the vehicle was back on deck with the batteries hooked up to the charger, and the tech team had turned in. The *Central America* had not moved for 130 years. For the night she would go nowhere.

Bob Hodgdon had chartered the *Seaward Explorer* to bring out more vehicle parts, pick up Don Hackman, and drop off two new deckhands. Early the next afternoon, the supply boat came alongside, Hackman transferred for his trip in, and the deckhands, Bryan Anderson and Tod Steele, boarded the *Navigator*.

"I told Tod that his primary duty was to take care of Old Dad," said Craft, "to make sure that I did not sweat! And he looked like he'd been kicked right in the face."

Tod looked like he'd been kicked in the face ever since the supply boat cleared Cape Fear. The sea had begun to rise, some rollers chopping into waves, enough to make someone not accustomed to the bounce feel like a rotten spot had sprouted midbrain and was spreading outward. Bryan Anderson was twenty-nine and a sailor. He had raced sailboats to Bermuda across blue water in seas so high that the boat foundered in the trough because the wind could not reach the sails. Tod was twenty years old and had never seen the ocean.

Tommy had needed two more tech crew members to handle deck lines and help launch and recover the vehicle, and he couldn't advertise. He had to find two men they already knew and persuade them to come to sea without telling them what they would be doing or how long it might last.

They thought first of Bryan Anderson, whom Barry had befriended at the University of Florida. Bryan had earned his bachelor's degree in fine arts with an emphasis on drawing. He was now a mechanic at a Gulf gas station. Barry had called Bryan because Bryan was good with his hands and he knew engines, and he had sailed on blue water. Plus, Bryan, who wore small, round, tortoise-shell glasses, had a calm way about him and was modest with a sense of humor. Perfect attributes for a shipmate.

Barry told him that a supply boat was waiting in Wilmington to bring parts and groceries out to a ship two hundred miles offshore, and if Bryan could get up there tomorrow, he could catch the supply boat out. Barry described the job only as "the opportunity of a lifetime." Bryan quit the Gulf Station, stuffed his go-to-sea duffle with T-shirts and shorts, grabbed his grandfather's blues guitar, and flew to Wilmington. Waiting for him at the airfield were Bob Hodgdon and Tod Steele.

Tod Steele was a polite young man, reared in the steel community of Youngstown, Ohio. His sister Paula had become the home office backbone of the Columbus-America Discovery Group, responsible for the payroll, the bills, the insurance, and the partnership filings. Tod had just finished his sophomore year at Ohio State and was visiting his sister at the old Victorian, when Tommy called from the *Navigator* looking for a deckhand. Tod thought, "Who would turn down a chance to go out on a boat?"

"Yeah," he told Tommy. "What should I bring?"

"An extra pair of shoes," said Tommy, "'cause your feet get wet."

Tod had four hours to catch the last flight out of Columbus that could get him to Wilmington in time to jump the supply boat. He cleaned out his dorm room, dumped it all on Paula's living room floor, packed a small duffle, and Paula drove him to the airport.

The first leg of the trip out on the *Seaward Explorer* was a two-hour excursion from Wilmington to the mouth of the Cape Fear River. Tod thought being on a boat was all right. At the mouth, they swung east, and for the first time, Tod saw the ocean. Then darkness fell, and about the time the *Seaward Explorer* passed the tower at Frying Pan Shoals forty miles out, that rotten spot in the middle of Tod's brain had spread to the outer reaches of his cranium, which coincided with him finishing a dinner of greasy pork roast and a salad with anchovy dressing.

They might be nice kids, Tod and Bryan, but they were not Craft's idea of deckhands. He wanted men who spoke the same language as he and who knew a scupper from a fair lead. But Craft had not put the team together, Tommy had, and Tommy had requirements other than at-sea experience: First, he had to be able to trust everyone involved with the tech team—no strangers. Next, he liked the men strong minded but not hotheaded; no ship was big enough. Although he might not consult with Tod and Bryan on technical decisions, Tommy wanted even the deckhands not afraid to think and not afraid to defend their thoughts. Mostly, he wanted them ready to get tired and dirty and stay tired and dirty for a long time.

So Craft's deck crew was Bryan Anderson, a blue-water sailboat sailor who first would have to unlearn everything he knew about sailing, because for reasons Craft could not explain, sailboat sailors spoke a nautical language entirely different from that spoken by the navy and the Merchant Marine; John Doering, a forty-nine-year-old treasure hunter whose idea of being at sea, Craft surmised, was leaving the dock in the morning with a peanut butter sandwich and returning before dark; and Tod Steele, who presently was barfing pork roast and anchovies over the side of the *Seaward Explorer*.

With Bryan, Craft had to wring out the sailor and let the seaman soak in. But Bryan understood tools and engines, and he learned quickly. Craft watched Bryan closely, and because Bryan was anxious to learn, he made few mistakes.

Doering was not a kid eager to please. He was a laid-back treasure hunter and liked everybody and saw no reason to make life any more complicated than it already was. He would jump on the crane, fire it up, swing it around, and pick something up. "You try that with a five-thousand-pound object," said Craft, "while you've got a three- or four-degree roll to the ship and maybe a little pitch, it doesn't work." Craft wanted him to put three slings on it and a tag line before it came off the deck. Doering would say, "Why do I wanna do that?" That was not the answer Craft was looking for. When Craft said, "Do it this way," he wanted to hear, "Okay. Is this the way you want it done all the time?"

Tod was by far the youngest man on the ship. After Craft had barked in his green face about making sure that Old Dad did not sweat, he asked Tod, "Do you know anything about lines?" Tod said, "No," and Craft motioned for him to take a seat, and for the next two hours he explained to Tod the evolution of rope and portage. He showed Tod a half dozen kinds of lines and told him the attributes of each and how they were to be used. "He was incredible," said Tod, "and it never ended. Every day he'd explain different things to me." Craft taught Tod how to tie a dozen good knots and then how to splice, until Tod was one of three men on board who could braid an eye at the end of a nylon line. Within two weeks, Tod could hang upside down from the towpoint off the fantail out over the water and tie a bowline with one hand.

When Craft ran his new deck crew through their first complete launch, it took them fifteen minutes to get the vehicle from the deck into the water unhooked. One week later, they had it down to two minutes.

LATER IN THE afternoon Tod and Bryan arrived, Tommy wanted Craft to launch the vehicle again, so they could get a camera down on Sidewheel. But problems with the crane kept the vehicle on deck until late that evening, and by then the weather had begun to fall apart. Craft was against launching, but Tommy saw it as an opportunity to learn more about deploying in marginal conditions. "That is gross stupidity," said Craft, "and I've told Harvey that many times to his face. If the weather is lousy, don't launch the vehicle! Keep the damn thing on deck!"

In forty years at sea, Craft had never had a sailor seriously injured. He respected the weather; when it started to break down, he had to know when to stop and when not to launch in the first place, or he would pay a price, maybe a steep one. "Harvey's version," said Craft, "is that any smart man can go out and do this stuff. Harvey's full of shit when it comes to that, and I've told Harvey that."

Craft figured that when the wind whistled up to twenty-five knots and the waves rose better than eight feet, a state-five sea, it was time to stay indoors and mend your equipment. Tommy figured it was time to explore the situation, to see how the vehicle moved through the bounce and roll of the air-sea interface, so he could add this to his store of information. "There probably were times where I would push Don and say, 'Don, yeah, there's seas here, but look what they're doing, this is maybe different.' Don's not really a research guy, whereas I viewed a lot of what we were doing as an R&D effort. Maybe somebody else would think it was really risky, but I am constantly trying to push the limits to figure out more about what can be done. It's part of my personality."

After heated words with Tommy on the back deck, Craft launched the vehicle a little after eight o'clock that evening, but it was on the bottom for only two hours when Tommy agreed they had to pull up. The seas and wind had continued to build, the weather fax confirmed more was on the way, and they needed two hours to get the vehicle back on deck. They recovered in seas approaching ten feet and the wind peaking at thirty knots.

For the next three days, the wind topped twenty knots and the seas ran to six feet with a confused swell, and for those three days the techs worked on the vehicle. With the ship heaving, they had to cover tools in plastic garbage bags and tie parts down with bungee cord and duct-tape coffee cans to the frame of the vehicle to hold the bolts. Waves sometimes washed over the low freeboard and curled around their ankles. Despite the weather those three days, they expanded the vehicle's capacity enough to operate the Mesotech and the video camera at once. Now they could find the wreck on the Mesotech again, then work their way closer until they could see it on camera.

By midmorning on June 17, the sea had dropped to gentle swells and the wind to light breezes, and Craft orchestrated a safe launch of the vehicle. As the cable reeled off the winch for a two-hour descent, they heard the drone of a single-engine plane. Their pilot had arrived with the first airdrop of the season, and everyone came out on deck to watch him circle.

The pilot had flown out from Wilmington, the "beach," and as silly as it sounded, there was something special about that. They would see him only through the windshield a hundred yards out, but in this lone-some patch of water on the far side of the Gulf Stream, he was their physical link to the outside world. The techs called it "touching home."

The pilot's name was Steve Gross, and he was the vital artery between shore support and the ship. A supply boat often took twenty hours to get out to the site; Gross could make it in two. With each day at sea costing Tommy twenty-five thousand dollars, every hour made a difference.

Gross was sixty-five and had been flying for forty-six years. In World War II, he had trained navy pilots. After the war he earned degrees in mechanical and electrical engineering at the University of Washington and worked at Boeing on electronics, radar design, and systems analy-sis. He was a fly-by-the-seat-of-his-britches barnstormer pilot, the kind you could picture flying in a little single-engine plane and suddenly throwing back the canopy and crawling out in the breeze toward the tail with a screwdriver clenched in his teeth because the dadburned rudder felt a tad too lively.

Tommy had bought Gross a forty-two-year-old Republic RC-3 Sea Bee, a single-engine seaplane shaped like a grasshopper, white and or-ange and black, with a big swale down its back and a wing across its head, and the pusher prop mounted backwards, above and behind the cockpit. The Sea Bee still had its original engine, and no one who knew airplanes seemed anxious to switch places with Gross, flying it across two hundred miles of open sea. The chief mechanic at the Wilmington airfield wondered out loud how Gross could fit through the door of the hangar with balls as big as his. Gross figured that if a good aviation mechanic thought he had balls that big, maybe they were inversely pro-portional to the size of his judgment.

Gross had been settled in Wilmington about a week when Tommy called Hodgdon and told him they needed a small bag of electronics parts flown out to the ship. This presented a problem: Contrary to their name, seaplanes don't land in the open sea, and the windows don't roll down or slide back; yet somehow Gross had to get that bag out of the cockpit. He decided he would simply force the door open the best he could and drop the bag into the ocean and let them pick it up in the Zodiac. All he had to do was fly the Sea Bee low to the water without stalling, keep both hands on the wheel, and nudge that door open against an airstream running eighty-five miles an hour, or about the same speed as the blow in a medium-size hurricane.

Bryan and Tod putt-putted out in the little Zodiac to wait for the drop, several white-tip sharks following them and circling. They watched the Sea Bee bank and level off at drop altitude, and just as it came even with the Zodiac, about twenty feet off the water, they saw the door crack open. Then a large package squeezed out the bottom of the crack, sucked straight up and back into the pusher prop, and exploded. Suddenly the air behind the plane filled with dozens of sparkling bubble-wrap packages of computer parts raining down toward the waves.

For another hour, Bryan and Tod looked for little packages of bubble wrap floating in the water; fortunately, the parts had been bound in sets, and they recovered most of the sets. Even with the mishap, Gross had saved Tommy one day's operation, so in its first flight the Sea Bee had paid for itself.

W HILE THE TECH crew watched Gross circling, the new vehicle had reached the bottom and now waited for them to continue searching for the ship. But when Moore turned on the vehicle's systems, the Mesotech sonar was not working. It had worked fine a hundred feet down during routine checks two hours earlier; now only the camera seemed to be operating. Trying to find something eight thousand feet below with a camera when you can see only twenty feet ahead in the dark is an act of faith. Tommy had experienced that the previous summer. Mike Williamson once compared the procedure to trying to snag small objects with dental floss from the top of the Empire State Building. But rather than spend hours to recover the vehicle, check it, and launch again, they decided to try searching with only the camera. From two o'clock that afternoon till ten o'clock that night, they made short runs back and forth with the camera, and the only thing they saw were tracks in the sediment where the camera sled from the previous year had careened off the bottom. When they recovered the vehicle at midnight, they tore apart the Mesotech. The weather was now optimum, light airs and smooth seas, and they had no time to waste.

For the next two days, with the Mesotech still not working, they ran nine more track lines with the camera, and they found nothing. They knew the ship was just below them only a few meters away; any moment the sighting could come; but watching the white glare of the vehicle lights reflecting off the barren ocean floor quickly became tedious. Doering's head sometimes hit the back of his chair in a dead sleep, but the rest of the techs left him alone until his snoring got so loud they couldn't think. The only excitement they had was when Steve Gross returned to the ship, tried to land in the open sea, and blew the windshield out of the Sea Bee.

No pilot with half a brain and ten minutes in the air would consider landing a small seaplane two hundred miles at sea, but Tommy had told Gross that morale on the ship was low and they needed a mailbag brought out, and Gross said he would try. From the air the ocean looked like a great millpond. At fifty feet and again at ten feet, Gross thought, "It still looks good." At five feet he saw the heavy roll, but it was too late. The ocean suddenly swelled beneath him and smacked the hull of the Sea Bee so hard it flexed the metal around the cockpit. "We thought for sure he'd bought the farm," said Brockett.

Tommy helped Gross realign the windshield and snap the rubber seal back into the groove, then Gross taxied out quickly before the sea grew higher. From the ship, they could see his wake and then a series of white splashes. "Pretty quick," said Gross, "it was only one tick and then the tail a little bit and finally I'm flying." The crew cheered. Gross circled the ship, tipped his wings, and disappeared into a dot on the horizon headed west.

A week now had passed since they imaged the ship twice on close-in sonar the night before Hackman left. Of that week they had lost three days to bad weather and most of another four to equipment problems, and in the twenty-two hours they had searched with the camera they had got not even a glimpse of the ship.

Early the afternoon of June 21, the *Navigator* was dragging the sled at a quarter knot, only the camera sending back images from the ocean floor. At 1:15, Burlingham called down to the control van and announced that the *Seaward Explorer* had arrived again and was standing off. He and Craft had to transfer parts, supplies, chill boxes of produce, and fourteen thousand gallons of fuel. Scotty asked them not to tie on until the *Navigator* had hit the end of the track line and begun to turn.

Thirty minutes later, Burlingham called the control van again and said he now was preparing for the *Seaward Explorer* to come alongside. In the control van, the glare of white sediment on the monitors suddenly turned to shades of gray with distinct lines and contours, and then right in front of the techs, like an apparition, appeared the wreck.

Moore yelled to Scotty, "Tell the bridge to fend off and back down!" Then he slowed the vehicle.

Scotty called Burlingham. "We've got a contact here! Can you delay?"

But Burlingham had already left the bridge. The crew of the *Sea-ward Explorer* had thrown lines and were snugging her rubber tire fenders up against the *Navigator*.

Inside the dark control van, the techs watched the stern of the ship glide by on the monitors. Scotty was trying to get as many acoustical readings as he could, but he needed a full twenty to thirty seconds of quiet, and he couldn't get it with the thrusters whining. Thirty seconds went by. Then a minute. Then suddenly they felt a hard thump, the control van jerked, and the wreck disappeared from the monitors. Once again they were looking at the whiteness of the barren ocean floor.

"I remember all the panic and the screaming and hollering," said Brockett. "'What the hell's going on?'"

Then they realized they were now tied up to the supply boat and the two vessels were rocking against each other in the swells. They felt more thumps. Moore raised the vehicle a few hundred feet, and except for Scotty, everyone left the control van and went out on deck to help off-load supplies and groceries. Scotty stayed behind, hunched over a pad of paper at the nav station, and tried to calculate where they had been when they saw the ship. But the geometry didn't work. "Scotty is generally pretty laid back, and it appears that nothing ever upsets him," said Brockett, "but this time he was white knuckled."

Once every hour or so, Moore or Brockett or Bob Evans would poke his head into the control van to see if Scotty had worked out the navigation yet. And every time, Scotty would say, "Nope." Bob's visits remained civil. Brockett got progressively sharper. Moore became so enraged he finally stopped coming in. He couldn't understand how Scotty could have the most sophisticated navigation system available and not be able to figure out where they were. "It can't be that difficult!" he told Scotty. Moore didn't realize that although Scotty had the numbers to work with, they meant nothing: The software was defective. Scotty suspected the software was defective but refused to mention it until he was certain.

The *Seaward Explorer* finished refueling the *Navigator* and departed at five-thirty that afternoon. The vehicle remained on the bottom, and the techs waited for Scotty to figure out where they had been. But as evening came on and then darkness fell, he still could not calculate where they had been when they saw the stern of the ship. At midnight they gave up and recovered the vehicle.

WHEN HE AROSE the next morning, now June 22, Scotty discarded the faulty data he had from the navigation software and started calculating his own grid, using whatever figures he could rely on: the position of the ship, the layback on the vehicle, some of the ranges he had recorded. At noon, he thought he had enough of an answer to try again, and they launched the vehicle.

For the first three hours, they hovered above where they thought the ship was, moving the vehicle a few meters this way and a few meters that way. The video camera was working well, and even the Mesotech seemed to be functioning. Then late in the afternoon, they acquired a target on the Mesotech. Scotty alerted Tommy on the intercom, and with all of the techs watching the monitors, he guided Burlingham closer and closer to the target, and once again the white sediment on the monitors suddenly filled with varying shades of gray and they saw the stern of the ship on camera. In the dark control room, everyone stared at the monitor for five minutes before they again lost contact. But Scotty now had a good fix on the site. He told Burlingham to come around and swing the bow of the *Navigator* to the northwest and proceed in that direction.

Bob had watched all thirty-six hours of video the sonar crew had shot the previous summer, and he had analyzed the two minutes when they had the ship in view; but even during that two minutes the light level was so low he could discern nothing except a few ship timbers and some anchor chain. His real understanding of the site had come from a suite of high-resolution work done with the SeaMARC, and in one image, right in the middle of the target, was a hump that looked like a paddle-box and a sidewheel.

"You just look at that sonar picture," said Bob, "and you say, 'Good lord, look at that!'"

Rising straight up just forward of that hump appeared to be the main mast, and aft appeared to be another mast, the mizzen. And if you studied that sonagram, really studied it, as Bob had, you could see spokes in the paddle wheel. That's what they were looking for.

Burlingham turned the ship and assumed a new heading to the northwest. With the camera running, the sled crept forward a few meters at a time. After an hour, they saw the stern again, but this time Scotty had put them on a bearing almost perpendicular to the last. Burlingham stopped and backed down, and Moore dove the camera sled just inside the starboard wall. For the next fourteen minutes, they followed the hull in a slow sweep, from the stern all the way to the bow.

Although the lights were not bright enough to see large sections at once, they could see small portions distinctly. The camera sled inched forward toward the bow, and the scenes once bright at the center of the screen slowly receded into the dark and new scenes appeared out of the same dark. They could see the ribs of the ship, and before they had gone far, they saw a long cable coiled round and round inside the hull. As the lights and camera passed over the coils, their shadows played across the ribbing. In the first several scenes they saw nothing but more ribbing and more cable, each scene virtually the same as the last. And then Bob noticed in the sediment just outside the hull what appeared to be two bottles. A little farther on they saw anchor chain trailing across the white floor into the darkness. Then the scenes returned to more ribbing and more cable, the only thing distinguishing one scene from another being a section of charred planking, as if part of the hull had once burned. They had no explanation for that.

No one spoke, but everyone wondered why the site looked so clean. How could Moore drop the sled inside and follow the starboard wall? Where was the decking, the superstructure, the stairways, the cabins? Maybe the scientists had been wrong, maybe the degradation process continued even at these depths. Maybe everything but the hull itself had simply degraded and dissolved and returned to the sea.

After several minutes, they approached midships, and based on their various target models, they expected now to get a glimpse of a sidewheel or the paddle-boxes or at least the shaft and some of the iron works of the engines, but the camera glided by, sending back images of an empty hull, a few more ribs, and planking. Where were the sidewheels, where were the boilers, where were the masts? Where was the coal? They saw not one lump of coal inside or outside. Yet the *Central America* sank with more than two hundred tons of coal in its hold, and coal wouldn't degrade; it wouldn't go away; even if the superstructure and the iron of the paddle wheels disintegrated, the coal would remain, huge piles of coal. But they saw nothing, nothing but the hollow shell of an old wooden vessel.

"You could see down to the rakes of the bilge," said Tommy. "You could see everything, and it was empty, completely empty." Before they had gone the length of it, he knew. "I had a feeling the minute I saw it that this wasn't the *Central America*. There wasn't enough debris. No cargo, no engines, no fastenings, no junk."

The fastenings alone on a steamer like the *Central America,* just the bolts and nails and brackets to hold together the engines and boilers and support the weight of the coal, would weigh close to one hundred tons. Even if the engines and boilers weren't there, the site should have been littered with debris, and there was none of that. Except for the wire coiling through the bilge, the hull was empty. "With a passenger liner, you'd expect a lot of anything," said Tommy, "dishes, personal effects—you know, stuff. And it's almost like this thing was full of cotton and the shipworms ate it all."

This was the first time anyone had "ground-truthed" the sonagram of a deep-water wooden shipwreck; the first time anyone had gone to the bottom and looked, then compared what they saw to what appeared in the sonagram. And what appeared in the sonagram of Sidewheel, what had impressed Williamson and his crew and Bob Evans and everyone else who had seen the sonar images of the wreck, wasn't there on the bottom.

Before they reached the bow, Tommy said, "Let's get out of here."

No one could explain the hump in the middle of the sonagram that had looked so distinctly like a paddle wheel. The hull was degraded and tipped slightly on its side; perhaps it had degraded in a way that left a big moon-shaped section that appeared on the sonagram as a hump, a misleading sculpture cast by the sea. Another trick: The faint straight lines they had seen on the sonagrams that looked like two masts standing tall corresponded with anchor chain they now saw trailing out of the hull and across the ocean floor.

The excitement and the expectation over that site had built for months, and since the techs first saw those white blotches painting the target's outline, perfect even from a hundred feet away on the Mesotech, that excitement and that expectation had risen right through the antsy frustration of waiting out bad weather and dealing with Mesotech malfunctions, navigation glitches, and teasing glimpses of the hull, and finally the big moment had arrived. So many people had convinced themselves and others that the suite of "Sidewheel" sonagrams perfectly portrayed a 280-foot sidewheel steamer that emotions now banged against an abrupt realization: The gold pea lay under another shell somewhere in the Atlantic. And the sea offered no clue.

With the possibility of competition appearing at any moment and the whole search map now to be reconsidered, they had no time to waste on a site with such low probability of being the *Central America*. After the camera crept past the bow and they had the length of the ship on film, Tommy called off the dive. Until they found the real site, they might use Sidewheel as a test site to build their database on deep-water wooden ships and to practice with their equipment, but Tommy had concluded that the site that so obviously held paddle wheels and massive boilers and huge engines and tall masts was no more than a sailing vessel that perhaps had burned before sinking and now lay hollow and empty at the bottom of the sea.

IN HIS FAILURE-MODE thinking Tommy had considered the possibility that Mike Williamson might be wrong in his analysis of the sonagrams. Williamson was the best, but no one had ever seen a wooden ship in deep water, so he had nothing to compare a sonagram to. If he was right, they had no problems. But what if Sidewheel was not the *Central America*?

After a tightly controlled effort to construct a probability map using the most sophisticated methods available, they had searched over 94 percent of that map and so had a 94 percent probability of having imaged the *Central America*. Even in Tommy's limit-your-risk world, that was an overwhelmingly high probability. The *Central America* had to be somewhere in their database.

Tommy assigned Bob Evans and John Doering to review the major targets on the optical disk. First, they analyzed the images in black-and-white, comparing the features of each to what they had seen in the Sidewheel images and what they had found at the Sidewheel site. Then they went to the color mode and experimented with drawing more information from the images. They analyzed several other anomalies. When they compared the sonagram of Sidewheel to these sites, they all appeared to fall into two categories: either a highly reflective rocket shape indicating hard metal, or a softer return indicating an old sailing vessel. Finally, they got to the sonagram labeled "Galaxy."

In 1986, the sonar team had imaged Galaxy only after Tommy insisted they stop running high-resolution passes over Sidewheel and finish running the broader-swath track lines through the lower-probability cells to the east. When Williamson imaged Galaxy, he studied the sonagram and pronounced the ship "a steel-hulled vessel with associated debris." But before Tommy would allow the sonar team to return to Sidewheel, he had wanted a high-resolution shot of Galaxy, and he kept them there until they had swept the target at closer range three more times. The debris field surrounding the core had intrigued Tommy; it swirled about the site like a galactical spray, so Tommy had named the site "Galaxy." It was the last wreck site they had imaged in 1986.

Bob and Doering now sat in the Elder van watching the software paint Galaxy in black-and-white, and before the lines had finished neither could believe what they saw staring back at them. Nearly filling the screen was an eerie caricature: a human skull with sunken cheeks and large, hollow sockets for eyes.

"Have you ever seen this picture," said Doering, "if you look at it from afar it looks like a skull, but when you get closer you see it's a woman sitting in front of a mirror combing her hair? This looked very similar to that. You could see a skull, and it looked like he was scratching his head."

At the core appeared to be a primary deposit, indicating the hull of a ship, then came a secondary ring of debris, and finally a tertiary plume of more debris. It looked like a ship that had twisted and turned on its way to the bottom, spilling its contents as it traveled down through the long column of water, the contents roiling around the hull as it descended, the hull crashing into the seafloor, and the contents scattering in a swirl around it. The reach of the arm up toward the top of the skull was part of that swirl.

Now that they had ground-truthed Sidewheel, they knew more about interpreting sonar images, and they saw quickly that Galaxy appeared to have a lot of hard targets at the center and thousands of tiny bright reflections in the debris field that formed the skull, scatters of what they thought might be chunks of coal.

"The more Bob and I talked about it," said Doering, "we went, 'Goddamn, this has all the earmarks of being the *Central America*!'"

They showed the sonagram to Tommy, and Tommy immediately saw the skull. He liked the site because it differed significantly from all the others. It appeared to have sufficient mass to be the *Central America* and to be long enough. And it was surrounded with plumes of debris, now so obviously absent from Sidewheel and the other sites. After studying the sonagram and comparing it to the others and talking to Bob, Tommy estimated that the site had at least a 50 percent probability of being the *Central America,* and maybe a much higher chance than that.

For two days after they made the camera run along Sidewheel, Bob and Doering studied sonagrams while the rest of the tech crew tested equipment and rendezvoused with the *Seaward Explorer* twice more for fuel. The weather was calm, the sea rising no more than a foot.

On the 25th of June, Tommy decided not to dive again until they had reconfigured the vehicle for better photo and recovery work. He wanted thrusters to position the vehicle, additional cameras to take color video and stills, and a manipulator to retrieve an artifact. As soon as the tech team upgraded the E-cube vehicle to the E-vehicle, he wanted to use Sidewheel as a test site to practice with the thrusters and the manipulator and to learn more about deep-water, wooden ships. Then he wanted to move on to Galaxy, study the site, select an artifact, get every-thing poised to file a claim in court, and begin to explore.

For the next five days, the tech team worked and ate, and slept once in a while. Scotty hid in the control van with microprocessors and a soldering iron. Moore and Brockett dismantled the camera sled and combined the old aluminum modules with new modules to enlarge the skeleton, then began to route hydraulic lines and electrical wires and computer cables for the new capability. The weather now was nearly optimum, light airs and small seas; they could work with both hands. On the 27th, a shallow storm front blew through, packing winds to forty knots and seas to twelve feet, but then the sea lay back down, and for the next three days the winds held steady at ten knots, the seas two to four feet. Throughout each day and long into the night they worked on the vehicle, making it bigger, heavier, more complex. And then just before midnight on June 30th, a clear night with stars shimmering, they saw a glow moving slowly along the horizon to the northeast.

THE BRIDGE HAD picked up the target on radar three hours earlier, a ship sixteen miles out. Burlingham disengaged the engines of the *Navigator* and estimated the other ship's speed at one to two knots. "Out here, you don't do two knots," said Burlingham. "There's no reason for you to do two knots." Cargo ships and tankers traveled through running ten to fifteen knots on a constant heading, and if you extended that head-ing in a straight line toward the coast it would intersect with a port in the United States, or it would lead across the Atlantic in a conventional shipping lane. The heading of this boat led nowhere.

From the same direction as the radar blip, they now could see the glow of her deck lights, faint at first but brightening, like the moon ris-ing slowly over the curvature of the earth. Tommy was napping in his bunk in the Elder van when Burlingham woke him up. "There's some-thing out here that looks like a city," said Burlingham. "I suspect it isn't a freighter."

On the bridge, Craft had a pair of binoculars trained on the glow, and he could see the mast. From the highest point gleamed a large red circle, and below that a white circle, and below that another red circle: RAM lights, Restricted Maneuverability, a signal that the ship was towing another vessel. But Craft could see none of the side lights or the stern light required on a surface tow. Whatever followed was submerged, and the only submerged vessel anyone would tow out there was a deep-water side-scan sonar.

Tommy no longer had to guess whether or how or when the competition might arise. It was now, that ship was who, and he had one advantage: He knew they were out there, but they didn't know he was; on their radar he was just another ship passing through. Quickly, he decided several things: Keep them at the edge of radar range until they depart the area and watch for them coming back; call Rick Robol; continue studying the sonar images; assemble and test the vehicle; stay away from Galaxy until we're ready to launch. As long as Burlingham kept the *Navigator* below the radar horizon, the other bridge couldn't tell who they were or what they were doing. It was a game of hide-and-seek, and the other crew didn't even know they were it.

All that night and throughout the next day, the bridge kept the *Navigator* just out of radar range, periodically creeping up to the edge to get a fix on the other ship, then dropping away. The ship moved slowly, sometimes erratically, so Tommy knew they were at the beginning of their search, still coordinating the boat crew and the tech crew and adjusting to whimsical currents. He wasn't sure yet in which direction they would strike out in long track lines, but he found it suspicious that they had begun their search within a few miles of the Sidewheel coordinate Robol had filed with the court. Maybe Sidewheel would occupy them for a few days, just long enough for Tommy to get his vehicle into the water and back up with an artifact.

The second morning, with the other ship straightening her track lines, Tommy had Steve Gross fly out, determine her heading, and take two coordinates an hour apart. When he got the second lat-long and had the bridge plot it on the maneuvering board, he realized that if the vessel continued on course, the sonar fish behind her could pass close enough on the first track line to image not the Sidewheel site but Galaxy.

Tommy watched them still miles from the target but moving closer to Galaxy on the radar. Burlingham, Craft, and the Nicor captain were with him on the bridge, along with Barry, Bob, and Brockett. The day

before, Tommy didn't want the other crew to know the *Navigator* even
existed; today that was less important than keeping them away from the
site. He told Burlingham to point the bow of the *Navigator* on an intercept course with the other boat. "Harvey was concerned about not doing
anything illegal and not putting the ship at risk," remembered Brockett,
"but he was adamant about not letting those guys survey the target."

The *Navigator* closed quickly, until an hour later they could just see
the other ship off to starboard. Craft watched her again through binoculars. Her lines were distinctive; she was all white with a broad stripe
fanning back off both bows, then arcing and plunging into the water.
Craft couldn't make out the color of the stripe, but he recognized her
lines.

"It's the *Liberty Star,*" he said. "She's out of Cape Canaveral."

Four years earlier, out of professional curiosity, Craft had inspected
the *Liberty Star.* She had a sister ship, the *Freedom Star,* both built for
Morton Thiokol, both operated out of Cape Canaveral, both booster
pickers for the space shuttle program. During gaps in the launch schedule, both were available for charter. Tommy wanted a closer look at her
back deck, to see what capability she had on board.

They had closed to about eight miles, the *Navigator* running to the
southwest on an intercept course with the other vessel heading south
southwest. Once they got within a mile or two of the other vessel,
Burlingham could do nothing to force or even encourage the captain to
alter his course. Burlingham was even uncomfortable with his bow
aimed at another vessel still eight miles away, because the other captain
had had them on radar for at least a half hour, and that other captain
knew their course and that that course would intercept his own within
an hour.

On the bridge, they suddenly heard on the hailing channel, "This
is the motor vessel *Liberty Star,* calling the unidentified Nicor workboat
on my port. Come back." The captain had recognized the Nicor Company's broad orange stripe around the wheelhouse.

Tommy told Burlingham not to answer. He had learned from the
lawyers that the law was unclear on what happened if an interloper
attempted to recover the same ship. If the interloper interfered in a
certain way, the court might split the award between them. "The idea

was to make sure they didn't get to that point," said Tommy. Even if the vehicle wasn't ready, he wanted to maneuver ahead of the *Liberty Star* and set up over the Galaxy site.

The other captain hailed them again, this time calling them the "unidentified gray-hulled workboat to my port."

No law required them to answer, but what Craft called "the Rule of Common Sense to Protect Your Own Ass" dictated that you let the other captain know what you intended to do, so he would not endanger you while you were doing it. "It's just good practice to answer such a request in a civil, courteous, and technically correct manner," said Craft.

They decided that Craft should be the only one to talk. He was older and wiser and had been around the sea for forty years. But before Craft responded, Tommy wanted to anticipate every possible turn the conversation might take. He wanted Craft to understand the importance of giving up nothing over the radio while at the same time not appearing to be evasive. They might ask this, they might ask that, what are you going to say? Do you feel comfortable saying it this way? What about talking about it like this?

Tommy called it forging an understanding, getting others to diverge quickly, present their best thoughts, and from their different perspectives distilling the best strategy. Craft called it chaos. "Until you personally stand in the wheelhouse of a vessel and watch and listen to the alleged decision-making process of Harvey, Barry, and Bob Evans, you do not understand what true chaos is."

The captain of the other boat hailed again. The two boats were closing now, the *Navigator* steaming at about ten knots, overtaking the other vessel traveling at one and a half knots. Tommy called down to the back deck and had Bryan and Tod throw tarps over the vehicle and arrange the tarps to make the vehicle look a different shape. He told Burlingham not to let anyone on the *Liberty Star* see anything other than the bow of the *Navigator*.

As Tommy stood ready with pens and paper to write Craft notes, Craft picked up the radio. "This is the research vessel *Nicor Navigator,* Whiskey, Yankee, Quebec, 7-4-5-8, calling motor vessel *Liberty Star*. Request you go to Channel 8."

When they had switched frequencies to get off the hailing channel, Craft said, "Hello, Skipper. You're trying to get ahold of us. What can I do for you?"

"We've got a deep tow behind us," said the other captain. He gave his course. "We were just tracking you here, wondering what your intentions are."

"We don't plan to cross behind your path," said Craft, "and we understand that you have equipment in tow."

In giving his course, the other captain had confirmed something the bridge of the *Navigator* had just realized might be happening: Either the *Liberty Star* had changed course or the bridge had slightly miscalculated her original heading. She now appeared to be on a bearing that would take her far enough west of Galaxy that the site would not lie within the swath of her sonar fish.

Even under Craft's Rule of Common Sense, he had provided all of the information the other captain needed to proceed safely. Then they heard the captain's voice again. "Say, are you boys out of Florida or Charleston or up Norfolk way?"

Tommy started writing notes to Craft, who made a face like all those notes were only distracting him. He got back on the radio.

"Uh, can I help you with anything there, Skipper? We got something to chat about?"

"No, that's all right, Captain. This is the motor vessel *Liberty Star*, out."

Burlingham kept his bow aimed at the other vessel, slowed, and let the *Liberty Star* continue on its track line until it disappeared over the horizon. Craft recorded in his log, "Liberty Star is rigged for deep water survey and may have a deep water ROV on board."

Tommy wanted to know the *Liberty Star*'s speed; her bearing; the length of her track lines; and how long she took to turn, realign on the next track, and begin again. He needed to understand her search area so he could predict where she might go, but he wanted to avoid making her crew suspicious. His greatest advantage now was that he knew what they were doing, and they still knew little about him.

During the day, Tommy insisted that Burlingham keep the *Navigator* within radar range of Galaxy, but at night the crew shut down the deck lights, blacked out the windows, and hunted for the *Liberty*

Star. Even before they captured her on radar, they saw her glow projecting far into the night sky, like a small city two hundred miles at sea. "Man, they were lit up," said Barry. "On a clear night, it was just dazzling." With his ship blacked out, Tommy crept up to the edge of radar range and plotted her, then fell below the horizon and came up again. Even if the other bridge saw something on radar, it was but a pinpoint along the rim. "We were just close enough to the edge that it would be inconsistent," said Tommy. "They wouldn't see us every sweep." The other crew had no way of knowing that the same ship they had encountered on an earlier afternoon now stalked them from just beyond radar range.

The military might call this maneuvering ECM, or Electronic Counter Measures warfare. Craft called it "a childish and asinine game of cowboys and Indians using radar instead of cap pistols." Craft went to sea to do a job, and he did it as efficiently and as safely as he could in a reasonable amount of time. "I didn't give a hoot who else came out there and looked," said Craft. "We had x number of places to look for the *Central America,* and the faster we did that, the sooner Harvey would be in a position to go to court and claim the ship. I have never been able to figure out his reasoning."

But Craft had never sat in a law office and talked to lawyers for hours about everything Tommy had to do to protect the site. He hadn't met with dozens of investors one or two at a time face to face and promised them he would use their money wisely. He hadn't spent ten years of his life trying to figure out how to do this. Tommy's decisions often stupefied people, until they learned more of the factors he had rattling around in his head at the time he decided.

One big factor was that the vehicle wasn't ready to launch, and if they sat on top of the site while they worked frantically to get it ready, they would be giving away their target, which they currently had no legal way to protect. Tommy surmised that the track lines of the *Liberty Star* would run twenty-five to thirty-five miles, which meant that on each track, she would drop over the horizon, sail beyond radar, and not be seen again for a day to a day and a half. He could use that to his advantage. If he knew exactly where she was, where she was headed, and how fast she was headed there, he could calculate how much time he had to get onto Galaxy, launch the vehicle, find out as much as he could, recover the vehicle, and get back over the horizon before the *Liberty Star* returned.

On July 2, the *Liberty Star* hit the south end of her track line, turned, and headed north, still keeping west of the Galaxy site. The following day, Tommy instructed Gross to fly out again, plot the position and bearing of the *Liberty Star,* then shoot a roll of film with a telephoto lens and drop it near the *Navigator.* When Doering developed Gross's pictures, they confirmed what Craft thought he had seen through binoculars two days earlier: Something shrouded in a tarp sat on the back deck, and it appeared big enough to be a deep-water ROV. Whoever was on the *Liberty Star* was prepared not only to run track lines and listen with sonar, but also to drop an underwater robot on anything that sounded interesting.

TOMMY COULD HAVE gone to court with an "artifact" and avoided all of these worries if he had followed one piece of advice Hackman had offered half in jest before he left the ship three weeks earlier: Fake the artifact.

"I always explore all possibilities whatever I'm doing," said Hackman, "and cheating is always a possibility. So I said, kind of jokingly, 'Let's go get some driftwood and tie it to the vehicle and bring it back up and say, "Here, we found it, we've retrieved a piece."' And his comment was, 'That's a very interesting possibility, but this project will be so scrutinized that I would never try to get away with anything.' We were out there alone and nobody would ever know what we did, but he was always looking at the big picture, so he will not chance anything like that. I'm probably the fourth most knowledgeable person about this project, and I have yet to see anything that's even remotely shady, illegal, unethical, or immoral."

Tommy would not break the rules, but he would forsake sleep to study every one, dissect it, examine the parts, and search for fresh interpretations. "We have to be smarter than everybody else in order to compete and still have integrity," he said. "We have to see every angle and think of all the options."

For three days now, they had charted the movements of the *Liberty Star.* Tommy had sequestered himself in the COM shack, studying the plotting sheets, and he could see the pattern of her track lines clearly now. He calculated that by the time she completed the track she was on and shifted east again for the next track, then headed back south, she would be aimed right for the western edge of the Galaxy site. He could stay away from the site only hours longer. Then he would have to pull up on Galaxy and sit motionless above it, a stationary blip in the path of the *Liberty Star,* and the confrontation would begin.

Since the *Liberty Star* appeared on the horizon, a supply boat had picked up Ted Brockett for another job, so they were short one knowledgeable head and two skilled hands. But the rest of the crew continued to work on the vehicle through the night, night after night, napping only when they could no longer function without sleep. At one stretch they stayed awake for thirty-six hours.

The vehicle had grown from a helpless creature that could barely see and hear, into a deep-ocean exploration system with five cameras for eyes, a sonar for ears, an eighty-pound telemetry unit to talk, seventy-pound thrusters to walk, a sixty-pound manipulator to pluck, a thirty-pound brain to command, hundreds of pounds of batteries for a heart, and hundreds of feet of hydraulic hoses and electrical cables, the arteries and nerves, to connect it all. But when they tested it on deck, circuit breakers blew, signals scrambled inside the cable, and the manipulator hung like a useless appendage. By the night of July 3, it still was not ready to go to the bottom, yet sometime the next day, the *Liberty Star* would be on top of Galaxy, and the only hope Tommy had of stopping it was to be on the site, doing the best he could with the system he had.

Under the glare of the deck lights that night, they dropped three acoustical pingers around the Galaxy site to form a subsea navigation grid, so Scotty could calculate the vehicle's position on the bottom. By two o'clock on the morning of the 4th, Scotty had the grid in place and calibrated well enough to read ranges. Eight minutes later, they launched the new vehicle. "We had no choice," said Tommy.

A little past three-thirty, the vehicle reached bottom; on the monitors, they saw the white sediment of the ocean floor. A half hour later, they began searching for the ship in ten-meter zigzags. For an hour they saw nothing. Then the cameras passed over mysterious six-foot rings

cut into the floor by sea cucumbers. In five minutes, they had drifted beyond the rings and once again saw only white sediment.

While the vehicle probed at eight thousand feet below, Tommy, Barry, and Bob huddled with Craft and Burlingham on the bridge. In a few hours, the *Liberty Star* would appear on radar about twelve to fifteen miles to the northwest on a heading that would cross the Galaxy site. This time Tommy did not want Burlingham to slide back over the horizon to avoid detection: Robol had advised Tommy that Admiralty Law required the one claiming a ship to stay on top of that ship; Tommy could rest his crew and repair his equipment, but he could not leave the site, or he risked losing control. On the bridge that morning, he told Burlingham that if the *Liberty Star* tried to enter the waters above Galaxy, he had to stand fast and, if necessary, drive her off. But Burlingham refused.

Burlingham had his own law, the sea captain's International Rules of the Road, and if another ship towing equipment needed to pass over the site, the Rules forbade him to squat motionless on top of it *with nothing in the water*. When the vehicle below lost power, they had to recover it, and the moment it hit the deck of the *Navigator,* the *Navigator* became the "burden" vessel and the *Liberty Star* became the "privilege" vessel. Then Burlingham had to give way, and the Rules didn't care if he was resting his crew or mending his equipment, or that giving way might compromise a recovery effort under Admiralty Law. If that vehicle was not legitimately in the water at the end of a long cable, Burlingham would allow the *Liberty Star* to run right over the Galaxy site.

Robol got on the phone with Burlingham to explain the legal ramifications, but Burlingham did not trust lawyers who were sitting on their behinds in their offices back onshore, when he was two hundred miles at sea in command of a ship and responsible for the lives of twenty men and another sea captain he couldn't predict. The Rules were clear: If you the captain of the give-way vessel encountered another vessel restricted in her ability to maneuver, you gave way, or you the captain lost your license. Maybe paid a fine. Maybe went to jail. Not the lawyers, you the captain. They didn't have to explain things in simple terms to Burlingham; he was as smart as any of them. He just went to college to be a sea captain instead of a lawyer or an engineer, and no lawyer or engineer was going to tell him to do something at sea that he as a sea captain knew should not be done.

With a confrontation unfolding at sea, Tommy's first thought was: Keep everybody on board this vessel from fighting, and especially don't alienate the captain of your own ship. "You can't say, 'The hell with your license, Burlingham. What the hell does your license have to do with anything?' You've got to be sensitive to that." But Tommy also had little time for tact. With Burlingham and the others on the bridge, he started talking like this: "We got these problems. Okay? You believe that way. I believe this way. Okay? Here they come. Now, what are we going to do?"

If Burlingham would not stand fast, the only way they could keep the *Liberty Star* from dragging her sonar fish across the Galaxy site was somehow to persuade the other captain to skirt their work area; the only way they could do that was by carefully stringing together words that might imply that they had a stronger position than they had but when laid out one word after the other still were true. They all talked, and Tommy laid out a scenario: what Burlingham would say, how the *Liberty Star* might respond, what Burlingham would say back.

Burlingham was willing to listen, but he told Tommy that if the two ships squared off, he would not lie to the other captain. Tommy could arrange the words in a way that perhaps left a few things unsaid or merely implied, but Burlingham would not lie.

Tommy didn't want him to lie; he wanted to tell them nothing, whether they had equipment in the water or out of the water or what kind of equipment it was or what they were doing with it. He wanted to reveal nothing that would arouse their suspicion any more than it had already been aroused. When he thought of oblique responses to questions they were sure to ask, he turned to Burlingham and said, "Is that okay to do that?"

Burlingham often responded, "No, I can't."

Sometimes Tommy then said, "Okay, if we can't do it that way, how about doing it this way?"

But once in a while, Tommy told Burlingham, "Well, we might have to." A couple of times tempers got so hot, they had to clear the bridge.

"Burlingham was in a tough situation," admitted Tommy. "You're not supposed to use the Rules of the Road to play cat-and-mouse games. It's just that if we didn't do what we were doing, we were going to get run over."

One thing everyone realized: They had to say something, because in about three hours the captain of the *Liberty Star* was going to see another vessel on radar, this time perched motionless in the middle of his track line. Then the captain would be on the radio asking what's going on, and Burlingham would have to respond. Tommy wanted the scenario to unfold another way; he wanted to take the offensive, be the one to break the silence and ask the captain to stay clear of their work area. That was part of his strategy. One thing he insisted on is that they decide on a plan and resolve all of their differences before they got on the radio; and once they got on the radio this time that only Burlingham speak.

While the strategy sessions continued through the early morning, the *Liberty Star* appeared on radar just twelve miles to the northwest. The vehicle was still down, moving slowly over the ocean floor, searching for the site; but after five hours the batteries were too low to continue, and the tech crew was exhausted from being up all night. They ended the dive at 9:30 and began bringing the vehicle back to the surface. During their first dive on Galaxy, they had seen nothing on the ocean floor but the curious rings carved by sea cucumbers.

AT TWELVE-THIRTY, the *Liberty Star* reached five miles out, and Burlingham broadcast a *sécurité* call, a message to all vessels within range, although he and Tommy knew that the only vessel within the sound of his voice was the *Liberty Star*.

"Sécurité, sécurité, sécurité," Burlingham began. "Research vessel *Nicor Navigator*, Whiskey, Yankee, Quebec 7-4-5-8. We are a gray-hulled research vessel working in the area bounded by latitude 31° 43' north, 31° 43' north, 31° 40' north, 31° 40' north, longitude 76° 22' west, 76°22' west, 76° 18' west, 76° 18' west. We are conducting underwater operations. We request all vessels transiting near us to please keep clear of the work area." The coordinates circumscribed a box three nautical miles north to south by a little over three nautical miles east to west.

A few minutes later, they heard on the bridge, "*Nicor Navigator,* this is *Liberty Star*. We're a research vessel to the north of you there. We're doing bottom surveys, and I reckon we'll pass a mile from you. Is that all right? Over."

Even if the *Liberty Star* cleared the *Navigator* by a mile, she would be inside the box, and the sonar fish following two and a half miles back could easily image the target.

Off the air, Tommy said, "Absolutely not, we've got gear in the water!" The vehicle might be on deck, insisted Tommy, but the three transponders they had deployed for the subsea navigation grid were tethered to weights on the bottom and hovered around the site. They rose no more than fifty feet, and the sonar fish was flying probably at about three hundred feet, but still the transponders were *gear in the water*. They just weren't connected to the ship.

"*Nicor Navigator* back," said Burlingham on the radio. "If you're going to clear our area from the coordinates that I just gave on the radio, that will be clear of us. If you're going to come inside these coordinates, our gear does extend in various directions and we'd appreciate it if you could stay outside our work area." He repeated the coordinates. But as soon as Burlingham had signed off, they realized they had made a mistake. They didn't know on which side of the box the *Liberty Star* would clear them, and with the currents running strong to the west, if she tried to clear to the east, the currents would still drive the sonar fish through the box and over the site.

Burlingham hailed the *Liberty Star* again. "Good afternoon, there, Skipper," he said. "The way it's looking now, I just wondered is it possible for you to clear us to the west, over?"

The voice from the *Liberty Star,* a different one this time, said he would have to refer to his charts.

Burlingham surmised that whoever had chartered the *Liberty Star* had not told the captain what they were looking for; that as far as the captain knew, the sonar techs were out there running bottom surveys; that, until now, they had the whole Atlantic Bight to themselves. The captain was probably dumbfounded that a confrontation appeared to be taking shape out in this lonely patch of ocean.

The *Liberty Star* continued toward the northwest corner of the box, and the radio on the bridge of the *Navigator* sat silent. Four miles, then two, then one and a half.

Watching from the bridge, Burlingham reminded Tommy, "If they push me, I'm getting out of the way. I'm not going to put myself in a close-quarter situation just to suit your priorities."

"Can we say it another way?" suggested Tommy.

The question epitomized what bothered Burlingham about Tommy. The man was always groping for exceptions. In Burlingham's mind the Rules were clear: The vehicle now sat on deck recharging; the tech crew was in the sack doing the same; we have nothing in the water; we cannot be the privilege vessel. But Tommy pushed. "If they try to come through, can we say we're preparing to put equipment in the water? Or I think we're going to have it in the water in fifteen minutes?" He suggested that maybe within the Rules were maneuvers Burlingham could perform if he thought about them ahead of time. "Like a game of chess," said Tommy.

That idea intrigued Burlingham. He wouldn't cheat and he wouldn't lie, but he would try to outwit. In specific situations, he had some leeway under the Rules to move and countermove his ship so that he was properly positioned to be the privilege vessel. "But it was a challenge for me," said Burlingham, "to do what Harvey wanted me to do without severely bending the Rules."

He got back on the radio. "Yes, sir," he said as politely as possible. "I just wondered, have you decided what your intentions are in passing outside our work area? The way we're working at this point we may be going anywhere unexpectedly, and so if you enter our immediate work area, the possibility of an *in extremis* situation may arise."

The phrase came right out of the Rules, *in extremis,* meaning If you enter this box, that will put our two ships at close quarters and we may collide. "If the situation were reversed," Burlingham said later, "I probably would have taken that as a threat."

This time, the captain of the *Liberty Star* responded much more quickly. "We'll be continuing on this course for a brief period of time," said the captain, "until we get to the northern limits of the area you specified, at which time we'll alter course to our starboard and lead you to our port hand. Over."

The captain still had not confirmed whether he intended to pass "a mile from the *Navigator*" or "outside the box," a big difference, because the box was over three miles wide. "He's trying to cut in as far as he can," said Tommy.

Burlingham was back on the radio. "Roger, I just want to make sure I understand; when you reach the limit of our work area, you'll be turning, I assume, down a course of 180 made good and skirting just the western boundary of our work area, is that correct?"

"We'll plan on leading you approximately one mile to our port hand, which should be about the western boundary." Still an evasive answer, but then he asked a question that helped Burlingham. "Do you have any tethered buoys in the water at this time? Over."

The sonar fish pulled by the *Liberty Star* flew too high to snag one of the $8,000 transponders forming Scotty's subsea navigation grid, but the captain did not ask how high off the bottom the buoys floated.

"Yes," said Tommy, off the air. "I think we ought to say yes."

"*Nicor Navigator* back," said Burlingham. "That's roger, we do have tethered items in the water at this time."

"Roger, are all these tethered items within the parameters of the box you gave me, over?"

"That's roger, Skipper. They're all within our delineated work area. Roger. I just want to check one more time," added Burlingham. "You are saying that you will skirt our western edge of the operation area?"

"That's affirmative," said the captain.

The *Liberty Star* was now no more than a mile from the northwest corner of the box. Bob had plotted her location every fifteen minutes.

TOMMY WENT BACK to the COM shack and called Robol. He had fans blowing on the SAT COM phone, but the COM shack broiled in the July heat, cooking the electronics, and Robol's voice kept fading in and out. Between the crackles, Tommy picked up that Robol thought the crisis at sea was not yet imminent, that he still wanted an artifact; he did not want to go to court and try to claim the site based only on the sonagrams. Tommy had interlopers towing a big sonar fish right toward the center of the site, and Tommy didn't trust them to keep outside the box that Burlingham had described. If Robol thought this crisis was not yet imminent, then before they got cut off, Tommy wanted a few words from Robol to convey in the strongest language possible their legal claim to the site, however uncertain that might be.

Robol replied, "You can say you've entered an action."

Tommy liked the sound of that; it was true, yet the words might stump someone else. "Okay, we'll use those words, 'entered an action,'" and he wrote down the phrase for Burlingham to repeat on the radio. But Robol was not finished. "And tell them you are rescuing a vessel in marine peril."

The words came right out of the law books. When someone discovered a ship and tried to recover it, it was called "rescuing a vessel in marine peril or distress," and it didn't matter if the ship was sinking or sunk or had been sunk for a long time, the one who discovered the ship was rescuing valuable property from the ravages of the sea and returning it to the stream of commerce. A sea captain may not interpret the phrase that way, but that's what the law said, and that was the beauty of the phrase. Barry was in the van while Tommy was talking to Robol. "I realized that if we could use the actual legal terminology and be absolutely truthful at the same time, it would send them a signal they wouldn't understand."

Tommy and Barry returned to the bridge to coach Burlingham on what to say, but Burlingham was reticent; he already thought that some of his comments to the captain could have been interpreted as an oblique threat. They told him that "rescuing a vessel in marine peril or distress" was the legal terminology; it could mean a lot of things. Maybe the ship was sinking and the cargo and tackle had to be saved before she went to the bottom; but maybe she had been sunk for a long time and continued to corrode and deteriorate due to the water and the pressure and the salt of the deep ocean.

Finally, Burlingham said, "All right, I can say that."

The *Liberty Star* approached the northwest corner of the box and began to alter her course to starboard, heading south. Tommy feared that after proceeding along the western edge for a short distance, the captain would suddenly cut inside the box, swinging the sonar fish close enough to image the Galaxy wreck site. Burlingham again hailed the other captain on the radio.

"I would just like to advise you of something," said Burlingham. "You may possibly want to copy it down." He began reading slowly. "We have entered a court action in the U.S. District Court for the Eastern District of Virginia. We are currently in the process of rescuing a vessel in distress. We advise you that any action you take which would endanger or interfere with our rescue operations could result in your being found in contempt of court. I am concerned for the safety of your vessel and ours."

Burlingham repeated the statement twice, until the other captain said, "Good copy."

They signed off, but within minutes the captain of the *Liberty Star* was back on the radio, asking for the "entity" on board the *Nicor Navigator,* the date the action was entered, the name of the vessel in distress, and the period of rescue.

"I guess I have a personal concern as a fellow mariner," the captain continued. "If a vessel is in distress, I would think the Coast Guard would be desirous of having that information passed along to us to assist in your search efforts. Over."

Burlingham caught the sarcasm; the captain was probably smiling at those gathered on his bridge as he said it. "Roger," said Burlingham. "The term 'distress' is a legal term with precedence behind it. We'll get back to you with those answers shortly."

When Burlingham called back a few minutes later, he identified the "entity" as "Columbus-America Deep Search Inc." They had filed on May 26, 1987. The vessel in distress was "an unidentified, wrecked, and abandoned sailing vessel." For the period of rescue, Burlingham answered, "Initial recovery is underway. This portion of the operation is partially dedicated to formulating a total and comprehensive plan. Dates and times will be filed with the court as the recovery proceeds."

Then Burlingham said, "We're just wondering what entity is on board your vessel."

"Stand by," said the captain, and when he came back, he said, "Our response to your last: the trustees of Columbia University."

GARY KINDER

The *Liberty Star* proceeded along the western edge of the box, by
Bob's calculation weaving inside the boundary and back out again three
times. By midafternoon, she had reached the southwest corner of the
box, where her course altered from due south to southeast, and she proceeded in that direction until evening, when she swung in a large arc,
east northeast, north, northwest, and headed back toward the box. At
ten-thirty that night, she passed within half a mile of the southwest
corner of the box and continued on course to the northwest, disappearing from radar the following morning.

UNTIL LONG AFTER the *Liberty Star* had turned below the box and
headed back to the northwest, Moore and Scotty worked on the vehicle.
At 8:15 on the morning of July 5, they launched for the second time at
the Galaxy site, and Moore stopped the vehicle a hundred feet down to
test the systems. The lights and the cameras worked, and the Mesotech
was operating again, but the thrusters were too weak to land, and the
manipulator joints too stiff to reach out and grasp. Moore ran the vehicle to the bottom anyway, for they had yet to find the ship.

By midmorning, they were on the first track line, with Scotty still
refining his navigation grid. For an hour and a half, he guided Moore
in zigzags, several meters one way, then several meters another way. Just
before noon, they were watching the black-and-white monitors in the
control room when suddenly small pieces of debris appeared, although
the lighting at that height was too dim to tell what it was. "You could
barely tell you were looking at anything," said Bob. But it was not white
sediment or sea cucumber rings or geology. After lunch Moore eased
the vehicle lower, and they ran a second track line and then a third. And
then a little after five in the afternoon they saw what they had been looking for, what Bob and Doering thought they had detected in the speck-
led sonagrams, what Tommy knew right away was missing from the
wreck site at Sidewheel: piles of coal.

Because the cameras could see only a twenty-foot patch of ocean
floor at a time, they could not tell how much coal littered the site, but
the coal fields looked extensive. The whole site seemed enormous, far
bigger than Sidewheel, and covered with debris. It was more like what

Tommy thought they would find. "We didn't have to look at it very much," he said, "to know that it had very good odds of being the *Central America*."

They ended the dive at six-thirty in the evening, and the vehicle had just begun its ascent when the *Liberty Star* appeared on radar sixteen and a half miles out. She came at them again from the northwest on a bearing roughly southeast, trolling at 1.6 to 1.8 knots and aimed at a point a few miles off the southwest corner of the box. Tommy told Craft to get the vehicle on deck as quickly as possible, put the batteries on the charger just long enough to get them up for another dive, then be ready to send the vehicle back down.

With the *Liberty Star* steaming toward them and the vehicle on deck, Moore stood topside, trying to adjust the joints on the manipulator. Each had to be oiled and finely tuned, but the pressure at eight thousand feet was two hundred times greater than at the surface, and the temperature was thirty-eight degrees instead of ninety-two. Joints that moved fluidly on the surface jerked and vibrated or froze at depth, and the only way Moore could tune them was to open the brain box on deck, adjust three tiny screws for each joint so it flopped like a noodle, then send the vehicle back down to see how close he got. It usually took several trial-and-error dives to figure all that out, but Moore didn't have several dives. He had one: a few hours to change to a thinner oil and tweak the screws so that below he could at least straighten the elbow and close the jaws. As a backup, Bryan and Tod mounted what everyone called the crab trap, a three-foot-wide metal screen that would hang from the bottom of the vehicle. If Moore couldn't get the manipulator to work, he would try to scoop something in the trap. Until they brought up an artifact from that site eight thousand feet below, they had little more legal claim to Galaxy than did the treasure hunters aboard the *Liberty Star*.

Six hours after Doering had pulled the vehicle from the water with the crane, he lifted the vehicle off the deck and lowered it back into the water. It was 2:40 on the morning of July 6. They now had the *Liberty Star* in sight about four miles due west of the box, and they watched her still angling slowly toward them. As the vehicle neared the bottom,

Moore, Scotty, Doering, and Bob sat in the control van for the dive, and Tommy, Barry, Burlingham, and Craft stood on the bridge, tracking the movements of the *Liberty Star*.

WHEN THEY SAW the huge debris field at Galaxy on the second dive, they had chosen the artifact to present to the court: a lump of coal. "Coal is a perfect first artifact," thought Bob. "It's fuel to the ship. It was only twenty dollars a ton or so, but nonetheless it had a defined value. And it's an object undeniably from a ship." Tommy counseled with Robol, who agreed: Legally, the fuel from an old steamship could serve as an artifact to represent the steamship. The coal was also plentiful, piled high, away from the core, not too valuable, and unbreakable. With a catch-as-catch-can scoop and an unpredictable manipulator sporting ten-thousand-pound crushing strength, Bob did not want the first retrieval attempt focused on a teacup.

Within two hours of reaching the bottom, Scotty had guided them closer to the site, until they had debris again on camera. With the thrusters still malfunctioning, Moore could not land the vehicle, so he would have to cock open the jaws of the manipulator and try to grab something on the fly. With the vehicle gliding slowly through the water four feet above the floor, Moore twice saw tall piles of debris, reflexively dove the vehicle into the pile, and tried to close the jaws as it hit. Sediment curled up before the cameras, obscuring the scenes. As the water cleared, Moore could see both times that the jaws hadn't closed, but he couldn't tell if anything had caught in the scoop.

For the next two hours, they continued running the monotonous track lines, seeing nothing more on the cameras. And then suddenly looming in front of the vehicle was a pile of coal strewn with pieces of pipe, perhaps from the steam engines. Moore estimated the pile was six feet high. Scotty radioed Burlingham to change direction slightly and increase his speed. Moore waited until the pile was only a few feet away, then he dove the vehicle as fast and as hard as he could. The monitors exploded with sediment roiling through the water, and Moore blindly tried to snap the manipulator jaws closed.

"We weren't sure if we got it or not," said Bob, "but Moore really plowed into that thing."

It was now almost ten o'clock in the morning. No one had been to bed the night before, and all four men in the control room were exhausted from staring at the monitors for the last six hours. Moore could see that the manipulator jaws were open and empty again, but he thought they might have snagged a lump or two in the crab trap.

Bob radioed Tommy in the COM shack that they were ready to recover, but Tommy said to bring the vehicle up only a safe distance from the bottom, then leave it deployed over the side. Three hours earlier, the *Liberty Star* had stopped in her southeast track line and for some reason had turned due south, where she now dawdled. Tommy didn't know what she would do next, but he wanted the vehicle in the water to give Burlingham an advantage under the Rules.

An hour later, they saw the *Liberty Star* stop her dawdling to the south and suddenly sweep an arc 180 degrees, to the east and then north, and then head back toward the box at a speed five times faster than when she had been towing. Within an hour of turning, she was less than a mile off the southwest corner of the box, and Tommy was on the phone again with Robol.

With a tape recorder running, Burlingham radioed the captain and advised him that the *Liberty Star* was approaching the box he had described earlier. He wanted to know the captain's intentions. The captain radioed back that his ship was not inside the work area and would not enter. Burlingham reminded him that the *Navigator* had equipment in the water and that unless the two ships remained at a substantial distance the *Navigator*'s equipment and the equipment trailing behind the *Liberty Star* could cross, become entangled, and break, perhaps destroying deck equipment or endangering personnel.

The *Liberty Star* continued north, hitting the southwest corner of the box, her speed now down to only twice the pace of normal towing. Burlingham met her at the corner, aiming the bow of the *Navigator* at the *Liberty Star*'s broadsides half a mile away, and as the *Liberty Star* steamed up the western edge of the box, Burlingham crabbed sideways, nimbly sidestepping with his dynamic-positioning system up the boundary, always keeping the *Liberty Star* just beyond the edge.

"I'm sure they found it hard to believe what we were doing," said Burlingham. "The *Liberty Star* was designed for the space shuttle booster retrieval system, so it was decked out to the nines. We were strictly a mudboat." But the mudboat was only camouflage; the *Navigator* could hover in one spot or inch sideways left or right; and her navigation system was far superior: The *Liberty Star* navigated on a system accurate to within hundreds of feet, sometimes only hundreds of yards; Burlingham at all times knew where he was within fifteen feet.

When the *Liberty Star* hit the northwest corner, she suddenly looped back to the southeast and feinted toward the inside like she would cut through the middle of the box, but there she encountered Burlingham's bow. She continued turning 360 degrees, dropping just inside, then went north again for a short distance, turned sharply east, and followed the northern edge of the box.

"It's hard for anyone to even understand the situation we're in," said Tommy. "We're supposed to be just sitting there doing our jobs and the other people interfere, that's the way the court likes to see it, but it became impossible to try to do that."

Tommy worried that if the treasure hunters on the *Liberty Star* persuaded their captain to nick the Rules just enough to force that sonar fish across the Galaxy site, then they might become co-custodians of all the gold on the *Central America*. "Yeah, they violated the Rules of the Road," said Tommy, "but now they know what we know about the site."

The vehicle still dangled in the water nearly eight thousand feet below. At two o'clock in the afternoon, Tommy saw the *Liberty Star* approach the northeast corner of the box. As the captain executes his turn, thought Tommy, Burlingham can keep the *Navigator* virtually still on the water, and that'll give Craft just enough time to get the vehicle on deck so we can see if there's anything in that crab trap. He told Craft to raise the vehicle and keep it just below the surface.

With the *Liberty Star* in the middle of her turn, Craft ordered Tod and Bryan out on the hero boards over the side of the ship with knives and marlin spikes, ready to transfer the weight of the vehicle from the cable to the crane. Then he pointed his index finger upward and ran it in tight circles, signaling to Doering to take in the line. Dipping into small waves with the roll of the ship, the vehicle broke the surface and came up snug against the tractor tires at the end of the boom. With the vehicle stabilized, Doering raised the boom, and the vehicle rose from the sea in a rush of water. The manipulator arm still stuck out, the jaws frozen open with nothing in them. As the vehicle reached eye level and the water coming off slowed to drips, they could see the crab trap underneath, and that, too, was empty. If Moore had scooped up anything by crashing the vehicle into piles of coal, it had washed out somewhere in the wait at the bottom or in the eight-thousand-foot ascent. No one was surprised, but everyone had hoped.

The afternoon was hot, the sky cloudless, the sun high. The *Liberty Star* had completed her turn and was pointed south. Doering was swinging the vehicle over the rail when the sunlight suddenly sparkled off something more than the droplets of water cascading from the vehicle. The sparkle caught Doering's eye. Before he set the vehicle down on deck, he yelled, "Get that!" Wedged into a corner of the vehicle's lower frame was something shiny and black. As the vehicle met the deck and the line slackened, one of the techs reached into the frame and pulled out a chunk of anthracite coal almost six inches across.

"How a piece of rock wedged in an aluminum frame could make it all the way to the surface without falling out with the vibrations and the air-sea interface, I don't know," said Doering. "But it did."

Bob dropped the tag line he held on the vehicle and examined the lump. It obviously had come from a ship at the bottom of the sea, for cemented to it were the white calcareous tubes of marine worms.

Headed south along the eastern edge of the box, the *Liberty Star* hit the southeast corner and turned west to run along the southern boundary. By midafternoon, she had circumnavigated the entire box, the *Navigator* inside and Burlingham nosing the intruders to the edge all the way around. The *Liberty Star* now turned back to the southeast along her original bearing and disappeared off radar late that night.

No one could explain how the lump of coal made it all the way to the surface tucked into a corner of the vehicle frame, but no one cared. They had an artifact from the ship, and now they had to get that artifact to the courthouse so the U.S. marshal could "arrest" it and the judge could award them the site and warn away the *Liberty Star*. But the courthouse was in Norfolk, three hundred miles away, and the *Liberty Star* circled the shipwreck site like a shark. They couldn't leave. A supply boat would take two days to pick up the lump and deliver it to Norfolk, and Steve Gross already had tried to land out there in the smoothest seas he had ever seen. He would not try it again. But Tommy had another idea for Gross. "It was a pretty crazy thing I was asking him to do," admitted Tommy. He wanted Gross to fly out in the Sea Bee and snatch the artifact from the air.

With the SAT phone still fading in and out, Tommy explained his plan to Gross back in Wilmington. The idea was simple: All Gross had to do was risk his life in a maneuver that inside territorial waters was strictly forbidden by the FAA. Tommy would have the artifact hanging off the stern of the ship on breakable line, and Gross would align the Sea Bee on a run parallel with the stern, ease his altitude down to a few feet off the water, drop his speed to just above stall, fly within a few feet of the ship, snag the artifact with a grappling hook trailing a hundred feet behind the plane, then goose the engine before he augered in. It was just the sort of low-tech engineering solution Tommy loved.

But as Gross listened to Tommy, his mind passed Tommy's and now he started thinking about what he was going to do with this thing hanging off the tail of the Sea Bee once he got back to shore. He couldn't fly to Wilmington and land at the airport, dragging the artifact along the runway at a hundred miles an hour. Even if the package survived, he couldn't stop in the middle of the runway, alight from the cockpit, and pull the thing in. He figured right away that the options narrowed to one: Fly back and land in the Cape Fear River. The problem was that the Cape Fear River in a lot of places was no deeper than six inches. To land in it, Gross had to read the charts and know where he was, and to do that he had to be able to see. He needed daylight to land in the Cape Fear River.

Darkness descended over Wilmington about eight-forty-five, but Gross needed more than the last gray shafts of evening to see where he was going. At the extreme, he had to be looking for a place to drop into the river no later than eight-thirty. It was now after three o'clock in the afternoon and he was in his shorts in his room at the Cricket Inn talking to Tommy on the telephone. If he hustled to the airport, he might be able to get the Sea Bee into the air by shortly after four o'clock. He started thinking, "So going out there at four, it's going to be six before I get there, and screw around an hour, it'd be seven, and get back at nine. That's too late to land in the Cape Fear River and know what I'm doing." But he told Tommy he would try, and he told Tommy to have everything ready when he got there so he didn't have to screw around for an hour. Then he raced to the airport and lifted off in the Sea Bee at 4:09 with an ETA at the ship of 6:05.

To give them a second chance if the first attempt failed and the artifact ended up back on the bottom of the ocean, Tommy and Bob decided to cut the lump in half. It was a hefty chunk that filled an outstretched hand; its two halves would weigh over a pound apiece. Bob took the coal into the workshop and found it so hard that only a hacksaw would cut through it. When he finished sawing the lump an hour later, he threw a couple of handfuls of Styrofoam peanuts into a large plastic mayonnaise jar, lowered one of the pieces into the jar, and packed it full with more Styrofoam peanuts, so it would float if it ended up in the water. Then he wrapped it in layer after layer of duct tape, incorporating a harness they could use to string the whole package onto a thin nylon rope. They would make a loop of the rope two hundred feet around, then stretch the loop from the ship a hundred feet out to the Zodiac, the top part of the loop taut so Gross could snag it with a grappling hook.

From four thousand feet up, Gross saw the light softening along the horizon to the east, and he worried about making the pickup in time to get back to the river. On the back deck of the *Navigator,* Tod and Bryan launched the Zodiac and Bob strung the jar onto the nylon loop. Doering tied the loop to the head of the crane with heavy fishing line, then straightened the elbow until the crane was as high off the deck as he could get it.

Gross arrived shortly after six, and Tommy talked to him on the hand-held, explaining how the maneuver would work. Then Tommy gave the radio to Scotty and got into the Zodiac with Tod and Bryan and twelve feet of PVC pipe to hold up the other end of the line away from the ship. As always, the white-tip sharks followed the Zodiac out and circled it, seven-footers, not particularly aggressive but unpredictable and known man-eaters. Tommy had to hold the nylon rope as taut as possible at the top of the PVC pipe while trying to stand up in the

Zodiac, which was rocking in four-foot seas. Bob wondered what would happen to the project now if a little confusion in the swells suddenly tossed Tommy into the ocean and the white-tips had him for a snack. Tommy was concentrating on a more immediate problem: Gross had arrived, but the nylon rope was snarled, and he was trying to swing it in large circles like a jump rope to straighten it out.

Gross had paid out on the grapple and had it twisting in the air one hundred feet behind the Sea Bee. He made one pass off the stern so Scotty could see how the hook was trailing, then he was ready for the pickup, but Tommy waved him off. Gross thought, If we can't succeed at the other end, we can't start at this end, and it's six-thirty right now. That's the limit. "If it's not ready," he told Scotty, "we need to abort."

Tommy was still standing up in the Zodiac, swinging the nylon rope round and round. Gross yelled at Scotty again, "I've gotta leave. Tell Tom I'll come back tomorrow." Then he had another thought; he would try to grab it even with the lines tangled, but that would be the only try.

Gross circled one more time and came in low between the stern and the Zodiac, the hook bouncing and twisting in the air a hundred feet behind him. Burlingham held the *Navigator* as still as possible. Bryan kept the Zodiac nudging away from the ship to keep the line taut. Tommy raised the PVC pipe as high as he could.

When the hook hit the rope, everyone on the ship heard the nylon sing, until the weaker fish line popped at either end. Gross was still horizontal, and the package now trailed at the end of a hundred feet of grappling line plus another hundred feet of nylon loop. It drooped toward the water, and suddenly it hit a wave, then another and another. Gross felt a hard tug, the nylon sang again, then it snapped, and the package skipped across the ocean for another quarter of a mile.

Bryan raced the Zodiac up the Sea Bee's trail, and they found the package still wrapped in duct tape, bobbing on the surface. Gross figured that the next time he would pull up sharply the instant the hook caught the rope. But there would be no next time tonight.

Back in Wilmington, Gross went to bed at 11:00 and set his alarm for 3:00 A.M. The following morning, he drove to the airport, gassed the Sea Bee, filed a flight plan, and was airborne at 3:52. For the first hour he flew in darkness.

The sea had settled overnight. When Gross arrived at the ship just before six, the Zodiac bobbed in gentle swells, and Tommy was attaching the fish line to the nylon loop and the PVC pipe. During the night, the *Liberty Star* had disappeared from the radar screen on the *Navigator,* but apparently she had turned back immediately, for she reappeared just before midnight. She was now about to nick the northeast corner of the box a mile away.

Gross took one pass at the stern so they could see if his hook trailed properly, then he circled, dropped the Sea Bee low, aimed his nose at the center of the nylon loop, and eased down to no more than ten feet off the water. This time there were no tangles. The hook caught the loop, Gross pulled back on the wheel and was climbing steeply even before the nylon stopped singing. When the fish line broke, he was up, gaining altitude, headed for the beach in a brightening sky. Trailing behind him in a stiff wind was a shipwreck artifact recovered from the ocean floor eight thousand feet below.

It happened fast, and everyone on board watched the performance. For the past four months, all of their efforts had been focused on one thing: recovering an artifact from the *Central America*. Now that artifact was on its way to the courthouse, and a whole lot of stomping and backslapping and yahooing erupted on the back deck of the *Navigator*.

GROSS LANDED IN the Cape Fear River, the bundle bouncing along the surface until he had stopped the Sea Bee. He opened the door, reached out with a boat hook, snagged the nylon loop, retrieved the package, tossed it in the backseat, and took off for Wilmington to gas the plane. Then he left immediately for Norfolk, where Robol met him at the airport.

Tommy had phoned Robol to tell him the artifact was on its way. When the courthouse opened that morning, Robol was at the door with an order for the judge to sign that directed the U.S. marshal to arrest the artifact and the ship it represented as soon as the artifact arrived at the courthouse. He also filed an amended complaint, changing the site from the coordinates at Sidewheel to those at Galaxy.

After Robol met Gross at the airport, he returned to the courthouse in early afternoon. He took the mayonnaise jar, still wrapped in duct tape, to the U.S. marshal's office and watched while the marshal unwound the tape, then reached into the jar of Styrofoam peanuts and lifted out the coal with the curious worm tubes stuck to it. The marshal was amused. He said to Robol, "This is the first time I ever arrested a lump of coal."

That afternoon, the judge reviewed all of the papers, including the arrest warrant and the amended complaint. He wasn't certain he had jurisdiction over a ship beyond the three-mile limit, but if no one challenged the Columbus-America group and therefore his jurisdiction over the site, then he would never have to address the issue. For now he assumed jurisdiction over the matter and awarded Columbus-America full recovery rights to the site.

At two o'clock the next morning, the *Liberty Star* reappeared on the *Navigator*'s radar screen, this time less than nine miles out. The mate on the bridge awakened Burlingham, who awakened Tommy, Barry, and Bob, and they all gathered on the bridge to watch the *Liberty Star* creep closer to the box.

Tommy knew that the court had awarded him the ship, but he had no proof of it, because the SAT COM phone was down again, and he could not receive a fax of the orders. "It was going in and out just driving me crazy," said Tommy. "I needed those papers to convince Burlingham and everybody else on board that they weren't going to lose their licenses, and that I wasn't just a wild man."

When the *Liberty Star* arrived at the northern edge of the box, Burlingham hailed the captain and requested him to stay clear of their work area. Once again the captain said he would neither enter the area nor interfere with the *Navigator*'s operations. Later that afternoon, Robol was finally able to fax the orders to the ship, and Burlingham hailed the *Liberty Star* a second time. He read the statement, quoting portions of the arrest warrant and the judge's order designating Columbus-America substitute custodian of the ship and all artifacts recovered from the ship. The captain acknowledged a good copy of the message and said nothing more.

The following morning, July 9, the *Liberty Star* turned at the end of another long track line, shifted about a mile, and headed back on a course that would slice right through the eastern half of the box. The captain radioed Burlingham and read the following statement:

"As of 9:32 EDT today, we are not in receipt or notification of any order or injunction issued by any court of competent jurisdiction that precludes us from surveying the open waters of the Atlantic Ocean. Nothing included in the filings of yesterday or previous days restricts our right to continue search and survey."

Obviously, the party on board the *Liberty Star* had talked with its own lawyer. "Accordingly," continued the captain, "we hereby advise you that we have started a survey transit line, which will pass through the area specified by you, and we request that you do not interfere with our passage along this survey line.

"We will be towing deep-towed hydrographic survey equipment at the end of a thirteen-thousand-foot cable at an altitude of approximately three hundred feet above the seafloor at a speed of approximately 1.6 knots over the ground. We will not be able to execute any sudden changes in track or speed, and the ship's heading may be substantially different from the track."

Robol had wanted an injunction ordering others to stay away from the site or face legal sanctions, but judges will not issue an injunction unless someone's life or property is in imminent peril. This was the imminent intrusion Robol had been waiting for.

Tommy told Craft to launch the vehicle immediately as scheduled. Then he telephoned Robol, and Robol helped them draft an affidavit for the judge: They began with the terse message from the *Liberty Star,* then continued, "The R/V Nicor Navigator is presently engaged in underwater recovery operations with a remotely operated work system deployed at a depth of approximately 9,000 feet on a tether of approximately 11,000 feet. The Nicor Navigator is showing day shapes ball-diamond-ball and is restricted in her ability to maneuver due to these underwater operations.

"At this writing, 11:30 A.M. EDT, the R/V Liberty Star is at a range of 8,600 yards, bearing 332 degrees true. On her present track, it is estimated that she or her tethered gear will pass within plus or minus 200 yards of the Nicor Navigator or its tethered gear in 150 minutes time.

This constitutes a direct threat and interference with our underwater recovery operations."

Barry signed it, dated it July 9, 1987, twelve noon, and faxed it to Judge Richard B. Kellam at the courthouse in Norfolk. By one-thirty, Judge Kellam had granted a temporary restraining order for ten days, and Robol had faxed a short statement out to Tommy on the *Navigator*. The *Liberty Star* had been heading toward them for the last three hours and had just entered the box at the northern edge.

Burlingham radioed the captain. "I've got something I'd like you to copy." He read the statement from Judge Kellam, repeated the coordinates for the four corners of the box, and concluded, "You are hereby notified that if you do not leave these coordinates immediately, you and all your principals may be subject to sanctions and penalties for contempt of court."

They expected to see the *Liberty Star* veer off, but she held her course. Tommy sent a telex copy of the statement to the *Liberty Star,* and still her course remained steady. Burlingham radioed again and said he was sending two men over in the Zodiac to deliver the document, but the captain refused to let them board.

Burlingham already had his stern aimed at the *Liberty Star*. He shifted the *Navigator* into DP mode and sat motionless halfway between the center of the box and the eastern edge. The vehicle remained in the water, and he flew the ball-diamond-ball of a vessel restricted in its ability to maneuver. That put him even with the *Liberty Star,* and under the Rules of the Road if the ships were even, the ship overtaking became the burden vessel. "I made them overtake me," said Burlingham, "and they can't cross my stern, because I'm towing."

As the *Liberty Star* approached, Burlingham started to creep forward ever so slightly, at the same time sidling to the west. The *Liberty Star* altered her course to the west, and Burlingham sidled farther to the west, the *Liberty Star* holding course long enough for the two ships to come within seven hundred meters of each other, less than half a mile, until she veered farther west and Burlingham had pushed her out of the box.

A few minutes later, the captain of the *Liberty Star* radioed Burlingham and read a terse statement that there was no temporary restraining order, because a $100,000 bond had not been posted, and that even if there were a temporary restraining order, it would have to be served either on their lawyers in Boston or by a federal marshal to them out at sea. Further, read the captain, the representatives aboard the *Nicor Navigator* had made "unfounded claims and veiled threats concerning possible harm to the M/V Liberty Star, its underwater survey equipment and personnel on board." Therefore, the captain of the *Liberty Star* would no longer honor radio communication from the *Nicor Navigator,* and all further contact between the two ships would be conducted strictly according to the Rules of the Road.

By five o'clock, Robol had secured the $100,000 bond. Tommy told Burlingham to radio the *Liberty Star* and call Dr. William Ryan to the bridge. Ryan had run the sonar survey for Harry John's X-marks-the-spot search for the *Central America* three years earlier. If Columbia University was involved, Ryan was on board: He headed the university's Lamont-Doherty deep-water sonar search team. With Ryan listening, Burlingham read another, more detailed, statement.

"We hereby advise you that the United States District Court for the Eastern District of Virginia has assumed jurisdiction over the wrecked and abandoned sailing vessel located within the confines of our work area, and a warrant for arrest has been issued under Admiralty Law. Furthermore, Columbus-America Discovery Group has been duly appointed the United States marshal's substitute custodian. This means that Columbus-America, together with the master of the research vessel Nicor Navigator, is legally empowered to act as substitute custodian for the United States Marshal Service. . . .

"You are hereby admonished to keep your vessel, the Liberty Star, and your sonar, recovery gear, and all other tethered items well clear of our work area. Should you choose to ignore this admonition, we will not hesitate to fully uphold our duty as substitute custodian and to enforce the court's orders, using all means necessary and proper under admiralty and international law."

Burlingham told Ryan that the bond had been filed and that the order was in effect. He added, "I would suggest you call your lawyer." Then he turned to Tommy and said off the air, "I better not be lying." The captain of the *Liberty Star* radioed Burlingham that he would stay clear of the work area for the night.

The restraining order was good for ten days. Robol now filed a motion to show cause why the *Liberty Star* and her personnel should not be held in contempt of court for having interfered with Columbus-America's ongoing recovery. He also asked for an injunction against the *Liberty Star,* and because the captain had described the entitiy on board his ship as "the trustees of Columbia University," Robol served each trustee. "That stirred things up a bit," said Tommy. "They were all subpoenaed to court." The hearing was set for July 15.

THE DAY AFTER they got the restraining order, Craft launched the vehicle again. For the first time, the thrusters operated so they could land, and Moore had the manipulator tuned well enough to reach out stiffly and pluck artifacts from the debris fields. Although crude compared to the capability Tommy had planned to take to sea that summer, this simple dive introduced a new era in deep-ocean technology. At nine thousand feet, they shot video and dozens of stereo stills, and Moore retrieved fragments of wood, several more lumps of coal, and a sampling of mud. Four times, he tried to pick up timbers or objects made of iron, but they were so soggy or rusted through, they collapsed.

The ship had deteriorated dramatically. Timbers protruded upward from the wreckage, and a little of the decking was still visible; they could see patches of the hull and maybe a piece of the bow. But mostly the site was covered with coal. In some areas, only where the coal seemed mounded could they identify what appeared to be the core of the site.

While the techs explored the debris, Tommy and Barry prepared for their trip in to Norfolk. Tommy had to testify in court, and the *Navigator* could not leave the site. The only "captain" Hodgdon could find to come out and pick them up was a teenage boy willing to make the four-hundred-mile round trip in his father's fishing boat with two of his buddies, a crew Hodgdon described as "hippie-type kids smoking pot." When they brought the boat alongside the *Navigator* on the evening of July 13, Tommy's brother-in-law, Milt Butterworth, was on board. He and Tommy's sister Sandee had moved from Chicago to Columbia, South Carolina, where Milt was now a professional photographer and videographer who taught film at the University of South Carolina. He had come out to help the techs document the site.

When he boarded the *Navigator,* Tommy and Barry climbed down to the small fishing boat, and it pulled away at 9:45 that night.

The boat smelled like dead fish. It was dirty, and Tommy suspected that lice hid in the beds. Worse than that, he had no telephone to talk to Robol or his suppliers or his crew back on the *Navigator*. He and Barry settled in below and tried to talk about the hearing, but they were too tired. Then they tried to sleep, but in the confused seas, the bow would rise and smack the next wave, sending shudders through the boat. The crew smoked a little weed, tied the wheel down to keep her close to on course, and crashed on the floor of the bridge. In the middle of the night, Tommy heard the wheel creaking against the tautness of the line, and the pounding of the hull did not stop for twenty-two hours.

The next evening, Hodgdon and Gross met Tommy and Barry at the docks and drove them to a small airfield south of Wilmington, where it was raining so hard water ran in the streets, and thunderstorms stretched far to the north. Gross had to stay with the Sea Bee in Wilmington to patrol the site and photograph intruders, so Hodgdon had found a young pilot with a little single-engine Cessna, and he rounded up a copilot to navigate and run the radio. Tommy and Barry tumbled into the two backseats, and the pilot taxied in the rain out to the runway and got them into the air just before the airport shut down for weather, and they flew to Norfolk that night in a rainstorm. Tommy had to be in court the next morning at nine o'clock, when he would be cross-examined by lawyers on the other side, and he had to talk to Robol and Kelly and Loveland before he took the stand.

WHEN KELLY AND Loveland had studied shipwreck cases back in Columbus, they noted the name of the lawyer who had argued each case, and repeatedly they saw the name of one man, David Paul Horan. Horan was a sole practitioner from Key West, a lawyer with an undergraduate degree in marine biology who loved only one thing more than practicing law, and that was diving. He loved diving so much that clients once offered to pay for a wind vane that would sit atop his office: When the wind blew, the words "Lawyer's In" would rotate into view, and when there was no wind, it would read, "Lawyer's Gone Diving." Horan had successfully argued Mel Fisher's case against the State of

Florida all the way to the United States Supreme Court, and in seven years of litigation, he had never lost a hearing, a trial, or an appeal. They hired him to ride shotgun in the courtroom next to Robol.

By the time Tommy and Barry got to Robol's office, it was well after midnight. "Everybody was about ready to fall asleep when we started," remembered Kelly. They told Tommy he would testify about what had happened at sea, that he would be the only witness from Columbus-America they would call. There was no time to coach him on demeanor, or rehearse, or warn him of traps the other side's lawyers might try to set for him. What they wanted to review with what little time they had was the story of what had happened at sea. When Tommy finished, it was four-thirty in the morning.

The story helped the lawyers understand the confrontation that had transpired topside in the sun, but the photographs sobered their intensity. Through the photographs they entered a world where nothing had changed for millions of years, until this ship had crashed into the bottom, and when the crashing had ceased and each piece had fluttered to the ocean floor, nothing had moved for over a century. In that world there was no confrontation, no deadline, no court date; there was no time, no season; there was no light, only the purest blackness in which the wreck sat mouldering.

"The first time I saw it," said Kelly, "it was like two o'clock in the morning. Tommy says, 'Look at these slides,' and he put them in the stereo viewer, and there it was, some ribs of the ship. It was eerie. You could see timbers along the side, and there was a deadeye right there. And they had pictures of funny-looking fish, too. One was this white— lord knows what it was, some sort of life form that probably nobody had ever seen before, totally white, you know, just swimming by."

W HEN THEY ARRIVED at the courthouse a few hours later, they saw sitting in the corridor not the trustees of Columbia University, but Tommy's real competitor: a renowned treasure hunter named Burt Webber, wearing his trademark safari jacket and a huge silver-and-gold diver's watch. Webber, forty-five, had gained fame and some fortune in 1978, when he beat John Doering and Seaborne Ventures to the wreck site of the *Concepción* on Silver Shoals off the Dominican Republic. Webber was the real client, the treasure hunter on board the *Liberty Star* who had organized the whole expedition with funding from a group called Boston Salvage Consultants. Three years earlier he had been involved in another project "connected" with the *Central America,* and, as he put it, he had been "tracking research diligently on it for years."

Presiding over the hearing was Richard B. Kellam, senior judge of the Eastern District of Virginia, Norfolk Division. Kellam was seventy-eight. He had been a federal judge for twenty years, and he was sharp enough to keep ahead of a courtroom full of lawyers and cut them off when he thought they were wasting his time.

"Good morning, gentlemen," said Judge Kellam from the bench. "Are y'all ready to proceed?" Not one minute had passed before the lawyers were debating whether Kellam had jurisdiction to listen to anything anyone said about a shipwreck two hundred miles offshore. Kellam decided that as long as Columbus-America could prove it had an artifact from the ship, then he had jurisdiction over the ship, at least long enough to hear the parties argue the issue itself. With that settled, he told Robol to call his first witness.

The lawyers had found Tommy a shiny gray suit, a white shirt, and a gray tie, but they couldn't find shoes to fit him. No one in the group had the same size feet as Tommy. He took the stand wearing the gray suit, the white shirt, and the shoes he had worn in on the fish boat, an old pair of maroon Reeboks.

Robol liked the shoes; they added a certain down-home look, a scientist with more important things on his mind than how nicely he was dressed. After having Tommy describe his work at Battelle, his early interest in deep-water wooden-hulled ships, and the technology his group had developed, Robol presented the judge with several sets of stereo slides of the ship, which Tommy looked at through a hand-held viewer and narrated for the judge. The first showed a piece of hull and the deadeyes used to rig sails. The next was at the edge of what appeared to be the coherent site, illustrating various states of degradation. The third set was of barrel hoops, with the vehicle's manipulator in the foreground.

When Tommy looked at the fourth set, he said, "This is in the bow area of the shipwreck, and of course in all these pictures you can see the proliferation of the anthracite coal, coal obviously being the most prominent component of this particular shipwreck site."

Next, Tommy narrated videotape of John Moore's recovery. "This is one of our manipulators," he began, "that is attached to one of our modularized sections, that is attached to our modular vehicle, that is picking up a lump of coal, that has been refrigerated in seawater, that is here on the desk."

On the television screen, the mechanical arm slowly picked a lump of coal from the pile and with jerky movements brought it up in front of the camera. Even in the poor quality of the footage, everyone could see something growing out of the top of the coal, a delicate, spongelike life form no more than an inch and a half high.

"God knows what the hell it was," Horan said later. "It looked like a tulip champagne glass: pure white, gelatinous—weird stuff. But that little white tulip just cooked their goose."

"The importance of this particular lump of coal," continued Tommy, "is that it has a sessile organism on it, which as you can see survives at these depths. We have refrigerated it, and it seems to be surviving so far."

On the desk in front of Tommy sat a large, wide-mouth bottle filled with cold seawater. In the bottle was a chunk of coal, and attached to the coal was the same wispy white life form, floating upward like a tiny hourglass.

"Where did the coal come from?" asked Robol.

"This coal came from the shipwreck site that we saw the video of, which is the same shipwreck site that this court hearing is about."

On cross-examination by the lawyer representing Boston Salvage, Tommy admitted that the *Central America* was his primary interest, that the *Central America* had a lot of gold on board, that he had disturbed the site but only minimally in the debris field, that the first thing he had recovered was on July 6, 1987. The lawyer wanted to know if he had recovered the items on that day just so he could file in court.

"I think it's fair to say," admitted Tommy, "that the fact that other people were in the area certainly sped up the process."

Next, the lawyer asked about the site, and Tommy described the size of the coherent core and the plumes of debris extending outward two hundred meters. He noted that he had not determined the ship's identity; that he had the present capability to recover artifacts from the ship, whatever it was; and that the confrontation had not been life threatening. That was the end of Tommy's testimony.

When Judge Kellam told Robol to call his next witness, Dave Horan shocked the other side by calling as a hostile witness a man named John F. O'Brien, the president of Boston Salvage. O'Brien and several of his current partners had also been involved in Harry John's search for the *Central America* in 1984. Loveland later described him as "a business promoter kind of guy, so he was a perfect contrast with Tommy."

With O'Brien on the stand, Horan wondered out loud why the *Liberty Star* had left the dock one month after the *Nicor Navigator* and gone straight for the center of the original ten-mile-radius circle Columbus-America had filed on for Sidewheel. But O'Brien said he knew nothing of Columbus-America's claim; he said he had never even heard of Columbus-America until he got a telex from the *Liberty Star* describing an encounter with a group calling itself by that name.

"Have you located any wrecks?" asked Horan.

"We have sonar data that details certain anomalies," said O'Brien. "We are not sure what they are yet, because we have not camera ID'd them yet."

"Unless enjoined by this court," asked Horan, "do you intend to conduct activities within the present injuncted area?"

O'Brien's lawyer objected, but Horan told Judge Kellam there would be no reason to issue an injunction if Boston Salvage agreed not to enter the box. Kellam overruled the objection.

Horan repeated his question. "Sir," snapped O'Brien, "our overall plan calls for a trapezoidal shape out in the ocean that is within that whole seven-hundred-square-mile area, and that box is scheduled to be surveyed. We have gone off that track on three occasions now. We are off that track because the court told us to. If the court tells us we don't have to, we are going to get back on our track, because that's what we contracted to have done."

"With regard to the vessel that you have now seen pictures of," said Horan, "that you have seen videotapes of, within the present injuncted area, what are your plans vis-à-vis that vessel?"

"I could not tell you," said O'Brien, "because I simply looked at a few photographs, which don't necessarily mean anything to me. I don't know if I'm interested in salvaging that. I'm not sure what it is. It hasn't been identified. I'm not sure if that stuff came off it."

"Are you willing to stay out of that present salvage area that is subject to the court's temporary restraining order?"

"I do not believe that I need to make that commitment now."

"I'm asking you the question," repeated Horan. "You can say that you either are or are not."

"We are not going to stay out," O'Brien shot back. "Unless the court instructs us to stay out, we are not willing to stay out."

"No further questions," said Horan.

Columbus-America rested its case, and the defense called the captain of the *Liberty Star,* Ed Sottak. Sottak, with his deep tan, starched white uniform, unfiltered Camels, and gravelly voice, looked so much the epitome of a sea captain that Loveland wanted to take a picture of him.

On the stand, Sottak recounted the running of his early track lines and then the various encounters between the two ships, emphasizing that he had tried to maintain a course that would keep his ship one mile from the *Navigator.*

Horan listened to the captain's testimony and figured, Forget the distance between the ships, that fish is two and a half miles behind you! You feint one way, cross over the other, the ships are still a mile apart; but that fish continues straight on ahead. "He knew where he wanted that fish to be," said Horan. "He just wanted to find out what we were doing underneath us."

Horan cross-examined the captain and got him to admit that even if he knew at all times the location of his ship, he never knew, especially when turning, the position of the sonar fish.

"We don't know how close your sonar fish came to the plantiff's salvage vehicle, do we?" said Horan.

"Exactly, I have no way of being that precise about it."

"And you don't know whether those transponders sticking up off the bottom over there were in fact endangered by that fish coming by with that thirteen-thousand-foot cable, do you?"

"No. I cannot give you a precise position of that cable under the water."

Horan finished his cross-examination of the captain, and Judge Kellam adjourned the hearing at 5:00, instructing the lawyers to be back in court the following morning at 8:45 for closing arguments. After hearing the arguments, he intended to rule quickly.

When the lawyers and Tommy and Barry arrived in court for the second day of the hearing, John O'Brien appeared in the courtroom with his lawyers, but noticeably absent from the corridor outside were Burt Webber, Dr. Ryan, and Captain Sottak. Judge Kellam gave each side thirty minutes for closing argument, then retired to his chambers. When he returned, he issued his opinion from the bench.

Listening to the testimony and the closing arguments, Kellam had become convinced of what Robol had written in law school ten years earlier and what he had told Tommy he foresaw as a likely progression: that American law, as flexible and embracing as it was cast over two hundred years before, had to keep pace with progress in science and technology. And if technology could now take us to the bottom of the deep ocean, then the law had to go there, too. "Though the items are small," said Kellam, "it would seem that all that is required is that it be identified as a part of a particular wreckage, and, thus, it has been done." Kellam relied on one of Horan's Mel Fisher cases to help him ground the intellectual leap in precedent: If he had jurisdiction over some of the artifacts, then he had jurisdiction over the whole ship, and it made no difference if that ship lay outside the territorial waters of the United States; the federal statute gave federal courts jurisdiction over all admiralty claims. "And if the court has jurisdiction of the wreckage and of the plaintiff," continued Kellam, "I would think that it has jurisdiction to protect that wreckage, to see that it is not destroyed."

As he spoke about the need to protect the site, he noted, "The area marked off is small in comparison to the amount of area of the Atlantic Ocean, some three miles by four miles, or twelve square miles. There isn't any question, in the minds of anyone, that the plaintiffs have located some sunken ship. Whether it's the SS *Central America* or not, I doubt if any of us know. They may have a good idea that it is, but the injunction in this case certainly should not, and does not, run to the vessel SS *Central America,* until it is clearly established that the vessel in question is that vessel."

As he neared the end of his opinion, Kellam said he could understand how long cables affected by currents and unpredictable in their travel might interfere with the salvor's subsea operation if dragged back and forth in the vicinity. "I just think that what damage the defendant might suffer, by not being able to drag the fish through this small area of water, is nothing compared with the risk that might result to those undertaking the salvage operation."

Once Columbus-America recovered items of value, Judge Kellam would decide who owned them. In the meantime, the judge enjoined "anyone in the world that knows of it" from interfering with Columbus-America's salvage operation at the wreck site. By 10:10, the hearing had concluded, and Judge Kellam had excused the lawyers.

As Judge Kellam spoke from the bench, the *Liberty Star* was already over halfway between Norfolk and the wreck site. She had left Norfolk in the evening of the previous day, just after court had adjourned.

"They sent the captain of the *Liberty Star* back out there with the ship while the judge is in court comparing final arguments!" said Loveland. "They figured, 'We'll get out there while these guys are still in court.'"

Burt Webber, Boston Salvage, and Dr. Ryan and his sonar techs still wanted a peek inside that little box, just an hour or two for their fish to fly over whatever had the Columbus-America group so excited. Loveland and Kelly surmised that their strategy had been to hustle out to the site before the court granted the injunction, add a couple of hours for the news to reach the *Liberty Star* and for Burt Webber and his crew to verify with the lawyers that, yes, there really was an injunction, and in that time sweep the box with their sonar and produce an image of the site for themselves.

But Tommy had already considered this scenario, and that's why he had spent twenty-two hours on a tossing fish boat and three more in the air in an electrical storm to get to Norfolk without taking his ship off site. When the *Liberty Star* cleared the radar horizon that afternoon, the *Navigator* sat there like a hen on a golden egg, and even the court now warned not to ruffle her feathers.

About five o'clock, Bob Evans called Robol to tell him the bridge had sighted the *Liberty Star* heading toward the box. This time she had nothing in the water and was moving at better than twelve knots. They watched her come within seven miles of the northern edge, just close enough for visual verification, then stop, turn, and slowly head away. They lost radar contact with the *Liberty Star* at fourteen miles, but they had not seen the last intruder.

On the
Galaxy Site

⚓

August 1987

Aᴼᴛᴇʀ Jᴜᴅɢᴇ Kᴇʟʟᴀᴍ rendered his decision, the Columbus-America Discovery Group issued a statement in the form of a news article. It was headlined "SHIPWRECK DISCOVERY OPENS NEW ERA OF DEEP OCEAN EXPLORATION," and the article began, "A shipwreck with many characteristics of the *Central America,* a sidewheel steamship that sank in 1857 with a load of California gold and the loss of 428 lives, has been found off the Atlantic Coast." With the find released, Tommy and Barry flew to Washington, D.C., for an interview with the *Washington Post,* and then on to New York, where they met with reporters from the *Christian Science Monitor* and *U.S. News and World Report.* When they returned to Wilmington a week later, they holed up in a motel, where the phone calls cost less than the ten dollars a minute on the SAT phone at sea. "We were hammering the phones," said Tommy, "calling as many people as we could about different things, engineers, partners, scientists, to make as many connections as we could."

Anyone on the ship would gladly have traded places with Tommy or Barry for a motel room that didn't move. Most of them had been at sea for nearly eight weeks with no break, working odd hours, often getting no sleep. They had to cling to their plates in the galley, or their food ended up on the floor. They had to shower and shampoo—remove the cap, pour the shampoo, and screw the cap back on, all with one hand—while their feet danced a cha-cha in the shower stall. To go to the bathroom they had to brace themselves just to sit down; then they couldn't flush without first filling a bucket in the sink and tossing that into the toilet. "Sometimes it'd take more than one bucket," said Tod, "and when you're in heavy seas and it's splashing out, it can get really nasty."

After two months of this, everyone had had to swallow a crawful of irritation. They were on there with people they wouldn't necessarily have chosen to have as their friends. "But you absolutely have to get along with them," said John Moore, "you have to put up with their irritating habits." Yet Moore would sit down next to Doering at the breakfast table, and Doering would say, "G'morning, John," and Moore wouldn't even acknowledge Doering was there. "He wouldn't even grunt," said Doering.

Doering himself was still grumbling about Tommy's leaving Bob Evans in charge of the tech crew. Bob was one of the youngest men in the control room, the only one besides Milt who hadn't spent hundreds of days at sea, yet he was coordinating every dive, trying to carry out Tommy's instructions for exploring the site, but he might rather have been back in Columbus with his new girlfriend, sorting out their relationship and playing jazz piano at night around the Short North.

On deck, Craft still fumed out loud over his frustration with Tommy. In the control room, Doering fumed over Bob Evans and his "goddamn mud samples," and Moore's verbal fits kept everyone tense. Even Scotty, probably the least judgmental and the calmest person any of them had ever met, ventured that he admired John Moore, "but he's impossible to work with."

To calm everybody down, about every ten days Craft unlocked the beer and a little hard stuff and had a deck party; not much by beach standards, but compared to the monotony and irritation of life on that 240-by-50-foot island, any celebration seemed like Mardi Gras. Craft started the beer chilling in the afternoon, then barbecued hot dogs and hamburgers that evening, and everybody gathered on the back deck in the still warm air of the Gulf Stream. Blue-black waves rolled by from as far out as they could see in all directions, the crew with a light buzz, bullshitting and grabassing, eating chips and dip, and listening as Bob Evans cranked up his keyboard and played jazz. He called it "music to dive by," and one of the mates sometimes joined in with a banjo, and Bryan picked some rhythm and blues on his grandfather's old Gibson. And they floated around in the Gulf Stream making as much noise as they could and no one cared.

"It sounds childish when you describe it," admitted Craft, "but after you've had three or four weeks with nothing going on, it's a welcome break."

Sometimes the parties lasted till long after midnight, the tension dissipating with the day's heat, the stars out and philosophy hanging thick on the deck. The crew liked the schedule with Tommy gone. Craft had them getting up at the same time every day and starting work at the same time and quitting at the same time. And pontificating at night under the stars was an opportunity for them all to beat up on Tommy for pushing them too hard, too far, when he was there. "He assumes that everybody is like he is," said Moore, "that they never get tired and that it doesn't matter whether it's night or day. He has trouble realizing that the other people here don't have the same kind of motivation that he does."

AT TOMMY'S DIRECTION, the tech crew was to film and photograph the ship. They were not to land, not to move in close, not to explore among the timbers and the silt and the piles of coal. Tommy did not want the site disturbed; all he wanted were thousands of photographs and thousands of feet of video.

From the day Tommy and Barry departed for Norfolk on July 13 until they returned three and a half weeks later, the tech crew launched the vehicle fourteen times, but only five of the dives lasted until the batteries ran low late in the afternoon. Sometimes they couldn't launch because the seas rose to seven or eight feet and the wind to twenty knots and lightning cracked bright and jagged all around them. Sometimes they had deployed the vehicle, but a squall line appeared suddenly along the horizon and blew in so fast they were forced to recover early.

Other times, the breeze settled at under five knots and the seas rose no more than two feet but the winch was down or they had electrical problems or the telemetry system failed to process the signals traveling up and down the cable. A major frustration became the vehicle's three thrusters, the eight-inch-diameter, two-and-a-half-foot-long impellers used to propel the vehicle around the site. At eighty-five hundred feet, the pressure would drive a pencil eraser through a hole no wider than two human hairs, .004 of an inch. The thrusters cost five thousand dollars apiece and were the best available, but the rubber seals couldn't hold back the salt water, which seeped through and shorted out the motors. On several dives, the techs surveyed the site for no more than thirty minutes before one of the thrusters stopped spinning. Then they had to rush the vehicle to the surface, tear open the thruster motor, and dry it out with a heat gun or a hair dryer, or throw it into the galley oven before it corroded. They ordered new seals, they machined their own seals, they redesigned the old seals, but nothing worked. The thrusters flooded nearly every other dive, and if one or two thrusters were out, they couldn't survey the site the way they needed to, and the site was massive.

The core seemed mounded at roughly the center of the site, and debris fields extended outward far beyond that, but at most points they couldn't tell where one ended and the other began. Shipworms, the sea, and gravity had turned the once stout timbers into soggy wood and collapsed them beneath their own weight. A thin layer of white silt covered the entire wreck, and fragile sponge life, like white tissue paper, undulated hypnotically in the gentle current. A few rattail fish, a foot or two long, it was hard to tell, crossed back and forth over the debris, and a small number of crabs and sea stars sat sprinkled among the wreckage.

The main video camera, the SIT, was black-and-white and the picture grainy, but it worked well in low light, and they could see parts of the ship coming into view far sooner than with the other cameras. Moore used that camera to guide the vehicle closer for the still photos. Doering worked next to Moore, sometimes snapping pictures from his own console and taking notes. "John Doering's seen more shallow-water ships than probably anybody around, and he knows what he's looking for," Tommy had said. "He's got a draftsman's eye, but no one in the world had ever analyzed a deep-water, wooden-hulled, historic ship. It takes a lot of people from different perspectives to interpret a ship like that."

When the vehicle came up, Doering took the film from the day's run and developed it in color strips. After the strips had dried, he laid them on the light table and studied each frame. Earlier, he had photographed the nineteenth-century drawings of the *Central America* and superimposed the drawings over the sonagram of Galaxy. "It looked good," he said. "And when we got down there, it still looked good."

For the three weeks he joined the crew that summer, Milt Butterworth worked with Doering in the photo lab, developing film and interpreting the photographs. "We used to invite virtually everybody in to look at the photographs, in case I missed something," said Doering. "But Milt was the greatest one. He'd look at the photographs and he'd say, 'Look at this shot, this is something,' and I'd look at it, 'Nah, don't worry, that's nothing.' 'Yeah, yeah,' he'd say, 'that's it, that's it!'"

But that was all they could do: study photographs of things underwater, bent and broken, decayed and rusted, encrusted with sea life, distorted by white sediment, subject to interpretation and misinterpretation. They couldn't land to confirm doubt or suspicion; they couldn't touch anything. First they had to understand the site, and when Tommy returned they would select areas to scrutinize and venture into slowly. Until then, they flew back and forth, collecting thousands of photographs, and continued to debate what they saw.

As Doering studied the monitors from his place next to Moore, and as he developed and analyzed more and more of the pictures, two things concerned him. One was the paddle wheels. Flying at thirty feet above the site, they saw nothing that looked like the iron-spoked behemoths that had hung from the sides of the steamer. "That was the big thing," said Doering. "Where are these damn paddle wheels?" The other was the size of the site.

The size of things seen underwater could be deceiving, and the ship was so deteriorated they still weren't sure where the bow began and the stern ended; but it seemed short. On one run, Scotty marked a reasonable point for the bow and the stern in his navigation system, Moore flew the vehicle from point to point, then Scotty calculated the slant ranges. "The vehicle is not traveling ninety meters," confirmed Scotty. "We're traveling a much shorter distance."

EARLY THE MORNING of August 8, Tommy and Barry returned to the *Navigator,* accompanied by Don Hackman, back for his second stint. The *Washington Post* and the *Los Angeles Times* and every other publication that had run articles on the Columbus-America Discovery Group had reported favorably on the group's interest in the science and the archaeology of wooden-hulled ships found in the deep ocean. Tommy seemed refreshed and encouraged by the judge's decision and the public's response, and he was ready to spend another two months at sea.

But Craft took Tommy aside and suggested that a crew rest would be, uh, desirable. He pointed out that for the past three and a half weeks Tommy had been on dry land, where the ground didn't move, where diesel fumes did not congregate and thrusters did not vibrate and one could stroll for more than a hundred feet and not see the same thing. A little R&R back at the beach for the rest of the crew would in Craft's words be "very desirable." When Tommy had had time to think about this, he figured it might also be an opportunity to invite some of the partners down to Charleston for a face-to-face progress report. Eventually, he not only agreed to head in, but even proposed flying the crew's families to Charleston and putting them up in a nice hotel. On the afternoon of August 12, the *Seaward Explorer* arrived to stand watch over the Galaxy site, and late that night the *Navigator* was bound for Charleston.

By 0815 a day and a half later, Burlingham had the *Navigator* secured to the Union Street wharf; families waited on the dock, and Don Craft and his wife, Evie, announced that the first party, lunch at Henry's, was on them.

Henry's was the oldest restaurant and bar in Charleston. It oozed with the redolence of the Old South, like the tables and chairs had all been marinated in bourbon. That afternoon, Henry's was stuffed with summer tourists tasting she-crab soup and sipping mortar-muddled fresh mint juleps. In the middle of the hullabaloo stood a piano, an old upright about chest high that the servers stacked with doilies, dishes, and silverware. The Columbus-America group had been seated for no more than a few minutes when Bob Evans disappeared. All they could see was his white straw Panama hat with the black band floating through the crowd. Evie watched him sit down at the old piano.

"He picks these doilies up and sticks them on top," said Evie, "just like he was in his own living room."

Bob hit a few keys and then a few chords and people started look-ing at him. The piano was not in tune, but it was close enough. Those near the piano noticed somebody tickling the keys, and the talking slowed in ever widening circles, until the room had grown almost si-lent, and suddenly the intensity of those weeks of sliding dinner plates and leaky thrusters all exploded out of Bob's fingertips into Scott Joplin's "Maple Leaf Rag," in a way, according to witnesses, that would have made Joplin weep for joy, and the good-time alarms set inside all of those tourists sipping mint juleps started going off at once.

"The place went crazy," said Evie. "He played the most gorgeous stuff you ever heard in your life."

Bob hit the keys on that old piano like a man exorcising his demons, that long blond hair shaking down his shoulders, those short thick arms pounding the ivory faster than the eye could see. Henry's piano and Bob's Panama and Scott's music and the ambience of Charleston all melted together that afternoon and whipped everyone there into a fine south-ern froth. "It was just absolutely perfect," said Craft.

When the yankee from Ohio skidded to a finish and punched that last high note, it wafted upward into the rafters, and the room erupted in applause and shouts of "More!"

"The entire room started applauding," said Craft. "They wanted him to stay on forever."

That was the beginning of three days of partying in Charleston for the Columbus-America crew and their families. Tod's sister Paula orga-nized a party the first night at the East Bay Trading Company, more lunches at restaurants in the historic area of old Charleston, and a catered picnic, where rain drove them back under a gazebo. Milt brought along

slides of the trip and projected them for everyone to see, and a few partners flew in from Columbus.

While the Columbus-America crew partied with their families
and the *Navigator* hugged the Union Street wharf, a boy of two lay in
the Intensive Care Unit of Charleston's Roper Hospital, suffering from
a respiratory attack caused by pesticides. From the window of the ICU
waiting room, the boy's father, a brick mason named Wally Kreisle,
could see the *Navigator* down at the wharf. He knew it was Tommy
Thompson's ship, and the sight disturbed him. For the past six years,
Kreisle had been in a dozen libraries up and down the East Coast researching the *Central America*. He had become obsessed with finding
the fabled steamer. Now he had investors and a ship, a deep-water
side-scan sonar and a contract with Steadfast Oceaneering, the deep-
ocean search and recovery company used exclusively by the United
States Navy and controlled by Bob Kutzleb. Kutzleb was one of the
two deep-ocean experts who had come to Columbus at the invitation
of an investor two years earlier to grill Tommy. The investor had given
Kutzleb Tommy's proprietary and highly confidential concept paper
on the *Central America,* including how to find it with the most recent
generation of side-scan sonar. Kutzleb had signed a confidentiality
agreement, but only after first crossing out some of the clauses. With
Kutzleb's sonar experts behind him, Kreisle was ready to launch his
own expedition. He had read in the paper about the court's awarding
the ship to Columbus-America, but he had gained too much momen-
tum to stop now; he was waiting only for his son to recover before he
set to sea to find the *Central America*.

Since the *Liberty Star* appeared on the horizon late one night in early
July, Tommy had been preoccupied with battling intruders and dealing
with the media. At the top of his priority lists were legal concerns and crew
concerns and strategy. He had spent little time in the control van, watching flyovers of the site and analyzing data. Although his purpose at sea
that summer was to find and identify the *Central America,* not until late
August could he concentrate on the science and the engineering and the
technology to accomplish his goal. While the crew partied in Charleston,
Tommy ducked out of celebrations early and stayed up late to review the

video and the photographs they had shot in his absence. He spent long hours talking with Bob. He now felt he understood the site well enough to begin forays into the debris to search for signs besides an abundance of coal that would confirm: This is the *Central America*.

Identifying the ship was crucial, because treasure occupied an area no bigger than a small closet; if you were on the wrong ship, you could spend months searching for something that did not exist. If you identified the ship, the search for the treasure could be focused and likely places explored and eliminated until you found it. Tommy wanted to verify that the debris on this small patch of ocean floor had arrived there 130 years earlier when the *Central America* came to rest. Then he could begin a methodical search for the treasure.

Early the morning of August 20, the *Navigator* was back on site and the techs had launched the vehicle. Coming in for their first landing up close to the wreckage, Moore eased the vehicle onto the ocean floor. The fine silt that had collected for over a century rose into a cloud.

No one knew what the *Central America* should look like after 130 years at the bottom of the sea: all its timbers and rigging and boilers and steam engines and cast-iron fittings and giant sidewheels and the attendant flotsam and jetsam of six hundred passengers and crew. Tommy and Bob had modeled the site every way they could imagine. They had models in which teredo worms had eaten all of the wood; models in which the iron had wasted away to ions and disappeared; models in which neither had occurred and the ship sat upright virtually intact; models in which both had occurred and the site was almost nothing but coal. And when the silt settled, that's what they saw. "We thought we were looking at probably what happened," said Tommy. "There was wood there and there was iron there, but most of the iron and wood was gone and mostly what remained was coal."

In the background they could see a few broken and rotted timbers, dark and flecked with rust, but so much coal surrounded and covered the timbers that even at this lower angle, they could hardly discern the line of the hull. Bob had studied the architecture of the *Central America,* but the crisp drawings from exploded views of the mid-nineteenth-century sidewheel steamer looked nothing like the worn, decayed, funny-angled, broken, silt-covered, coal-strewn scene in front of them.

What struck everyone was the violence evident at the site. "At Side-wheel," said Tommy, "we couldn't see the violence. It was just sitting there like a rowboat, with nothing in it. On this site you get more of a sense of, Wow, this really was a disaster."

Tommy had one rule in the control room: No one was allowed to say out loud what he thought he was seeing. As soon as somebody said, "It looks like an anchor," then everybody else would see an anchor. Perceptions produced better results if one wrote, "wine bottle"; another wrote, "whistle"; another figured, "piece of pipe"; another thought, "It looks like a ladder"; another wondered, "Is it a lantern?" "If people get to talking about it," said Tommy, "they all arrive at a conclusion, and the conclusion has more to do with the social dynamics in the control room than it does with the reality of the situation."

A black timber embedded in the silt might look like a hole, or a hole might look like a blackened timber. Colors and shapes and textures could all conspire to make a collection of small coal pieces and bits of wood appear to be a decanter or an old iron plate or a mottled bar of gold. "A number of times we thought we saw gold mixed in with the coal," said Tommy, "but it wasn't obvious it was gold." They couldn't be sure of anything until Doering had developed the film from the higher-resolution still cameras and they had studied the pictures and gone back for a closer look.

Craft was in the control room when they saw the first piece of evidence, other than coal, confirming they were on the right ship. John Moore had landed the vehicle a few feet from what they thought were remnants of the superstructure. "We looked at the plates that the shrouds of the masts were secured to on the side of the ship," said Craft, "their configuration, the spacing, the general location of each set of plates; and they matched the *Central America* diagrams."

A few days later, Moore lassoed an anchor and brought it to the surface: one thousand pounds of forged iron with curved flukes taller than a man. It matched the size and configuration of one of the anchors the *Central America* had carried.

As he piloted the vehicle on the bottom, Moore watched two monitors, each at a different camera angle. To understand depth on his two-dimensional screens, he studied the shifting of shadows and the crossing of objects in the landscape. "If you want to see genius in action," said Craft, "watch John Moore take a manipulator arm with metal fixtures and pick up a piece of ceramic. The jaws are strong enough to crush your arm, yet there was not one piece of pottery that came aboard that had any evidence that Moore had even nicked the glaze."

During the weeks Tommy and Barry had been ashore, Moore had worked to refine the performance of the manipulator and gradually had coaxed it to precision. On the first dive after Tommy's return, he retrieved a bottle and an ironstone plate. On the next dive, he picked up two earthenware jugs. On another dive, they spied in the debris field what appeared to be a tiny porcelain jar with a lid. Moore landed near it, reached out with the manipulator, and tried to pick it up, but the lid slipped off. He retrieved the jar first, then went back and recovered the lid. When they got the vehicle to the surface, they saw that the porcelain jar was only an inch high by three and a half inches in diameter. Inside was a grayish cream. "What in that day," said Craft, "would be the equivalent of a lady's cold cream." In the cream were two things: a fingerprint and a strand of chestnut-colored hair.

As the artifacts came aboard, Bob studied them to determine when they were made. He kept several reference books in his lab, and he often called Judy Conrad, their historian back in Columbus, to track down trademarks, periods of design, pottery styles, and registry dates.

The *Central America* had sailed for only four years, from October of 1853 to September of 1857: Among the artifacts they recovered were two white ironstone mugs marked with the symbol of John Maddock, who made shipboard dinnerware from 1842 to 1855. They found an ironstone dinner plate by John Wedge Wood, who had produced his dinnerware from 1841 to 1860. The design of an oval Elsmore and Forster serving platter they recovered was first used in 1853. During one dive, Moore recovered a child's white mug engraved with pictures and maxims from Ben Franklin's *Poor Richard's Almanac*: "God Gives All Things to Industry" and "Diligence Is the Mother of Good Luck." Franklin dishes for children were popular in the early and middle nineteenth century. Moore plucked from the debris two two-gallon jugs hand painted with blue floral designs. One was pear shaped, which was popular in the 1850s and earlier, and the other was cylindrical, a style that began in the early 1850s, as potters began to mass-produce their wares. "Two different items of stoneware crockery," noted Bob, "and the 1850s was the transition between the two." The Edwards family of Burslem, England, manufactured much of the dinnerware aboard American ships, and Moore brought to the surface an Edwards saucer with a British registry mark and a registration date of July 18, 1853.

The period artifacts impressed everyone but Doering. He didn't even care about the anchor or the shroud plates. While Tommy was back on the beach, Doering had decided that because there were no paddle wheels, this could not be the *Central America*. Since Tommy's return and their closer look at the wreckage, he had seen nothing that changed his mind.

Tommy did not doubt Doering's eye, but even Doering had not seen a ship at these depths. And Doering was a treasure hunter, quick to pass judgment. Tommy was an engineer with a scientific bent and intent on following a reasoned course. He had researched how fast the iron of the paddle wheels would corrode in seawater, and experts had advised him that for each year the iron lay on the bottom, so many mils of thickness would return to the water as iron ions and drift away; multiply that times 130 years, and most of the iron would be gone. Already they had found rusted chunks of iron that were little more than deep-sea mirages; when Moore reached out to touch them, orange and brown smoke filled the water and the chunks vanished. How long ago had the deep ocean reduced the engines and the sidewheels to a thin shell that soon disappeared in a briefly roiling cloud?

If the paddle wheels had not corroded into hollow wisps of orange and disintegrated, perhaps they had spun off—on the way down, when they hit bottom; nobody knew. But all of Tommy's research and experts would not dissuade Doering.

Not only did the absence of paddle wheels bother Doering, he still thought the ship was too short. Tommy knew there was some question about the length. "But we didn't have enough information to make any valid judgments," he said. "We couldn't even see the site." If they could, without doubt, identify the bow and the stern, and if Scotty measured the distance from one to the other as considerably greater than 280 feet, then Tommy would be worried about length. Little could account for a target being longer than the *Central America,* but many theories could explain a shorter ship. The stern castle of the *Atocha* had ripped loose and floated off with tons of bronze cannons aboard and swept nearly ten miles across a shallow sea. The *Titanic* had snapped in two on the way down, and it had a steel hull. The *Central America* could have broken in half and the engines fallen out. It could have broken in half and the engines remained in one half, and that half could have gone straight down and the lighter half floated on downstream. Maybe one of the smaller targets they had seen with the SeaMARC was the rest of the ship. Hackman had another theory: that the ship had descended to the bottom at a severe angle and plowed into the ocean floor, sticking half upright in the bottom; then it had degraded straight down, making the ship appear shorter.

"So that wasn't negative proof," said Tommy. "All that it was saying is, We don't know that the length of the site is a significant issue."

Negative proof would be a plate dated later than 1857, or a stanchion not used in ship architecture till after the Civil War, or a bottle whose design they could trace to later in the nineteenth century.

"Yet artifact after artifact kept going period," said Bob. "'This is the right period. Look at this thing; this is the right period again!' How many period sites could there be with women, children, iron, coal, and a wooden hull? We even did a probability analysis that said there was a less than ten percent chance of that not being the *Central America.*"

But by September, one thing bothered Bob: The coal seemed too plentiful. The *Central America* had left New York with a full load of coal. The big sidewheel steamers burned about fifty to sixty tons of coal a day, and Bob estimated they kept a reserve of about two days. The engines stopped four days short of their destination, so that left about three hundred tons on board. The Galaxy site appeared to be covered with far more coal than that, but with the limited viewing range of the cameras on the E-vehicle and no way to move some of that coal to see what lay underneath, Bob couldn't be sure of anything.

"I had all these doubts," said Bob, "and yet was pretty sure that we still had it. But I couldn't figure out the form of the site well enough to even decide where to begin digging. I've got this big pile of coal down there, but where's the gold?"

WALLY KREISLE WAS a big man, about 260. He suffered from intense migraines, lightning bolts exploding inside his head, and he was in no mood to be trifled with. Tommy Thompson might have an injunction, but Kreisle wasn't about to let something as piddling as a federal judge's decree stand in his way. Before leaving port in late August with Kutzleb's sonar crew from Steadfast Oceaneering, Kreisle announced in the Georgetown, South Carolina, paper that the Columbus-America Discovery Group was on the wrong wreck, and that he was going to recover the real *Central America* in water one hundred feet deep about twelve miles off Cape Hatteras; he had already seen her sidewheels, and he had brought up pieces of steam pipe and copper sheathing from her hull. Then Kreisle left port aboard the *Cameron Seahorse* with a side-scan sonar capable of imaging a wreck at eight thousand feet and sailed straight for the Galaxy site two hundred miles out.

Rick Robol knew that Kreisle was mounting an expedition and had served him with notice of the court's injunction before he left the dock. When the *Cameron Seahorse* arrived near the site on August 29, Burlingham radioed her captain and described the box that no one else could enter. Then Burlingham asked for the name of the charterer, but the other captain refused to say. When Burlingham asked to board the *Cameron Seahorse* to serve the papers himself, the captain replied, "I don't see any necessity of you coming over here. I understand the boundaries, and if we at any time come into that area we would be more than happy to receive those papers from you." Then the *Cameron Seahorse* set up five miles to the east and began running track lines north to south, each track creeping closer to the eastern edge of the box.

A few days later, Burlingham notified the other captain that the *Nicor Navigator* would be temporarily away from the site to rendezvous with a supply boat and warned him not to enter the injuncted area. The following morning, the *Navigator*'s radar picked up the *Cameron Seahorse* angling inside the eastern boundary of the box, then heading out and continuing on her track line.

Three days later, the 5th of September, Burlingham's first mate was patrolling the injuncted area along the northern border. Overnight, the weather had risen, the wind coming in at twenty-five knots and the seas cresting at nearly ten feet. With seas too rough to launch, the tech crew planned to leave the vehicle on deck and run about eight thousand feet of cable off the winch to unkink the "assholes." Earlier, the *Cameron Seahorse* had passed two miles to starboard, steaming north; when she reached a point three miles off the northeast corner of the box, she swung 180 degrees to port, now headed south, and appeared ready to skim the eastern boundary. But suddenly she altered her course to the west and was now angling toward the box, headed for its center, and flying the colors and shapes of a vessel with an object in tow.

Burlingham had nothing in the water. He had assumed that no one who knew anything about running deep-water sonar searches would have a tow fish in the water in such weather. "Their data had to have been shit, as far as I can tell," said Burlingham. With the *Cameron Seahorse* only one thousand feet from the boundary, he radioed the captain and asked his intentions. The captain replied, "I intend to hold my course and speed."

Burlingham told him that if he continued on his present course, he would enter the injuncted area, and that if he entered that area, he would be held in contempt of court.

"Roger," the captain replied.

Two minutes later, the *Cameron Seahorse* entered the box a little over a mile to the northeast of the *Navigator*. Burlingham turned to the southwest, took up position over the site, his stern aimed at the *Cameron Seahorse*, and cleared the bridge of everyone but Tommy. Then Burlingham heard Tommy instruct Craft to hook an empty aluminum cube to the cable, throw it overboard, and run up the RAM colors and shapes. Burlingham was furious. "You can't change it! Once you have achieved status, you maintain that status!" Tommy argued that unkinking the cable had already been planned for that morning. "We have no obligation to reschedule our work requirements to conform to another vessel's movements!"

This time, Craft sided with Tommy. "I got in a real argument with Don," said Burlingham, "and he put that sonofabitch over anyways, and he knew I was pissed."

The *Cameron Seahorse* had closed to within two thousand feet of the *Navigator,* which was directly in her path. The captain radioed Burlingham and asked for his intentions.

"Harvey wanted me to get in his way," said Burlingham, "physically get in his way. Hell, he wanted me to almost ram 'em if I could."

Burlingham told Tommy, "I'm not gonna hamper 'em. That is grossly against the Rules, and that's what they're gonna look at with me. They're gonna say, 'It's in black-and-white, here's the book!' I have to stay out of his way."

He stood until he could stand no longer, then he radioed the *Cameron Seahorse.* "Be advised, I'm going to peel off to the west now just to ensure no entanglement of the equipment." He reminded the captain of his promise to allow Burlingham to serve the papers if the *Cameron Seahorse* entered the injuncted area.

"I said that with sincerity," replied the captain. "However, I've been advised not to allow anyone to approach this vessel."

Burlingham was livid that the captain had lied to him, and that he had so blatantly defied the court order. The *Cameron Seahorse* veered slightly to the east away from the *Navigator,* and Burlingham crabbed to the west to get out of her way. Both crews stood on deck watching the other through binoculars as the two ships passed no more than three hundred feet apart. An hour later the *Cameron Seahorse* had exited the box on a southerly bearing and was nearly two nautical miles from the *Navigator.* Burlingham watched the *Cameron Seahorse* slowly begin a wide sweep to the west and knew that the captain was getting ready to swing back in a crossing pattern that would cut an X through the box.

Burlingham waited until four o'clock that afternoon, just as the *Cameron Seahorse* was coming out of her turn, to broadcast a sécurité call requesting that all vessels in the area give him a berth of over eight thousand feet.

Ten minutes later, he called the *Cameron Seahorse* and asked that the charterer and a representative of Steadfast Oceaneering be present in the wheelhouse. "As it appears again that you are going to enter the injuncted area," he announced, "I'm going to read you a copy of the order granting preliminary injunction."

Suddenly, a different voice from the *Cameron Seahorse* broke in. "*Nicor Navigator, Cameron Seahorse*. This is the charterer, Wally Kreisle, K-r-e-i-s-l-e. Be advised that this vessel and the crew aboard are working under my direction. Also be advised that we are going to search the area that we are entering now and the area that you are describing."

"Roger, understand," said Burlingham. "I am now going to read to you the order granting preliminary injunction." He read the first paragraph. "How copy?" he asked Kreisle.

"Copy that just fine," said Kreisle, "but I don't understand any one part of it. I don't know what else to tell you, bud."

"Roger," said Burlingham. He read the next paragraph from Judge Kellam's order. "How copy?" he asked again.

"Copy just fine, don't understand any part of it," said Kreisle. "I've explained to you what this vessel intends to do and what I've paid for this vessel to do, and I expect that to be carried out to its fullest."

Burlingham signed off. Fifteen minutes later, he got back on the radio with the captain. "Look," he said, "I'm not recording this call. This is between you and me. You are going to be in deep shit if you come into our injuncted area."

He had the *Navigator* idling on the southwest corner, her bow pointed to the northeast, waiting for the *Cameron Seahorse*. The cable was still in the water, and this time Burlingham agreed to fly the colors and shapes of a vessel restricted in its ability to maneuver. He could stand fast or move left or right, and the other captain had to steer clear.

The *Cameron Seahorse* came in from the west about midway up the box on a bearing almost due east, her bow rising and falling in heavy seas. As she began to close, Burlingham sat motionless, and the other captain did not deviate. Tommy and Bob were in the wheelhouse with Burlingham, when Robol called. Bob went aft to the COM shack to take the call. Except for pitching in near ten-foot seas and a twenty-five-knot wind, the *Navigator* sat motionless in the water. Bob could look out beyond the heaving stern and see the bow of the *Cameron Seahorse* aimed at the *Navigator*. "He's come around," Bob told Robol, "and he's about ready to run right through the injuncted area again."

"I understand," said Robol.

"I don't know if you do understand," said Bob, "but I certainly hope you understand. This guy is running right up our ass!"

"I understand," Robol said again. He told Bob that even though it was the Saturday of Labor Day weekend, he had a hearing set for that afternoon and the judge was aware of Kreisle's actions.

"Good," said Bob, "'cause right now, I'm looking out the back window here, and I can practically read the name on his boat." Bob hurried back to the bridge.

The *Cameron Seahorse* continued to close, and Burlingham decided it was time to make his move. "Okay," he said, "which way do you want me to take him?"

Because of currents and the layout of the wreck site, Bob said, "I'd rather you take him to the north."

"No problem," said Burlingham.

Burlingham angled to port, forcing the other captain to deviate from his intended track. The other captain radioed Burlingham, "C'mon, can't you do anything else?" Burlingham said, "No, sir," and continued sidling to port, and the other captain had to veer away.

About six o'clock that evening, Burlingham pushed the *Cameron Seahorse* out the north end of the box. The tow fish, they estimated, followed two hours later.

Not only was it Saturday night of Labor Day weekend, it was also Judge Kellam's fortieth wedding anniversary. Phoning the judge's home, Robol apologized profusely for the intrusion and explained that the judge's injunction had been blatantly violated at least three times by a second group of interlopers and that a confrontation at sea was approaching crisis. That night Kellam granted an order to show cause, demanding that Kreisle and all other parties aboard the *Cameron Seahorse* appear in court Tuesday morning. Immediately, Robol faxed notice of the order to Kreisle's ship and to Kreisle's lawyer. When the *Cameron Seahorse* arrived in Norfolk, a federal marshal and a private investigator met Kreisle at the dock and confiscated all of his navigation and sonar records and turned them over to the court under seal. On Tuesday, Judge Kellam held Kreisle and everyone else on board the *Cameron Seahorse* in contempt of court.

* * *

At the end of the previous summer, Tommy had negotiated another option with Mike Williamson: If for any reason, Tommy needed to do more sonar work at the end of the '87 season, he could exercise a first right of refusal on the SeaMARC in September.

Tommy now had three reasons to exercise the option: One, he wanted to protect the injuncted area from Kreisle, who was unpredictable and already had defied one court order. Two, since Galaxy was in the last track line they had run to the east, he wanted to know what both Kreisle and Burt Webber had been finding out beyond Galaxy; he worried that one of them might find a cultural deposit and claim it in court just to operate near the Galaxy site and pounce on Galaxy as soon as Columbus-America left. Three, he didn't know how much of the *Central America* had broken up and scattered as it sank. "It was hard to interpret what we were seeing at Galaxy," he said. "To the east could have been chunks of the ship."

By now it was almost mid-September, and the weather window in the Atlantic was about to close. Burlingham moved the *Navigator* to a shipyard north of Charleston to beef up the winch, load twenty-one thousand feet of new co-ax cable, and pick up Williamson and his sonar team. From September 12th to the 26th, they searched another five hundred square miles of deep ocean to the east and south of Galaxy.

During the search, Tommy hardly showed his face outside the Elder van where he had his SAT phone and his bunk and where he took most of his meals. Over the entire summer, he had been able to study the wreck site at Galaxy for only two and a half weeks, and in many ways, he knew little about it. Although the vehicle surpassed anything tried in the deep ocean, it was still the emergency vehicle they had sailed with back in June. It was not designed for intricate or heavy salvage. "If we'd had a more advanced vehicle," said Tommy, "and we hadn't had all the interference from the competition, I think we would've found gold." But they didn't have a more advanced vehicle and the interference had continued nearly all summer, and they had found nothing of commercial value. Tommy was already planning for the '88 season.

They had spent all summer at sea, from one side of the weather window to the other. And after learning with their own eyes that the sonagram of Sidewheel had deceived them; after discovering what a shipwreck with a debris field really looked like; after all of the fighting and tension and maneuvering at sea; after all of the fighting and tension and maneuvering in federal court; after proving they could go down eight thousand feet to the ocean floor and work for long periods of time, which no one else in the world had ever done, and find coal in abundance and artifacts from women and children and ceramic dishes and jugs and bottles from the mid–nineteenth century; after four months at sea, they had found no gold, and the site had yielded nothing that confirmed or denied that all of the battles in court and all of the battles at sea had been fought over the *Central America*.

Columbus, Ohio

Fall, 1987

TOMMY HAD NOT wanted to go to sea in May. He had wanted to stay in Columbus, let Moore and Hackman and Scotty produce a vehicle that could do all of the wonderful things he envisioned, drop that vehicle in the water for the first time in August, test it at sea for weeks if necessary, then head for Sidewheel. If Sidewheel wasn't the *Central America,* and he had projected it might not be, they would search the sonagrams again and head for the next most promising site. When they found the *Central America,* he wanted to study the ship, film it, photograph it, understand it, and bring back a significant amount of gold. He would finish the recovery the following summer. Instead, the rumor of competition had forced him to sea long before the vehicle was ready, and the reality of competition had kept him there, fending off interlopers instead of concentrating on the science and the engineering and the technology. He had depleted his funding by having to rush production of the vehicle, having to fight the legal battles in court, having to pick at the site with a vehicle that could not document and recover the artifacts as they needed to.

"Tommy was tired," said Buck Patton, "and the project was seriously short of money."

That season had cost Tommy most of the $3.6 million allocated to the entire Recovery Phase. He had saved some of the funds from the Search Phase the year before, and when he combined that with what remained in the war chest from '87, the partnership had an emergency fund of almost a million dollars. But $4.5 million was gone, and in some of the partners' eyes Tommy had recovered nothing but a bunch of dishes and one infamous lump of coal. Some said, "That may be the most expensive piece of coal in history." "Will it burn?" asked others.

"A lot of coal jokes," remembered Patton.

Tommy knew that some of the partners, those close to the project, the ones who helped him with advice, realized the significance of what they had achieved that summer. Two milestones: First, they had done things with technology on the bottom of the deep ocean that no one had ever done before; second, a judge had accepted their novel interpretation of jurisdiction and granted them the exclusive right to work at their site; and when other intruders challenged that right, the judge had enforced his injunction with a contempt order. But no matter how extraordinary the accomplishments, the partnership also wanted to see gold, and there was none.

"You would think that after our performance and after what we won and how we persevered and how well our team worked and how we beat these guys," mused Tommy, "you would think it would just be the most heroic situation you could ever imagine, and yet that doesn't necessarily translate to an investor. The way the typical partner looks at it is, 'Wow, I didn't realize we were gonna have all this legal trouble. It's starting to look more risky.' So we got back that fall, totally exhausted, spent physically and mentally, and now the hard part starts. Now we're in an uphill battle to raise money to build the full-up vehicle."

The most important thing Tommy could do was communicate with his partners, but he would have to word his letters carefully. At sea, Tommy had had little time to communicate with anyone. He had sent each partner a couple of brief missives that he worried had confused them even more, because he couldn't tell them everything that was happening. "And it's hard for them to identify with you," said Tommy, "if you haven't been able to tell them the strategies that went into everything." He wanted to convey the struggle they had experienced at sea, so the partners could relive some of the adventure; he wanted them to understand the difficulty. But he had to be careful not to make the problems seem insurmountable; and without making promises, he had to tell them what he planned to do the following season.

For insight into the mind-set of the partners, Tommy talked frequently with Wayne Ashby and a few other partners with whom he had developed a rapport. When he wrote a letter, Tommy showed it first to Ashby and asked for his comments. "We worked together on a lot of them," said Ashby. "He'd call and ask me if he was explaining it the right way, should it be expanded, does it need to be mentioned at all."

Two ticklish issues needed explaining: How a world-class sonar team had, with great confidence, selected the wrong site; and why he needed more money to build the full-up vehicle to penetrate all of that coal and prove that Galaxy was indeed the *Central America*. Tommy wasn't surprised at the error on Sidewheel, because no one had ever done this before and he understood the difficulties. Anyone in the deep-ocean community would understand. But would the partners? If he tried too hard to explain why it was so difficult to do anything on the bottom of the deep ocean, the partners might think, Well, if the experts don't know, who does? And if the experts don't know, should I put more money into it?

In October, they mailed a ten-page letter to the partners, reviewing the setbacks caused by Tommy being forced to implement the E-Plan and the major achievements despite those setbacks. Based on the encouraging sonagrams of their original target, Sidewheel, they had expected to find a relatively intact steamship with sidewheels and hatchways. Instead, wrote Tommy, they found "a scene of great corrosion and abundant life eating away at the shipwreck, attacking the wooden hull walls until they slowly collapsed." Moving on to what they then realized was a far more promising target at Galaxy, they had recovered several artifacts from the debris field, but with the limited capacity of the E-vehicle, they could not dig through the coal.

"Although we were not equipped to perform major excavation in the main portion of the wreck where the gold is located," explained Tommy, "the importance of our progress in developing a range of capabilities with the emergency system cannot be overstated.... This will be of immense value over the winter as we complete the development of the full-up recovery system and prepare to embark on our intensive recovery effort next spring."

In November, Tommy held a partnership meeting at a downtown hotel, where the team displayed the artifacts they had recovered from Galaxy: ceramic dishes, bottles, and jugs. In one large aquarium a lump of coal supported the child's mug etched with the Ben Franklin maxims from *Poor Richard's Almanac.* The jar of lady's cold cream with the chestnut strand of hair sat under glass, preserved by the smoke of dry ice. Striking color photography hung on the walls, and television monitors ran tapes of the manipulator arm plucking the mug from the sediment and recovering the anchor. In his presentation to the partnership, Tommy discussed the 1987 operations at sea, elaborating on their technological and legal accomplishments, and he showed the partners stereo slides of the site and video of the vehicle working on the bottom. After the meeting he summarized the state of the partnership in another encouraging letter. "We have carefully analyzed the massive amount of data gathered during our operations last summer," he wrote. "With the operational problems well in hand, there remains one major factor that could work against our ability to recover the gold in 1988: raising the funds in time to let contracts to build the 'generic' recovery system." He told the partners they would need at least another $3.5 million to go forward with the next season. But few were quick to sign another check.

For partner Art Cullman, the summer of '87 was everything but what he had expected. When Sidewheel looked nothing like a sidewheel steamer and Tommy had to move on to another site, Cullman got worried. "By then," said Cullman, "most of us were a little bit shaken."

Partner Mike Ford saw two problems, one compounding the other. "Every-one was following like ducks in a line," he said, "until there was equivocation on the target. Then I think some people were momentarily dis-illusioned, maybe thought, 'My idol has failed me.'" At the same time, their idol was asking them for more money. But Ford viewed it philosophically. "In entrepreneurial ventures you always have tests; you have the pro forma and the actual, and invariably something happens in the actual which tests the mettle of the managers. It's how they handle that that really determines what the group is made of."

Others agreed. They had done time on the hot seat themselves, fending off competitors, dealing with surprises and crises, and generally watching their whole operation hang in the balance. So they could see it from Tommy's perspective, and they knew Tommy would either get through it or he wouldn't. "Business people are used to letting somebody else run with their football," said Jim Turner. He and others thought, Tom, you work it out and let me know what happens. But still the partners talked.

"When something starts to go wrong," said Buck Patton, "the phone rings off the hook in this city. 'What do you think? Are you going into the next phase? I think we need to have a meeting, or maybe get together for a drink.'"

Late that fall, eight of the partners gathered at Patton's office to talk with Tommy about their concerns. Money was one issue, but they also worried that Tommy tried to do too much, that he wouldn't delegate. "He feels he has to do everything," said Fred Dauterman, "otherwise it won't get done right." Patton wondered if they could persuade Tommy to limit the number of things he was trying to juggle: law, publicity, contracts, accounts payable, taxes, bookkeeping; let someone else handle those things.

But Tommy was launching a company that had a long start-up phase. There was no revenue, nothing sold, no transactions. Nothing yet could be systematized. That's how Tommy perceived entrepreneurism: Plow new ground, learn new ways. But to do that you had to lead each step along the new path yourself. In time, Tommy could systematize the functions, then develop a managerial team, and delegate fully. Until then, he had to oversee everything.

Bill Arthur, who had seen a thousand start-ups, was frustrated with Tommy for other reasons. "We were outta money!" he said. "And Tommy had that wide-eyed openness of the true scientist, like, 'So? What's to worry? Everybody knew we'd be outta money.' We were having all these wonderful scientific papers advanced, but the investors don't really give a sh—— care about his technical prowess. The reason we have investors, solely and exclusively, without any fear of contradiction, is gold!"

"Right now, Tommy, you have a failure of confidence," said Arthur. "We've spent all the money, and no gold." But Arthur saw a solution, and he had a proposal: Spread the risk; take all of that technical capability and go after several ships. "Let's parlay this into a business," said Arthur, "the recovery version of General Motors. Then let's take it public, sell stock in the thing."

Although that reasoning often worked well in the search for oil and gas and in similar ventures, it was the antithesis of Tommy's methodical approach. He had watched treasure hunters make that mistake: throw in four or five more wrecks to sweeten the pie. But that didn't spread the risk; it increased the risk by wasting money on these unlikely pursuits, hoping they would get lucky, then running short just shy of the good project. Long ago, Tommy had seen the wisdom of focusing on one promising wreck with a documented high value that they knew had not been disturbed. But he listened to Arthur. He always listened.

"I'm not calling the shots here," said Arthur. "But I think we have a case of gold fever on the money side and giddy scientific pursuit on the other."

Arthur was a smart man, a prominent professional, but Tommy had to separate the man from the idea; he had to assess Arthur's inclination to package and sell. Arthur had said it himself: "My business is structuring deals. I'm really good at that." His idea might be a good one—it might even avert a financial crisis—but it was not the right time to start a public company. Tommy would consider the idea, but he would have to find a way through the current crisis without mortgaging the future.

"Tommy's tough," observed Patton. "He's smart and he's tough. He listens to what I say and he listens to what others say, and then he decides what he's going to do, and he goes and does it. I've seen some real tough guys, and he's got my highest respect."

In the off-season, Tod reported to the Victorian on Neil Avenue and worked around the office. He saw Tommy frequently. "He lived in the office practically," said Tod. "Every hour of the day he was working." Tommy would stroll into the kitchen, rubbing the palms of his hands together in half circles, that Harvey grin on his face, his eyes bloodshot, to see if the coffee was on. Then he would disappear again behind the sliding oak door to the formal dining room to talk on the phone and figure on a legal pad. Tod found it curious that with all of the problems Tommy had to work out, he never seemed down. "I've never even heard him say anything negative about anyone," said Tod, "or about anything. Usually he's just burned out." Tommy stuffed himself with supplements and vitamins and sometimes went for thirty hours at a stretch with no sleep. It seemed like he never slept. "He's just so intense," said Tod. "He just goes and goes and never quits."

Tommy was worried, but he liked to worry. "The more you worry, the more you think," he said, "and the more you think, the more you know, and the more you know, the better you're able to deal with the situation. I'm not the type to worry to the point of dysfunction. The more I worry, the more energy I have."

Besides worrying about money and his partners, Tommy also worried about technology and his engineers. At the partner meeting, he had told the partners nothing about the problems his engineers were having trying to design the "full-up" recovery vehicle. To be able to explore and work through the coal at the Galaxy site as he needed to, he was pushing his engineers, especially Hackman, to do things with this new vehicle that either no one had ever been able to do, or no one had ever even thought of trying. He had designed the system on paper the year before, then had to settle for the E version. Now he had more knowledge about the site to incorporate into his thinking, and he used that knowledge to run thought experiments with Hackman and the other engineers.

One of the most troublesome problems with deep-ocean vehicles was their limited reach. Manned and unmanned vehicles had manipulators, but often the joints were too stiff to operate or too clumsy for a smooth maneuver. Even if a manipulator moved smoothly and operated at its maximum flexibility at the shoulder, elbow, wrist, and hand joints, it typically reached out three feet and operated inside an envelope four by six inches. The jaws moved barely this way, barely that way, slightly up, slightly down. Maneuvering any submersible in close enough to use the manipulator sometimes proved impossible. Even an operator as skilled as John Moore sometimes nudged an artifact two inches, which was enough to put it out of reach. Then they had to shut down, lift the vehicle, move it over, and wait for the silt to settle. With such limited range, they often worked for no more than ten minutes before they had to lift off and relocate.

Tommy had pondered the problem for years, and he figured that one giant leap for mankind's working presence in the deep ocean would be a robot that swiveled. He asked Hackman to think about this: What if they took the whole vehicle and put it on top of a rotating base, like a steam shovel or a crane? Hackman's first thought was, "A crane's got huge ball bearings, and those bearings and the gears and the hydraulic motors and the shafts will all be down in the mud." He reminded Tommy, "We have a flying vehicle that we can pick up and rotate any time we want to."

"I know," said Tommy, "but it takes too much time." He wanted Hackman to calculate how big a rotating base would have to be. In a few days, Hackman reported back. "It's going to weigh five thousand pounds."

"It can't weigh that much," said Tommy. "See what you can do."

Hackman had learned never to say no to Tommy; he had a different way of dealing with Tommy's ideas: He calculated exactly what it would take to do what Tommy wanted to do, then let Tommy decide for himself if it was worth it.

Hackman also had been around Tommy long enough to know that Tommy had a reason for suggesting an idea, and although that reason might not be apparent to anyone else, it was founded on something deep inside Tommy's head that had gotten there through some combination of his intellect, his experience, and his unique gift for viewing the world upside down. And often, after Hackman had stood on his own head and fiddled with the figures, he began to see things he hadn't seen at first. Sometimes he surprised himself when he reexamined old assumptions and found ways he could make Tommy's seemingly crazy ideas work. "This happened many, many times," said Hackman.

But when Tommy told Hackman to look for a way to put the vehicle on a rotating base, Hackman thought, "If he wants to see on paper how ridiculous that idea is, I'll show him."

In all of his days working on top-secret ocean projects, Hackman had never heard of anyone even thinking about trying something so outrageous underwater. However, Tommy was right about one thing: If they could make it work, it would increase the effective reach of the vehicle a thousandfold.

Hackman started thinking about the problem the way Tommy would think about the problem. Why does it have to weigh five thousand pounds? Well, he figured, it has to weigh that much because I'm assuming we have to have those big ball bearings and some of these other things. Are those valid assumptions? he asked himself. Maybe not. Maybe we wouldn't need those big ball bearings, after all; maybe we could put it on an oscillator, but then the whole vehicle would be sitting on a single shaft. A few quick calculations on bending loads told him that the weight of the vehicle would snap the shaft in two. He'd have to think of something else. But that was a start.

When he told Tommy how far he had gotten, Tommy thought out loud, "Maybe we can look at it more like the wing of an airplane, and instead of assuming the shaft would be rigid, let's assume it's going to be flexible." Then they ran a thought experiment to narrow the probable range of flex.

"That's how Hackman and I work really well together," said Tommy. "He would tell me the problem very clearly, and then I'd just put that in my subconscious and think about it and some new idea would pop out, and we'd talk some more."

In one of these sessions, Hackman realized, "If we're going to make this thing flexible, I can put a little ring of rollers out here to take the bending loads, and I can go to a smaller actuator to rotate it."

That made the whole base much lighter and more resilient. Hackman was pleased that they had solved a problem he had thought impossible to solve, but when he took the sketches to Tommy, Tommy wanted to know how much it would weigh now, and Hackman said he had reduced the weight to about half his original estimate. Tommy wanted to beat the original estimate by an order of magnitude, get it down to one-tenth, or 500 pounds. Hackman went back and pondered his sketches, and he saw new things. He sandwiched one ring into the other ring. Then he turned the rotary actuator upside down. Then he moved the hoses from the base to the vehicle. "I just kept working out the problems one by one," said Hackman. By the time he had finished, he had the specs for a rotating base that weighed 450 pounds and allowed the vehicle to rotate 360 degrees.

But Tommy wasn't through. Most manipulators were anthropomorphic, modeled after the human arm, but with only five to seven degrees of freedom. The human arm has twenty-seven. But even the human arm has only two segments, upper and lower, plus an "end effector," or hand. Tommy wanted three upper arms and five lower arms, and at least twenty-seven degrees of freedom, and he wanted to do things the human arm could never do. He wanted to extend the reach by telescoping the shoulder, the equivalent of a human arm changing the TV channel without a remote. Just shoot the whole arm across the room. Many ocean engineers had tried to do this, but every solution Hackman had ever heard of was complicated, heavy, and expensive, and most increased the reach no more than a few inches. One group had created a cage on rollers that weighed five thousand pounds and slid forward like a barn door yet extended the reach less than two feet.

Even if you extended the reach, you still had to see out there, or the new reach did you no good, because you can't work on what you can't see. Humans can't tighten the little screw at the hinge in a pair of glasses if they hold the glasses and the tiny screwdriver out as far as their arms will reach and point the screw away. But Tommy wanted a vehicle that could. He wanted to pop the eyeballs out with the arm, zoom in with the cameras, and backlight it so the scene was crisp, and he wanted to zoom and backlight from several angles.

In thought experiments with Tommy, Hackman conceived a simple solution: Inside a plastic-lined aluminum sleeve, he put a hollow aluminum shaft that slid like a long drawer. Cheap, light, low tech. All of the stresses and flexes worked for Hackman on paper, and when they built it, it weighed three hundred pounds and extended the reach of the manipulator and the light booms by five feet.

"That's what Harvey brought to it," said Hackman. "He had a concept of what he wanted to do and how he wanted to do it, and then I'd design something, and he would just look at it. 'Will it do the job I want it to do and will it fit in the place I want it to fit?' And if it wouldn't, he'd say, 'Let's keep working on it,' and so I'd go back."

To raise fresh capital, a partner had set up a meeting between Tommy and the investment banking house of Drexel Burnham in New York. But when Tommy flew to New York with Ashby and the other partners to meet with the Drexel management, he felt uncomfortable with them. "They really weren't our kind of people," he said.

"Money is money, as far as I was concerned," said Ashby, "but Tommy thought that working with them would destroy the entire project, the confidentiality. They wanted to get into his organization and talk to his people and talk to his researchers, and Tommy would never have permitted that."

Another partner offered to finance the building of the new vehicle and lease it back to Tommy. He also would arrange for a million dollar credit line. In one stroke, Tommy could have surmounted most of his financial problems, but then the partner would have owned the vehicle. Tommy had to find a way to deal with the current crisis without creating future problems. After three months of considering options, he finally decided he would have to raise money "the old-fashioned way," which meant sitting down face to face, one on one, with old partners and new investors and explaining the recent accomplishments of the project. Ashby and the lawyers advised framing it as a new offering, Recovery Phase II, and including enough units in the subscription to raise $7.5 million. The only problem was that every unit in the partnership was gone, 100 percent sold, so they would have to dilute all of the current units to make room for the new ones. And on this the partners differed.

"Some partners thought that we had accomplished a great deal," said Ashby, "and that the dilution should be very low; and others thought, Yes, we've accomplished a great deal, but these new shares should get a lot."

Earlier, Ashby had broached the idea of a 25 percent dilution with some of the partners, but everyone he talked to had bristled. He pulled back to 15 percent, and some of the partners agreed they had to do it; others didn't like the idea but said they wouldn't fight it. Most had been diluted in other ventures, so they knew why it happened, and in the end, if not harmony, at least there was agreement. After they finished the arithmetic, Recovery Phase II opened with 150 smaller-percentage units that sold for fifty thousand dollars a unit.

When Ashby talked to partners now, he emphasized the court injunction and how the vehicle had performed on the bottom of the deep ocean. "Wayne was a bulldog," said Dauterman. But as the fall of 1987 turned to the winter of 1988, money only dribbled in, and Tommy had to spend much of his time trying to predict when it would come; then based on that and a carefully conceived priority list, he cautiously ordered the next critical piece of equipment or line of expertise.

While Tommy allocated the funds sparingly, he pushed his engineers on the vehicle without their knowing that money was a problem. He confined his talks with them to concepts and specs and the problems of dealing with suppliers. Partners called him on his cellular phone at the warehouse wanting to know why he was not further along and why he had to improve what they had last year, anyhow.

Tommy never revealed to one group the problems of another. The partners may have had nightmares about their investment and the need for a second round of financing and a higher ante, but they didn't go to bed worried about the seemingly insurmountable technical problems ahead of Hackman, Moore, Scotty, and the engineers. Hackman, Moore, Scotty, and the engineers went to bed with those nightmares, but they didn't go to bed worried that the next time they called a supplier in Goleta, California, they would hear that Columbus-America was sixty days in arrears. Tommy went to bed, when he slept, with everyone's nightmares.

Columbus, Ohio

⚓

Winter and Spring, 1988

IN COLUMBUS WINTER begins to relinquish its grip in March, but only reluctantly. The battle with spring takes place under gray skies, and tree branches with pinhead green sprouts rattle in the wind, and something falls from the sky that is not quite snow, not quite rain, not even sleet, but frozen droplets that tinkle when they hit, almost like the tumbling of thin seashells. On one of those days, Bob Evans sat in his office on the second floor of the Victorian on Neil Avenue, reviewing the sonar records Mike Williamson and his crew had shot in the summer of 1986.

In the ongoing round tables, Tommy, Bob, and Barry had concluded that although there was a 90 percent probability that Galaxy was the *Central America,* they also had to plan for it not being the *Central America*. Everything at the site had been so encouraging and yet it was frustrating; they had found the ship, they had found the period dishes and the cup and the jugs, they had found the coal; but because the E-vehicle was so limited, they could not get under the coal to see what was there.

"It seems like we have the *Central America,*" said Bob, "but in science you're never sure about anything; you're 97 percent sure, or 99 percent sure. We were all putting a real happy face on, and at the same time, there's this doubt in our own minds about what we've got, or at least a willingness to accept an alternate hypothesis."

Part of Bob's responsibility during each off-season was to systematically examine all of the new data gathered at sea and reevaluate the data from previous years. Tommy was particularly interested in synthesizing a new understanding of deep-water, historic ships; with the updated information, they planned for the next season. That winter Tommy told Bob to reanalyze the sonar records for the entire fourteen hundred square miles searched in 1986 and catalog everything. Bob had two things to help him: the experience of having ground-truthed Sidewheel and Galaxy, and new-generation software that would enable him not only to turn the sonar data into bright, contrasting colors, but also to scan, zoom, mask, and otherwise enhance the information.

Bob had begun reviewing the records from day one of the search, and as he went through each file, he marked the anomalies. "I was recreating the voyage, starting with the first record, so I was reliving what they had gone through. I was saying, 'Okay, there's an anomaly here; here's another anomaly,' I think I called them small, medium, and large, just checked everything; I wanted to have a whole record of that."

Each time Bob entered a new file number, thin blue lines zipped from left to right and right to left across his monitor, marching upward until seven hundred of them filled the screen. "It's the way it comes up originally at sea," said Bob, "so you can watch it just as if the SeaMARC was moving along underneath you." When the blue lines finished painting, Bob had on the screen two miles of ocean floor three miles wide.

The light blue lines indicated a flat bottom of sediment, but among the blue, Bob sometimes saw flecks of green or various shades of red or black, tiny anomalies no bigger than a pixel on the screen, or one to two meters on the ocean floor. When he had noted every anomaly, he entered another file number and again the blue lines began to zip back and forth across the screen.

Williamson and the sonar techs had dismissed most of the anomalies in the original sonar search as too small, too hard, or too round and shapeless to be the remains of a 278-foot, wooden-hulled, sidewheel steamer. For Bob's review of the records, Tommy told him to note everything: big ships, little ships, geology, shipping containers, submarines, sailboats, every anomaly right down to the suspected oil drums. Two anomalies along the western edge of their search area had been dismissed as geology, but being a geologist, Bob thought, "They'll be interesting to look at."

For three days Bob had done nothing but review sonar records. So far, he had analyzed over a hundred. The first forty-nine records were short and unreliable because the weather had been stormy and the equipment not working properly. At file 0120, Williamson had switched the SeaMARC to the one-thousand-meter swath for the high-resolution passes on the promising targets. Bob had studied those images many times before on the master optical disk, Williamson's hit parade. "But I looked at them again," said Bob, "because there's some interesting ships."

In the middle of these promising targets were the high-resolution images shot along the western border of the probability map. "I was fascinated with what I was doing now," said Bob, "because this was the geology marked in the log book, and I was looking at these strange, long, linear features." He spent nearly a half hour studying 0126, then he went to 0127 and let it paint its way up the screen.

File 0127 was the target in the southwest corner of the probability map that the sonar techs had had trouble finding again on the high-resolution run. They had made three passes trying to relocate the target. On the first pass, the watch leader had written in the log, "No joy." On the second pass, the navigator had written, "Contact 200 meters port—geology?" and the watch leader had written, "Large geological feature." On the third pass, the watch leader had concluded, "Target believed to be a geological anomaly with no cultural value."

When Bob finished reading the comments from the watch leader and the navigator, he glanced up at the screen, and the thin blue lines running quietly back and forth had been interrupted by a large image only partially painted but already filling with green and red and black. "I saw this, quote, 'geological' anomaly coming on," said Bob, "and it got about a fourth of the way up, and I started getting weak in the knees." The target was long and thin at the core, "Like a central area of mass," thought Bob, "but surrounded by a galaxy of debris." He checked the log again to make sure it was shot on the one-thousand-meter swath. He measured it quickly. The core seemed significantly longer than Galaxy. He turned off the screen. "I thought about it for a minute . . . and I turned it back on . . . and I looked at it again . . . you know? . . . I'm like . . . you know? . . . I just . . . I thought . . . you know? THIS IS A SHIP! THIS IS NOT GEOLOGY! IT'S NOT ONLY A SHIP, BUT IT'S MUCH BIGGER AND THE RIGHT CONFIGURATION!"

What had distinguished Galaxy from the other shipwreck images were the fields of debris obscuring the contours of the ship. Every other image had looked like a ship, pointy and thin, but Bob now knew that the image of a wooden-hulled ship full of coal after a century on the bottom should appear rounded and amorphous. "The sonar team are the top sonar experts in the world," said Bob, "but this was the first phenomenon they had seen that looked like this, so I can understand why they called it geology. But I've spent all summer ground-truthing something else called Galaxy that looks like geology, too, only I know it's a ship. And this new target looked an awful lot like Galaxy. If this was the *Central America,* we had spent 115 days at sea on the wrong ship."

Bob rocked in a confused emotional sea, waves of frustration and waves of elation washing over him at once. They all had worked so hard that summer and had found nothing; yet he was staring at dramatic information that fit their new understanding of sonar images of large, deep-water, wooden ships. "I'm suddenly thinking that right here in our very own data might be the answer to our problem."

Despite a chilly wind rattling the bare branches of the trees, Bob left the office and started down the sidewalk, head bent, passing one old

Victorian after another. "I was questioning my scientific method. This anomaly had an area roughly mid-ship that was standing some thirty feet proud off the bottom. This corresponded, of course, with engine works. It had two rather bulbous, hard-looking objects sitting athwartships on either side. These, of course, could have been paddle wheels. I'm thinking about all this stuff, and I'm going, 'This is it!' I was drawing conclusions well beyond what I was allowed to conclude, so I was trying to slow myself down. I can't just unequivocally say, 'This is the *Central America*!' That's ludicrous. However, I was very quickly swept into the idea that no longer was Galaxy the only bet we had."

When Bob returned, Tommy was in his dining room office deep into a conversation on the phone. Bob went up to the third floor and found Barry also on the phone. He returned to his computer on the second floor. The screen was blank. Bob turned it on, and the anomaly appeared again on the left side, a cloud of color at the core against a neutral ocean bottom. "Maybe it is geology," thought Bob.

On the upper right of the screen sat another anomaly he surmised was geology, a ridge, thirty to forty meters wide and miles long. He studied the two, using everything he knew as a geologist and everything he had learned about comparing shipwrecks on the bottom to their sonagrams. He went back and forth, first one, then the other, then the first again, finding correlations and differences and seeing much more in the new anomaly of what he now thought he should be seeing. "It was obvious to me," said Bob, "that that strong 'geological' anomaly was significantly different from the other geology."

Late that afternoon, Barry came down from the third floor, tired and wanting a break. Before he could open Bob's door, Bob walked out.

"He looked terrible," remembered Barry, "like he was sick."

Bob had his head down. He saw Barry, but he said nothing; he half turned like he was going to walk back into his office. Then he stopped and stood still.

"What's the matter?" asked Barry.

Bob took Barry into his office and closed the door. He turned on the screen, and there was the image of the new target. "I've been looking at this for the past four hours." He measured it for Barry: It seemed longer than the Galaxy site. He measured it again to be sure. He held a small aerial diagram of the *Central America* next to the image on the screen.

"It's much bigger," thought Barry. "You can see the outline, and you can see the sidewheels on either side."

"Those are sidewheels, aren't they?" he said to Bob.

Bob dropped into a careful refrain that he would use for months. "All I can say is this site fits the target model of the *Central America* as we now know it. This looks like a large shipwreck with coal on it, wooden hull, with engine works in the middle. That's what it looks like."

The next day, Bob arrived at the Victorian, went upstairs to his office on the second floor, and closed the door. "I was very nervous about this piece of information," he said. Alone in the quiet of his office, he again brought up the image and measured every dimension, listing the figures; next to those he wrote the comparative figures for Galaxy. Then he experimented with his new software until he could bring Galaxy on to his screen at the same time as the new target and adjusted both to the same scale. Colored sprays radiated from the core of each. But the sprays forming Galaxy were colored consistently, and Bob knew they were mostly coal. Among the sprays surrounding the new target he saw varied colors and shadings, indicating a much more complex site.

Using different color schemes, he shot four pictures of the screen with a Polaroid camera. Then he took them downstairs to the dining room and slid open the pocket door. As usual, Tommy was on the phone, but he was on hold. Bob walked over to his desk. "In the data from '86," he said, "we've got another anomaly." He handed Tommy the pictures. "This is Galaxy, and this is the new target."

Tommy studied the pictures and recognized the new site as the one Craft had dragged him out of his bunk to see about five o'clock one morning. Craft had called the site "Geo." Tommy raised his eyebrows, nodded as though he found them interesting, unlocked his desk, and slipped the pictures into a drawer. "He immediately became very jealous of that information," said Bob, "the same as I had done."

Later that morning, Tommy went up to Bob's office, and Bob had the two shipwreck sites still on the screen. Tommy gazed at them, then looked at Bob with an expression that indicated he saw immediately what Bob had seen.

"It had these coal-like characteristics that Galaxy had," recalled Tommy, "and that's what we were looking for, comparative data, ground truthing. The more we looked at the image, the more we'd go, 'Hey, look at this feature here, look at that feature there—this could be a sidewheel steamer.'" But he was cautious; the experts had been fooled before. They had now ground-truthed a few deep-water ships and were building a database, but understanding sonagrams was still tricky. "What was obvious to me," said Tommy, "was that, Well, it's no longer a 90-something percent chance on Galaxy; we've got new information." Because the new target so resembled Galaxy, and because the targets shared the same differences from other targets, Tommy named the new target Galaxy II.

The two of them sat at Bob's monitor, scanning, expanding, and false-coloring the image of Galaxy II, pulling from it as much information in as many ways as they could. Bob tried every program he could think to run on his new software. They scrutinized the target "a zillion ways." They talked about the colors and developed theories on what they might mean. There appeared to be so much debris that at one point they even hypothesized that one ship had landed on top of another. "That's a tough one," said Bob, "that's a tough one to reconcile."

Tommy said, "If that's a second ship right on top of another one, I quit."

That was one of the reasons Tommy had moved toward deep-water ships, because they weren't piled on top of each other, confusing the artifacts and the identity of each. The odds of that happening in this area of the ocean at that depth were so little they could hardly calculate them. Whatever was out there had to be part of the same ship. "I was excited that we had produced more data," said Tommy. "The next question was, Well, what are we going to do about it?" How would they form an operating plan for the '88 season with two targets, each requiring a different approach? They might need to excavate in the coal, they might need to move heavy wood, they might need to run more sonar surveys. Tommy decided to put Galaxy II on a parallel track.

"It was forty miles closer to shore, so it became obvious to me that instead of testing our winch and doing all that stuff off of Jacksonville in deep water, we'd go all the way out there and test our stuff at that site. If nothing else, it was going to help us understand what we're looking at on Galaxy."

BEFORE TOMMY SET to sea in 1988 he needed a ship. He could lease again, but he wanted to build stability into the project, to have a vessel he could use to protect the site if he needed to. With his own ship, mobilization would be much faster, the crew could be at sea in a few days, and they could stay at sea as long as they needed to; and they didn't have to mount a new winch and a new crane and a new deployment arm and get used to a new deck every season.

"We were in all kinds of trouble financially," admitted Tommy. But that was a short-term problem, and as always he had to juggle the short term with the long term, and long term he could see they needed a ship. Already he had counseled with some of his partners about the wisdom of buying their own ship, and he had assigned Craft to survey ships for sale, to find out how much the owners wanted, how much it would cost to convert them, and how soon he could get them to Jacksonville and have them ready for a June 1 departure.

In December, a partner named Gil Kirk called Tommy with a proposition: I buy the ship and you lease it from me at a nominal day rate; that way you've got a ship whenever you want it, and I have collateral; that leaves you whatever cash you can raise to spend on other things. Tommy liked the idea. Wayne Ashby liked the idea. Tommy called Craft and told him to speed up the search for a ship.

By February, Craft had located the *Arctic Ranger,* a thirty-year-old side trawler built for the Fisheries Research Board of Canada as a vessel for scientists to study fish stocks on the Grand Banks. It had no working deck aft and only a small foredeck, but it had a wet laboratory and ample cargo storage and comfortable bunks for thirty-two crew and scientists. About the only thing Craft didn't like was the trawler's size. He wanted at least 200 feet in length, a big open deck, and a 50-foot beam to give them more stability. The *Arctic Ranger* ran 180 feet stem to stern, 33 feet abeam.

Craft flew to Newfoundland, drove north to Goose Bay, beat the owner down to $167,000, and brought in an icebreaker to cut through ice two feet thick so Burlingham could get the *Arctic Ranger* to a shipyard in St. John's. There, they hoisted her out, sandblasted and painted her, and gave her a quick overhaul, just enough to get the vessel through the Canadian steamship inspection, temporarily flagged in the United States, and down to Jacksonville, where they would begin the real conversion.

At the end of the first week in April, Burlingham sailed into Jacksonville, up the St. Johns River, and had the ship on shore power dockside, Pier 7, in Green Cove Springs the evening of April 9. He and Craft and a crew of carpenters, electricians, and day workers now had two months to transform the *Arctic Ranger,* old, frozen, Canadian fisheries research vessel, into the *Arctic Discoverer,* technologically unequaled deep-ocean recovery marvel.

Burlingham took charge of the ship's crew, while Craft prepared for the conversion. For weeks Burlingham's crew blasted and sanded the temporary paint job, and for another month they primed it all with a ruddy compound. Craft's crew ripped twelve tons of junk out of the bowels, including old hydraulic units and generators, old wiring and fishing equipment, and much of her three-inch-thick concrete, which had to be jack-hammered out in sections. Craft hauled the bulk of it to the junkyard. He converted the electrical system, air conditioned the ship, replaced the generators, mounted the SAT COM unit, installed the deployment arm, installed the winch, installed the crane. For a control room, he stripped the fish lab down to the bulkhead, installed a new deck, new paneling, electrical connections, painted the whole inside black, and created a small electronics shop adjacent.

The conversion continued every day through April and into May and then into June, as the *Arctic Ranger* slowly became the *Arctic Discoverer*. By June they had seven day workers, then thirteen day workers, still cleaning and blasting and painting, ripping out and throwing away, retrofitting and realigning and upgrading, preparing for the techs to arrive and begin mobilizing. They worked on Saturdays, they worked on Sundays. They painted the entire ship, everything from the bridge to the foredeck and forecastle, the aft deck and the hull, all the way down to the waterline, in bright white. They added touches of blue along the cap rail running the length of the ship, on the ladders, on the top of the short fore and after masts, and as a wide stripe around the stack. When they had finished, the new *Arctic Discoverer* looked shiny enough to be a hospital ship.

While the conversion continued in Jacksonville, Moore, Scotty, and the rest of the techs worked on the vehicle in a warehouse back in Columbus. Except for the low roar of a welder's torch and the whine of a hand band saw cutting channel aluminum, the warehouse was quiet. And still. No wind, no sea, no pitching and rolling, no tools sliding, no water roiling at their ankles. They could work with both hands at the same time and leave parts lying about rather than having to put them in coffee cans duct-taped to the vehicle. In the evening, they could return to a comfortable room and sleep all night, so they had "some constancy of biological rhythm," as Scotty called it.

By July, the conversion down in Jacksonville was complete except for the ship's thrusters, two shafts twenty-two feet long, each ending in a six-foot bronze prop that would keep the ship dynamically positioned above the site. A manufacturer in Houston who had built thrusters for the U.S. Navy had convinced Tommy he could produce a set in time for the *Arctic Discoverer* to set sail on June 1. By mid-July, he still had not delivered.

The delay allowed Burlingham and Craft to continue their conversion into mid-July, but by then they had nothing to do but wait. They couldn't leave without the thrusters, and everyone was getting tense. "We all started getting a little crazy," said Burlingham. "I'm serious. We were all definitely going off the deep end."

But back in Columbus, the delay provided more time for the techs to build the vehicle Tommy wanted. Moore called Burlingham frequently to find out the status of the thrusters and how much more time they had to work on the vehicle, and they continued to improve the lighting, the optics, the flexibility, the maneuverability.

Besides constantly working with the engineers, Tommy still met face to face with potential investors. Gil Kirk's offer to buy the ship and pay for the conversion had freed up much needed cash. In May, John F. Wolfe, the most well-known of a small group of business leaders who got things done in the community, wrote Tommy a check for a million

dollars. But a million dollars and a ship would barely get them to sea
with a vehicle that worked; they still had operating expenses. "We were
really getting into dire straits," said Tommy. "I was raising money and
talking to people all the way up to when we left." Wolfe's contribution
made it easier for Ashby and others close to Tommy to persuade the
partners to kick in another $1.5 million, and new investors had added a
million to that. By August, Tommy had $2,500,000 to add to Wolfe's
$1,000,000 plus his reserves, enough to get him through the season, which
already had been shortened by two months.

On THE 10TH of August, the techs crated everything for transport down
to Jacksonville and joined Burlingham, Craft, and the rest of the crew
aboard the *Discoverer*. The entire operation now depended on the thrusters, which finally arrived on the morning of the 14th. "The problem was,"
said Tommy, "we always thought the thrusters were going to work. The
guy'd say, 'Just one more week,' or 'We got to machine these parts,' and
then they'd come back and they were machined wrong. It just went from
one faux pas to another with those things." The crew had problems with
the diesel engines that ran the thrusters' prime motors and problems with
the hydraulic pumps. They couldn't get one of the thrusters to turn over
because the engine was full of water.

"I'm just livid," recalled Buck Patton, "just livid. They send him
this piece of shit, and they're late on delivery, and when it gets there
it's all wrong and Tommy has to rebuild it. You know? Everybody
was upset and rightfully so, because they had a very short weather
window left."

Just after lunch on August 19, a tug pulled the *Arctic Discoverer* from
the dock in Green Cove Springs and dragged it downriver. The tug cut
them loose just above Jacksonville, and they headed for a marina at the
mouth of the St. Johns, where they remained for another seven days,
working on the thrusters, checking for hydraulic leaks, trying to get
them to talk to the dynamic-positioning computers.

From the marina, Tommy called Dean Glower. "These investors
are not that patient," Glower told him. "You're a great guy, it'd be nice
to see a good guy win, but you're not there yet. You're gonna lose your
investors if you don't come up with some real cargo." Tommy finally

decided they could wait no longer; they had to go to sea with the faulty thrusters and continue trying to work out the problems.

Craft waved good-bye from the dock. He wasn't going on this trip; Burlingham could direct launch and recovery, and Craft had been with the *Discoverer* seven days a week since he first saw her dockside encrusted in Goose Bay ice back in February. It was now the end of August.

Aboard the
Arctic Discoverer

———— ⚓ ————

Late Summer, 1988

T HE NIGHT OF August 28, a tug nudged the *Arctic Discoverer* from the dock, guided her into the river, and set her free to find her own way the short distance to the mouth of the St. Johns River. With the sea buoy abeam at eight o'clock, Burlingham took her into the Atlantic and set his course for the first test site at Galaxy II. Tropical Storm Chris had just blown through, freshening the air and leaving the sea still high but dropping quickly.

The *Discoverer* traveled through the night at just under ten knots, the breeze riffling the white canvas tarp stretched across the foredeck, the air ducts on the forecastle spinning into a blur. Her bow cleaved the incoming seas, rising slightly and falling to rise again in fine fashion. The techs liked her seaworthiness. She was solid, and because of her rounded hull, her ride was far smoother than the *Navigator*'s; the techs watched the shelves of computers in the control room and saw nothing quaking.

All day and throughout the night on the 29th they steamed northeast toward Galaxy II, the winds now light and the seas small. Midmorning on the 30th, Burlingham halted their advance to test the new dynamic-positioning system. This time, the long shafts of the thrusters oscillated in sync, and the wide props spun at varying speeds, and the *Discoverer* held station.

In the afternoon, Burlingham and the deck crew hooked a spare aluminum cube to the end of the cable, launched the cube with the crane, and ran ten thousand feet of cable off the winch, allowing it to stretch and unkink. Then they recovered the cube and rewrapped the cable even and tight. By midnight they had secured the forward thruster and again were underway the last few miles to the test site at Galaxy II.

The following day, they drifted near Galaxy II in calm seas, working on the vehicle and continuing to fine-tune the thrusters. Burlingham kept his eye on the weather, which so far had been benign. "The last day of September you ought to be shutting down, getting ready to head in," said Burlingham. "So we're a bit behind schedule here on the first."

With Don Craft gone, Burlingham would be directing the deck evolutions during launch and recovery, and he wanted everything safe and smooth; as safe and as smooth as anything could be at sea. For the next three days, they prepared and tested deck equipment for the first launch.

The biggest equipment on the *Discoverer*'s deck was the port-side deployment arm, a stout T-head ten feet tall and weighing four tons. The deployment arm arced out over the water, and one morning Tod's job was to climb to the top and install a four-hundred-pound block, or pulley. While Doering fired up the crane to lift the block, Tod started up the rungs, but halfway, he realized he had forgotten the teasing wire and climbed back down to the deck. What happened next conjured nightmares in him for months. He had walked no more than five steps toward the workshop when suddenly the four tons of steel shot down toward the water, blew a piston, and flipped backwards onto the deck, blowing every light below and knocking plaster off the ceiling. Topside, there was a fresh divot in the steel deck. The whole incident was over in less than three seconds.

That was what worried Burlingham. You could prepare for a thousand contingencies at sea, but the sea had a thousand and one ways to beat you, and the last always occurred when you were carefully watching for one of the others. It could happen even on a calm day, with the sun shining, the wind at rest, and the crew performing routine work, like hanging a block off the deployment arm. Burlingham discovered that in the salt air, the lever that controlled the arm had rusted and stuck in the down position; when Doering switched on the battery pack for the first time, the power came up suddenly, and the arm dropped to obey the command of the rusted lever, blew apart with the force, and crashed back onto the deck. If Tod had not forgotten the teasing wire and climbed down to get it, he would have been crushed under eight thousand pounds of steel.

To withstand the harsh demands of the sea, equipment had to be heavy and the lines to move that equipment had to be strong, and still the rusting, the rotting, the pitching and pulling of the sea strained that heavy equipment and those strong lines sometimes beyond the breaking point. And when the equipment failed or the lines parted, just like the crashing of the deployment arm, it happened suddenly. John Moore had seen parted lines kill men on deck so fast it cut their last thought in two. On a derrick barge in the North Sea the crew was handling tonnage over the side, and one sailor was standing where he shouldn't have been, and a line parted and zipped him up the middle between the legs, cut him in half the long way. Deck lines stored so much energy that when they snapped, they hit you before you could hear them coming. "It's over," said Moore, "before you even know it's happening."

Sometimes it was a surprise; sometimes the danger was there and you knew it but could do nothing about it; sometimes you saw the potential ahead of time and prepared for it. The next near miss on the *Discoverer* came during a controlled experiment. Tommy had had half a dozen new blocks specially designed and built for that season, and he wanted Burlingham to test them. The heavy new blocks were the pulleys over which the lines would run to launch the vehicle. Although manufacturers rated the blocks to a specified load, the manufacturers were sometimes wrong, so Tommy wanted Burlingham to run tests at sea. Rarely did one fail, but if a block was not going to hold under extreme force, Tommy wanted to know it before he had the block, the vehicle, and half the crew in a tense situation.

With each new block secured near the stern and the men safely out of the way, Burlingham ran the line through the testing block and torqued it tighter and tighter until the line was as taut as it would have to be during launch and recovery; then he tightened it more for a safe margin. On one test, a block exploded, and the energy in the line that was suddenly released launched the block like a ball bearing in a high-powered slingshot. A yellow blur shot past the foredeck and over the forecastle, so far out across the ocean not even the watch on the bridge saw it land. If the block had hit one of the deckhands, the impact would have turned him to jelly.

Now into early September, still drifting near the test site, they continued to experiment with the vehicle on deck, tracing glitches in the electronic and hydraulic and computer systems they had created for a robot that would perform the way Tommy had originally envisioned. No more emergency vehicles. This would be what Tommy called "the full-up vehicle," a robot that could stay on the bottom for a long time and carefully explore and document and record, and later selectively recover and store and bring to the surface. And as they worked to ready it for the first test dive, they basked in sunshine and bobbed in friendly seas. Bob Evans wrote to his new wife, "We continue to have technical difficulties and haven't had the sub in the water yet. The weather has been beautiful and I hope we don't use it all up fixing and readying equipment."

At noon on the third day of floating above the test site, the winds rose from light to fifteen knots and the seas from calm to three feet. Then the gale hit.

On September 5, the winds shot to thirty-five knots and the seas to ten feet, and all the next day the winds continued at gale force, with furious squalls soaring to fifty knots, blowing the crests of the waves into spindrift, as the seas nearly doubled in size. Burlingham stood in the wheelhouse twenty-five feet off the waterline and saw walls of water cresting just below his ankles.

With four tag lines, they had lashed the vehicle to the deck under a heavy tarp, but the pounding of the foredeck twice ripped it loose, and they had to venture onto the pitching deck to resecure it with a dozen heavy canvas straps. Twenty-foot seas exploded off the bow and crashed onto the forecastle, then flooded down and across the deck in gray-green sheets.

With jolts and spasms, fall in the Atlantic already was arriving, and each new preview of the season to come would hit a little harder and stay a little longer. After two days, this one began to subside. By noon on the 7th, the wind had dropped to twenty-five knots and the seas to twelve feet. But even with the storm passing, the seas remained far too rough to launch. During the next three days, the seas gradually diminished to three feet and the winds to less than ten knots. Tommy had about three weeks left to find the *Central America* before the weather dropped and did not lift again till the following summer.

To RELOCATE THE Sidewheel site the previous summer, Bob had had the navigation numbers from fifteen high-resolution passes to use for his calculations; later for Galaxy he had had the numbers from three passes. But because the sonar techs in 1986 had thought Galaxy II was geology, no one had pinpointed the navigation. Bob had a single high-resolution image shot from one direction, and even that was skewed because the *Pine River* had already passed the target before the crew started to run the navigation records; no one knew if the sonar fish was even behind the ship. Now he had to provide Scotty with a coordinate close enough to the actual site that they could fly the vehicle within twenty feet of it after no more than a few days of running track lines.

Since he had rediscovered the Galaxy II image in March, Bob had wondered how he could relocate that target, and for months he had worked at the problem, taking the few numbers he had to work with and carefully assuming where he had no choice but to assume. It was high school mathematics, mostly geometry and trig, in pencil on graph paper, the way Bob did all of his calculations. *Speed down slope, about 1.3 knots; speed up slope, maybe 2.1 knots. At slower speeds the sonar techs probably had less cable out so the fish would sink; when they picked up speed, they probably had more cable out so the fish would keep close to the bottom; so the distance the fish dragged behind the ship was probably proportional to the speed of the vessel.* He calculated that when the fish imaged Galaxy

II, it was 1,835 meters, or a little over a mile, behind the ship. He calculated further to allow for the fish just coming out of a turn, and when he finished he had a coordinate. Then he worked it all another way, and then another way, each time reweighting his assumptions. And each time, he produced coordinates that fell within 160 meters of each other, creating an elliptical probability area for the target.

Except for Tommy, Bob, and Barry, everyone on the ship thought they were stopping at another site to test equipment on their way to the coal pile at Galaxy. "We didn't want to start working on the coal pile until we were sure we could stay there," said Moore. But Moore also sensed that Tommy's interest in this new site was far greater than it should be for just another test site with a burned-out, eroded hull on the bottom.

At seven o'clock on the morning of September 10, they were still working on the vehicle. The winds had all but died and the seas were down to a foot or two. "May try to launch today," Burlingham wrote in his log. But they didn't drop the thrusters and crank them up until twelve hours later, and by the time they got the ship holding station over the site, it was eight o'clock in the evening, and Tommy decided he did not want to launch. Two hours later, Burlingham disengaged the DP system, and they began drifting with the thrusters still down.

At midnight Burlingham recorded in his log, "Standing by as before."

*S*EPTEMBER *11, 1988: 131 years ago today, Addie Easton had huddled in the dining saloon of the* Central America *and watched Ansel remove his coat and join the other men in the bailing lines. Seawater in the hold had risen and doused the fires beneath the boilers, sending hot steam through the engine room, burning the firemen and the haulers, extinguishing the lamps, and silencing the big steam engines forever. The steamer had slid into the trough of the sea, and her giant paddle wheels had slowed and then stopped and hung motionless at her sides, like Ferris wheels in winter.*

At eleven-thirty that morning, the techs launched the vehicle without incident and went to the galley for lunch while they waited an hour and a half for the vehicle to reach the bottom. After lunch, they began to collect in the control room, Moore in the pilot's seat, Scotty behind him at the navigation computers, Doering to Moore's left in the copilot's seat, Milt Butterworth just to the right of Scotty, scanning the audiovisual monitors. Burlingham sat next to Scotty, learning Scotty's intricate navigation system. Bob Evans perched behind them against the wall, where he could see all of the monitors at once, and Tommy sat in a chair in the middle, his toes touching the carpet and his knees pumping.

The control room was small with a low ceiling of acoustical foam. The air was kept cool for the computers, no higher than sixty-five degrees and often five degrees cooler. On dive days you could tell the tech crew from the deck crew because the tech crew wore jeans and sweatshirts to keep warm. On three sides, stacked floor to ceiling, computers, monitors, and digital displays filled the room. At the electronic heart of the control room was Scotty's new logging system, which would keep track of all the computers. It would correlate navigation information from topside with information from the subsea grid. It would record every photograph and every video frame and store the crew's comments on each. Using the logging system, they could easily return to an artifact seen in the debris on an earlier dive.

On one TV monitor, they could see the cable still feeding off the big drum out on deck and Tod sitting in a chair next to the drum watching the level-wind. Cool white fluorescent tubes lit the control room until Tod alerted them that the vehicle was within two hundred feet of the seafloor. Then they turned out the lights, and the room was dark except for the glow of monitors and readouts from the digital displays. At 13:33, they had visual contact with the bottom.

By a little after two in the afternoon, they had checked all of the vehicle's systems and begun the first track line, but less than a half hour later, Scotty noticed that the *Arctic Discoverer* had drifted off track. One of his computers registered zero for the forward thruster. Scotty ran up to the bridge and found that someone had laid a book on top of the primary on-off button. He removed the book, reset the button, and the ship crabbed back over to its position and continued on line, traveling at about half a knot.

The Mesotech sonar swept into the darkness a hundred meters out, far beyond the range of the cameras. A target would show first on the Mesotech, then about three minutes later, the vehicle would pass over it, and to the rear the target would appear on the SIT camera in a flood of light.

The previous summer on Sidewheel, they had all squinted for three days and a dozen track lines before they had finally glimpsed the stern of the ship fifteen feet away. This year, they hoped that the brighter lighting and more sensitive cameras, and Scotty's improved navigation system, would help them locate the ships much faster. But no one was prepared for it to happen as fast as it did.

Not a half hour after they got the ship back on track, Moore saw on the Mesotech some small targets off to port. Scotty was briefing Burlingham on how they had set up this track line when Moore called out the heading of the vehicle and the bearing of the targets, and Scotty turned to record those in the navigation log. As he was completing the notation, Moore called out more targets, with heading and bearing, and Scotty again was writing as fast as he could, archiving the data, so they could find their way back to this spot. Suddenly, Moore was calling out more and more targets. "I'm getting sonar action," said Moore, "and now I'm getting more sonar action! We have got some really major action coming here!"

Scotty was trying to write it all down as fast as Moore called it out, and at the same time he and Burlingham were trying to watch what was coming up on the TV monitors, which now everyone was watching. Moore was talking fast, describing all of the targets showing up in the sonar sweep, and then he yelled, "Whoa! We've got a biggie here! I mean it looks like we've got something SERIOUS!"

Into the glare of light now glided the first of the smaller targets Moore had seen on the Mesotech a few minutes earlier: three white artifacts in the sediment.

"Cultural deposits here," said Moore. "Yep, we got a plate, it looks like."

Someone else said, "A bottle maybe."

"Looks like a bottle or something there, yeah," said Moore. "Got coal. Fiiirst ruuun," he said, laughing, "and we're coming across something. This is a huge area, too."

"Can you see out beyond a hundred meters on there?" asked Tommy.

"Not really," said Moore. "This is sixty to seventy meters. I've turned it up to make sure I don't miss anything."

The big object he had seen on the Mesotech would be coming under the SIT camera on the port side, but just as the vehicle reached the target, it started to twist to starboard. Moore talked to the vehicle. "I don't want you to rotate that way. Over the other way, over the other way."

"Remember that one last year, John?" said Bob, referring to the early dives on Sidewheel. "Where we went over the stern and barely saw it?"

And suddenly, a huge bulbous shadow began to grow at the lower left corner of the monitor. Bob had just said, "and barely saw it," when someone yelled, "Look at this! Look at this!"

"Whoa!" yelled Doering. "WHOA, WHOA!"

Bob looked up at the monitor. "Whooaaa."

And then Tommy said, "Oh, my God!"

IT HAD HAPPENED so fast, no one was prepared.

"Here we go!" yelled Moore.

"Oh, you know what that is!" said Doering.

"You know . . . ," started Moore, and like a chorus everyone in the control room shouted, "WHAT THAT IS!"

A huge paddle wheel was nudging into view, lying in the silt, the bottom portion partially buried, metal spokes fanning out from the center, a twisted pile of iron lying on part of it, all casting dark shadows onto the sand.

"You know what that is!" Doering said again.

"NO SHIT!" yelled Moore.

They had hardly had time to realize what they were seeing when Bob suddenly called out to Moore, "You better get up RIGHT NOW!"

"I'm pulling up right now," said Moore.

Bob had studied the sonagram of that site for months. He knew it better than he knew his closet. If his calculations were correct, the vehicle was only seconds from colliding with something that cast a long sonar shadow on the sonagram.

"In fact," said Bob, "you're over about a ten-meter high spot, so I'd GET IT UP!"

Just as Moore raised the vehicle, a forged-iron crank gear the size of a file cabinet suddenly turned white in the glare of the vehicle lights not two feet below the vehicle. It perched at the end of an iron shaft that ran thirty feet up from the center of the collapsed wheel.

"Oh, shit!" yelled Moore. "No sh . . . !"

The drive shaft of the starboard paddle wheel, snapped loose and sticking up at the precise center of the ship, had nearly snagged the vehicle. On the ocean floor, looking like the cage off an antique electric fan, lay the starboard paddle wheel of an old steamer, the iron spokes still radiating from the center and only the wood of its paddles missing. As the vehicle drifted slowly by, the wheel cast a weblike shadow onto the ocean floor that danced in the lights; rusticles dripped from its undersides and sea stars lay draped across its spokes.

Every man in the room knew that no matter how precisely Scotty could tune the navigation system, or how meticulously Bob could render geometry and trig, or how skillfully Moore could fly the vehicle, or how clearly Milt could focus and light the cameras, no one ever, ever, ever hit a deep-ocean target on the first track line with a camera. Ever. They had a greater chance of winning the Ohio State Lottery than they did making a shot like that in the deep ocean.

"The best excitement was when we went over the paddle wheel," said Moore, "because, I mean, it was totally unexpected!"

Tommy was trying to process this sudden information and all of its ramifications at once. He didn't know if they should even be seeing sidewheels. After the perplexing exploration of Galaxy the previous summer, when the search among the debris teased more than taught, when the progress proceeded in loops instead of along straight lines, Tommy had prepared himself for asking a lot of new questions and doing a lot of poking, trying to understand this interesting anomaly that Bob had discovered. And then fate had dealt him a scene that looked about the way a schoolboy would envision a sidewheel steamer would look after 131 years at the bottom of the sea. Big rusty sidewheels. Yet in Tommy's analytical mind, even this discovery only increased the odds; it did not prove this was the *Central America*. He knew that another sidewheeler had sunk in the Atlantic Bight and that by his and Bob's calculations, it might be within their probability area. That, too, had to be checked out. "All that's racing through your mind in about half a second," said Tommy, "and you're just going, 'Oh—my—gosh.'"

After a few minutes to consider what had to be done first, Tommy said, "We might have to put into effect those procedures we talked about." Burlingham was the only one in the control room who was not part of the tech team, and after a few minutes of seeing a few pieces of what might be the *Central America,* Tommy already was getting concerned about security. Nothing against Burlingham, just that to do his job as captain of the ship, he did not need to know what the site looked like on the bottom. "Need to know" was a phrase Tommy used often.

"They had mentioned something about it beforehand," said Burlingham. "It was in an oblique manner but obviously solely directed at myself."

Burlingham took the hint. "I'm outta here," he said.

With Burlingham gone, Scotty called up to the bridge to reverse direction, and as the ship sidled back along the same track line, they waited for twenty minutes, called up to the bridge again to stop the ship, then repositioned the vehicle directly above the center of the site. With the new lights, even at that height, they could make out the big pieces gnarled in the midsection and decide where to explore from there.

"We'll turn around . . . ," said Moore, as he ran the small thrusters on the vehicle and the vehicle slowly rotated to the right to reveal another huge wheel on the port side, "and we'll have a look over at this big bad boy, standing upright."

The starboard wheel they had first seen had tumbled outward onto the ocean floor; the port wheel had started to collapse but still clung to what was left of the side of the ship, the gears holding it up not yet corroded enough to release their grip.

Tommy watched the screen and said little. "These pictures are incredible."

"I must say," laughed Moore, "I'm rather impressed myself."

"The lights you got on here," said Tommy, "are also incredible."

"That's because they're where we put them," said Moore, "and not where most people put them."

Mindful of the starboard wheel shaft poking up, Tommy told Moore to lower the vehicle slightly for some closer flyovers. With the tech crew staring at the monitors, Moore flew the vehicle around the site, ten meters north, ten meters east, ten meters south, ten meters west. And for the next four hours, the vehicle waltzed above the ship.

AMIDSHIPS LOOKED LIKE a country junkyard in winter, quiet, serene, buried in white; like piles of old tractors and old cars surrounded by grayed and weathered fence posts, all under an inch or two of snow. But in these piles were old engines and boilers and water tanks and gears, all under an inch or two of deep-ocean sediment, like the resting place of a once proud steamer that had succumbed to the fury of the sea and accepted her fate with dignity and grace. Her sleekness, her blackness, the yellowed patina of her decks, the broad red stripe running stem to stern along her lower wale had all crumbled and turned to blue ash. Her spiderwebs of shrouds and the majesty of her sail and her real muscle, the enormous steam engines with piston strokes ten feet down, had disappeared or lay in disarray. Thin lines of blue and gray now ran at odd angles through the whiteness, long pipes perhaps, things of metal in odd shapes, some rectangular, broken, crisscrossing in piles.

As the vehicle flew slowly above the site, the black-and-white SIT camera relayed images topside of things that the tech crew the previous summer had only hypothesized might be found at the final resting place of the *Central America*. Since no one knew what a wood-hulled steamer should look like after nearly a century and a half at the bottom of the sea, the piles of coal and period artifacts at Galaxy had seemed more important than the absence of sidewheels and boilers and pipes. Galaxy was convincing until something more convincing appeared. Now that had happened in a rush: huge piles of angular metal strewn between two giant paddle wheels at an enormous site that they already had seen held at least a few of the trappings of a large passenger ship. Bob Evans's instinct the moment he saw the site on the sonagram had proved correct. After months of probing on Galaxy, they couldn't say for certain it was the *Central America,* only that there was nothing they had seen on the site to tell them it was not the *Central America*. Within seconds and after but a glimpse, they knew that Galaxy II was no longer a test site.

By the end of the dive, calm had settled over the control room. The vehicle performed as they had hoped, the DP system functioned perfectly, and they were seeing the remains of what they now were almost certain was the SS *Central America*. A few hours earlier, they had wondered if the vehicle would work, if they could find the site, and how long they would stay there before they moved on to Galaxy to dig in the coal. Suddenly, everything had changed. When they had completed the four-hour survey at twenty-five feet, Tommy ended the dive. "That was really convincing seeing the sidewheels," he said, "but I'm thinking, 'How do we explain this to the world?'"

Tommy had examined every possible scenario, and he had contingency plans for each, but with all of the pressures of disgruntled partners, thruster problems, vehicle design, competitors, buying a ship, money dribbling in, wondering where the gold was on Galaxy, wondering if the gold was not on Galaxy but on a new site, he hadn't spent a great deal of time wondering what he would do if on the first dive they landed on top of cast-iron sidewheels the size of a farmhouse. He had to guard that information carefully. If this was the *Central America,* then somewhere on that site lay hundreds of millions of dollars in gold, and that much money made people think crazy thoughts and do crazy things, and Tommy didn't have time to diverge on all of the scenarios that could arise if they returned to port and twenty men knew everything that was happening in that control room. He knew from working at Battelle it was easier to control the information itself. He decided that not even Bryan and Tod would be allowed inside the control room during a dive.

"I went out on deck and talked with them," said Tommy, "and I explained to them that we couldn't have everybody on the ship know, that we had to draw the line somewhere. I told them they would have to share that responsibility with Burlingham, and they did a good job, although I know it was difficult for them."

Day after day, Bryan and Tod and the ship crew would see nothing except the same faces on the same deck and the same water stretching to the same horizon. And only feet away, the rest of the tech crew would be looking for gold on a ship from another century. But the delicate balance between Tommy's need for secrecy and his desire to share the story with those who had helped make it happen had to be tipped in favor of security. The experiences in the control room would remain in the control room, and as difficult as it was for Tod and Bryan and the others to be on the outside, the secrecy would be an even greater burden for those inside; for they could share with no one the extraordinary sight they were about to see.

AT SIX O'CLOCK the next morning they were up working on the vehicle, a stiff breeze coming in from the northeast. With the *Discoverer* rocking in moderate seas, they worked on the vehicle all day, preparing for a more intensive study of the site. They wanted to explore and photograph the engines, examine the anchors and foredeck, and drop five-foot measuring sticks at selected points.

They launched again at eight-thirty that evening, the vehicle arriving at the bottom near ten o'clock. During those night hours on that day, September 12, 131 years earlier, the *Central America* had crashed into the seafloor. The miners trapped on board the steamer had died long before she hit, and their remains long ago had become one with the sea, drifting away slowly with the current. The only ghost left was the ship herself, alone in the dark, her captor quietly destroying her. When contrasted with the raging and relentless storm, the breaking apart, the final explosion and roar, the screams of that night in 1857, it was difficult to decide which was more surreal: the silence and monotony of the world in which she now lived, or the computerized world from which she now could be seen, with the monitors casting blue and the digital readouts glowing orange, and the steel umbilical running off the back deck down to a two-and-a-half-ton aluminum robot that illuminated the darkness of the world below.

By ten-thirty they had located targets on the Mesotech and had video contact with the site. Careful to avoid the starboard wheel shaft rising thirty feet above the core, Moore began flying the vehicle over the engine works: five meters northeast, five meters north, ten meters north, ten meters north, ten meters east, as the vehicle cast its light from only six meters above the site.

When it hit the ocean floor, the *Central America* was as big as a four-story building, three hundred feet long. Now, most of that building had collapsed in on itself, and they were looking down on it from above. The starboard side appeared to be collapsed onto the port side. The two remaining masts had fallen. Most of the deck was gone, the thick pine planking long ago eaten through. The supporting beams had also disappeared. Only the hanging knees remained, the braces for the supporting beams that shipwrights carved from the trunk of a solid oak tree and its first arcing branch, and they, too, were worn and riddled. Even the thick iron hog strapping that once girded the hull had broken loose and lay flat at the sides of the ship.

At the center, corrosion and time had buckled part of the cast-iron engine works, the broken shapes now further distorted with dripping rusticles. As Doering watched his monitor, he sometimes turned to Bob, pointed at the screen, and said something like, "That's the, uh, three-quarter deck." And then he would say, "Now, look at this thing here, that looks like one of those . . . ," and he would name some part of the ship they had discussed earlier. "I wouldn't see it at all," remembered Bob, "and after I'd talked to him, I couldn't not see it."

As Bob directed the dive, Moore fired scores of still shots, and Milt recorded everything on video. They searched the midships area for shapes that would help them identify the engines. That was a key element to Bob, for the ship was either the *Central America* or another sidewheel steamer that reportedly had sunk a hundred miles to the west, and the only way he could distinguish the two steamers was by their engines: Although similar in size to the *Central America,* the other steamer had a single "walking beam" engine that Bob described as looking like "the drinking ducks you see pumping oil out of the ground." The *Central America* had two oscillating engines. As they studied the engine works, Bob thought he was seeing the pieces from two engines, and he saw no drinking ducks.

From the engine works amidships, they gradually moved west and south toward the bow of the ship. At midnight they were hovering above an area they guessed had once been a little fore of the pilot house, near the foremast that Second Officer Frazer and Captain Badger had chopped down at the height of the hurricane. Among the silt and decayed hanging knees that once supported the cross beams under the deck, they saw two straight pipe stalks each with two large cylinders branching off in opposite directions. The stalks were horizontal, but rounded, so little silt had collected on their surfaces. Bob recognized them as the ship's whistles. At one time, the pipe stalks had led down to the boilers in the engine room, and the steam rising in the pipes, when released, had whistled out, like water boiling at tea time.

Moore eased the vehicle to within three meters, and they were studying the whistles, when Milt said, "What's that?"

Everyone searched the gray-and-white landscape of the foredeck but saw nothing except more silt and odd gray lines angling through it.

"It looks like a bell," said Milt.

"The year before," recalled Doering, "Milt kept seeing gold coins and gold bars and paddle wheels and all kinds of stuff, so it was like, 'Oh, Milt, you're seeing things again.'"

"No, here," said Milt. He walked over to Doering's monitor and pointed to it on the screen.

A gray hump nestled in the silt, something that appeared to be rounded with perhaps a slight flange. But with the scene only in black-and-white, and pools of silt collected on so many of the timbers, and shadows crisscrossing the silt, it was difficult to tell. They looked and they debated and they drifted around it at different angles, until some agreed with Milt: It might be a bell.

"But it could be like a spittoon," said Bob.

They drifted a little more, and Bob got a different perspective on it, and he too began to see what Milt was talking about. "It might be a bell," he said.

Moore eased the vehicle closer, and they studied the artifact for twenty minutes. Moore tapped lightly at the forward thruster for a gentle wash of water to tumble the silt from the flange. When the scene cleared, Bob and Doering both agreed with Milt: Not only was it a bell, but it was also inscribed, although it was so mottled they could not read the inscription. Even though the inscription could prove that this was the *Central America,* Tommy would touch nothing until he understood the site and could discern safe places to land. They left the bell half buried in silt.

For the last hour, they continued their flyover of the fore area, stair-stepping up the ship, first ten meters south, then ten meters west, ten meters south, ten meters west. When they arrived at the bow, they couldn't locate where the short bowsprit had pointed onward, because that whole portion of the ship had cleaved open like a book and crumbled.

At 3:00 A.M., they ended the survey and retrieved the vehicle. The following day, they launched again and explored the site for another seven hours. They placed a measuring stick close to the bell, another stick near the engine works, and another off the port bow. Between dropping the measuring sticks, they continued their survey, still hovering, touching nothing, documenting the site on video and with hundreds of stills.

ONE SPRING AFTERNOON in Columbus, Bob's curious mind had wandered back to that final scene: The men standing on the deck of the *Central America,* tossing coins into the sea and flinging gold dust about the cabin. That gold dust, Bob surmised, probably became part of the hydrodynamic plume, the cloud of debris that began heading down as the last waves washed over the decks of the *Central America*. With the ship plummeting at its center, the cloud had descended through the long column of water, expanding until it sprinkled onto the floor in plumes surrounding the ship. Bob hypothesized, *If this* is *the* Central America, *in those plumes one might find gold.*

Historical references agreed that apart from the gold shipments, the passengers themselves had carried about a million dollars in gold. At twenty dollars an ounce, that would be about fifty thousand ounces of passenger gold. Much of that would be in coin, but Bob had read many accounts of passengers stashing gold dust in pouches and treasure belts, which they upended in the final hysterical moments. Let's take a SWAG, he thought, a Scientific Wild-Ass Guess, and say that 10 percent of the passenger gold was dust. That would mean there were about five thousand ounces of dust on board. The grains of gold would be tiny, maybe half a millimeter on a side, he figured, so a cubic centimeter would be twenty grains times twenty grains times twenty grains, or eight thousand tiny pieces of dust for every cubic centimeter. Then he calculated how many cubic centimeters five thousand ounces of gold would occupy and multiplied that times eight thousand and came up with a number upward of a hundred million grains, and those grains would have been distributed throughout the pteropod ooze at something like one part per trillion. Whether four or five hundred miners could find, pluck, and store so many grains of gold seemed unlikely, but he concluded that in a fairly decent sediment sample one still might expect to find gold.

"But I don't think anyone else believed it," said Bob. Even Tommy had laughed when he mentioned the idea.

At the next launch, on the 16th of September, the dive objectives were to complete the photo coverage of the collapsed bow, survey the debris field off to starboard, and inspect the foredeck for a place to land near the bell. From midafternoon till well after midnight, they roamed, filming and photographing, and landing three times to retrieve four bottles and a plate. Each time they landed, Moore powered the arm out toward the artifact, and the shaft supporting the arm slid along the bottom, packing sediment into the shaft.

When the vehicle returned to the surface at 5:00 A.M., Bob removed the bottles and plate from the artifact drawer, then carefully scooped the mud from the shaft and carried it to his lab. He placed part of the sampling in a petri dish and dripped hydrochloric acid over it to dissolve the tiny skeletal remains of plankton that formed most of the sediment. Then he studied the petri dish under a microscope, poking through the remains with a pair of tweezers. Most of what remained were fragments of wood, bits of copper, coal, and iron, and even a few minuscule drops of solder from the vehicle itself. But just as he was about to quit his poking, at the edge of the dish he saw the tiniest flash of color. Everything else was brown or gray or black or all three, sometimes with a slight orange cast. This little pinpoint under the microscope, almost invisible to the human eye, was bright yellow.

Bob didn't know if it was gold, and even if it was, he thought, "You can't go to your investors with a piece of gold the size of a grain of salt and go 'Gold!'" Still, if the speck was gold, that was one more bit of evidence they were on the right ship, and if they were, that same particle of dust sitting in his petri dish had once winked back from a miner's pan along a stream in the Sierra Nevada.

He left the microscope on, locked the door to his lab, and went up to the foredeck, where some of the crew were casting for dolphin. He watched for a while, then fished for a while. When he saw Tommy and Doering, he asked them to come down to his lab, where he showed them the speck under his microscope. They decided to test it the same way James Marshall had tested the first gold he pulled from the American River that cold January morning in 1848, the original nugget that had started the California Gold Rush: Marshall had tried to break it by mashing it with another rock. This flake was too tiny to mash with another rock, so Bob left it under the microscope, angled the sharpest edge of a spatula against it, and pressed down. It didn't snap or crumble, but gave way softly. "Which confirmed," said Bob, "that this little fleck really was a fleck of gold."

Later, Bob parceled the rest of the sediment into petri dishes and gave each dish an acid bath. When he searched the cultural remains, he found a dozen more grains of gold, one stuck with a tiny plankton snail. He placed all thirteen in a vial, the original fleck marked with a groove from the spatula.

P ASSING INTO THE latter half of September, the late-season weather held steady at just shy of marginal, with winds bucking up to twenty knots and seas cresting at six feet. But any day, the winds would whistle up to thirty or forty knots and the seas swell to ten or fifteen feet and neither come down for a week. Regardless of threatening weather, their continuing exploration of the site had to be methodical: Finish filming and photographing from high altitude, then drop lower and continue to explore and analyze the more interesting parts of the ship, then drop lower again, and again, until the outline of artifacts sharpened and their bluish cast filled with color, and they had documented the site and knew where to begin their search for the gold.

But for the next five days, the vehicle was in and out of the water with a series of equipment problems and aborted dives. Not until the 22nd did they get the vehicle to roam the site again from early afternoon until late evening.

As Moore flew the vehicle over the collapsed foredeck, Milt fired the still cameras, one photograph after another from the captain's quarters up to the pilot house, and from there up to the bow. But the captain's quarters were gone, as was the pilot house; and the stump of foremast left behind by Frazer and Badger had disappeared. On the monitor, beams of light trailed across the soggy and riddled remains of thick planking once trod by hundreds of passengers, but now jutting only a short way into the wreckage before tapering into blue shadow. Among the shadows, they saw the shapes of bottles and the contours of ceramic pitchers and washbasins and hints of other artifacts, but at fifteen feet above the site, the picture on the monitors was too muted to see detail.

In one of the aborted dives of the previous few days, they had investigated an iron box lying in the silt forward of the bow. From a few feet away, it looked like a small rising oven mottled brown and orange from rust. Milt had photographed the box several times, then Moore had extended the manipulator slowly, paused, then moved it closer, then paused again. When the jaws reached the edge of the box and touched it lightly, the monitors suddenly filled with an orange cloud, and minutes later, when the water cleared, the box had vanished. "Just poof," said Tommy. "All we had left were the images we'd taken."

When they had filmed and photographed the foredeck, Scotty guided them through the darkness to a patch of sediment off the port bow, where earlier they had seen what appeared to be boxes made of iron. "There are bits of the historical record that talked about an iron locker in the hold," remembered Bob, "and there's a natural tendency to think *safes*." Moore landed the vehicle, and for a half hour they studied the boxes, some of them ten feet on a side. But most of them had a metal pipe jutting out one side, or fit within the architecture of the ship in a way that indicated to Bob that they were probably tanks for storing fresh water.

After filming the port sidewheel for the next two hours, Moore flew the vehicle over the debris field about fifty feet out. As they passed over the debris, someone noticed a solid geometric shape lying alone. It appeared to be another box, only much smaller than the ones they had photographed earlier. Moore guided the vehicle over and hovered above the box. As he dropped lower, they realized it was a leather suitcase, and sitting about six inches away, straight up as though waiting to be sipped, was a white teacup.

In all of his days exploring sunken ships, Doering had never encountered a scene so eerie. "It looked like a . . . like a train platform in foggy London," said Doering, "like somebody was sitting on his suitcase drinking his tea, and his train came in, and he set the cup down to catch the train and forgot his suitcase."

They were always interested in boxes. "In the old-fashioned, storybook concept of treasure," said Bob, "you have a chest about the size of a desk, and it's overflowing with emeralds and diamonds and gold." But a desk-size chest filled with treasure would weigh thousands of pounds and no ten men could lift it. Whereas a hundred pounds of gold would fit neatly into something smaller than a loaf of bread.

Bob thought the suitcase next to the teacup was small enough to contain valuables, and he told Moore to get as close as he could. When Moore eased the vehicle down, a huge cloud of sediment enveloped the suitcase. With the forward thruster trained on the scene and spinning slowly, a gentle stream of water pushed away the cloud, but when the scene cleared, the white teacup was gone and the leather suitcase had tumbled into the distant blue, where it sat, upright once again.

Moore brought the vehicle closer. A white feathery coral rose from the brown leather, and large pink anemones clung to the top and sides. Beneath where the handle had been was a name plate, but the letters were covered with sediment. Moore trained the forward thruster on the suitcase and spun it a few revolutions for a light wash. When they still couldn't read the letters, he ran the thruster a touch faster, and suddenly the suitcase opened like a clamshell, hung open momentarily, then slowly closed again. It was packed with shirts neatly folded behind the webbing that covered each compartment. "The very notion that fabric could survive at such depths in these conditions," said Bob, "opened up some interesting possibilities." They photographed the suitcase several times from different angles, then left it once again to lie alone in the darkness of the debris field.

Topside, the days dragged on, the weather rising, then lying down, then rising again, and each time, it rose higher than the time before and remained longer, and the periods of benign weather grew shorter. Worse than the weather, the monotony of being at sea had frayed thin the nerve endings of the ship crew. In similar circumstances in the Merchant Marine, the first mate had seen a cook throw a mayonnaise jar and split open a seaman's head; he had seen seamen hack each other up with fire axes. Tommy's security policy only made the crew more surly, but his security measures remained in place. He was responsible for protecting the investment of his partners, and if that caused problems with morale, he would have to deal with those, but he would not relax his security. "It isn't a matter of trust," he said, "it's a matter of being responsible for the kind of information you have."

When the vehicle came up after a dive, Tod and Bryan winched it under the awning, so the artifact drawer was facing forward away from the bridge, and placed a tarp over the front end. Then everyone had to leave the deck and clear the bridge except for the watch. Tommy even instructed the watch to step away from the windows until the artifact drawer could be emptied. Bob then opened the drawer, removed the artifacts, and took them to his lab to study and catalog.

On the eight dives so far, two had aborted after only minutes on the bottom. In the remaining six, they had spent most of their forty hours on the bottom, roaming above the ship, filming and photographing, each time dropping closer as they found some areas more interesting than others. Milt had bought film in 100-foot strips, then cut it and wound it into his own 250-frame cassettes, one for each camera. Using strobes to brighten the scenes, they shot as many frames as they could on every dive before the thrusters failed, or the topside software died, or a power glitch erupted in the subsea computer. Sometimes the strobes misfired or the apertures stuck.

After each dive, either Doering or Milt developed the film from the day's shoot, cut the frames into strips, sponged them off, hung them up to dry, and laid the strips on a light table to study with a loupe. The pictures from these still cameras were closer, sharper, more detailed than what appeared on the monitors during the dive. The crew understood the site only by analyzing these close-in stills.

The ship had started at the surface as a mass of cultural objects in a narrow capsule, 278 feet long, 40 feet wide, about 30 feet high. Bob estimated that the site now covered over ten acres. "It's an incredibly complex object that has an incredibly complex series of processes evolving upon it. You've got corrosion and biology and gravity and a little current all acting upon what originally arrived at the bottom, and that's gone on for 130 years, so that further confuses the issue, Where is the gold?"

As they flew the vehicle lower and lower, each new wave of photography revealed another world, clearer, more vivid, more detailed, no longer a country junkyard dusted with snow, but a desert oasis created by the flotsam of civilization and inhabited by the creatures of the deep ocean.

Because water bends light, the colors of the rainbow washed out before they reached the cameras: Red went first, then orange, then yellow and green. By the time the light had traveled twenty feet, only the purples and blues remained. Photographs from the early dives came back in dull shades of blue, which masked many of the artifacts and most of the life forms. "The best example of that," said Bob, "is a feathery life form called gorgonian coral. It's about a foot long by two inches across. You can't see them flying over, yet they're strewn all over the site."

As they worked the vehicle lower among the timbers and their shadows, the techs could distinguish more artifacts littering the site: small chips and splinters of wood, broken crockery, bottles, soap dishes, iron gratings, plates, vases, washbasins, copper sheathing, chunks of coal covered with rust. And the closer they got, the more color reached the cameras: On trunks and under beams, anemones quivered in the faint current a fiery red-orange or a yellow brighter than canary. Shiny black urchins and green sponges and blue and purple cucumbers perched among the wreckage. The white gorgonian corals stood in small forests, with tiny white crinoids stuck in their branches. In some areas, on slow patrol, swam a needle-shaped fish three feet long with a sharp face and translucent fins sweeping up behind its head.

The site teemed with another deep-ocean creature they never saw: the tubeworm. For over a hundred years, tubeworms had infested the site, boring holes into the wood until stout timbers had collapsed into pieces and disappeared. Tubeworms still lived all through the riddled wood at the site, but looking at a knee or a beam from the outside, no one could tell if the worms had bored it almost hollow, or if the beam stood firm.

THE BELL STILL lay among the rotted timbers of the foredeck. Although it was the single artifact that could prove they were on the *Central America,* Tommy wanted to leave it alone until they could devise a way to recover it without destroying other artifacts.

With the wind still holding below ten knots and the sea under three feet, they deployed the vehicle again on the afternoon of September 23. For several hours, they ran flyovers, dropping closer to the ship to film and photograph among the timbers in areas that in the higher surveys had looked promising as high-value cargo sites. Then Scotty guided Moore through the darkness to a clear patch of sediment off the port side of the foredeck.

For this dive they had removed the rotating base from the vehicle and mounted a small winch up inside the framing. At the end of the winch cable hung a grappling hook that held a basket snug against the bottom cube. Moore lowered the basket onto the clear patch, then flew the vehicle to hover a few meters above the bell. Half buried in silt, it looked small.

Doering had stood in front of the ship's bell on the *Queen Mary,* dumbfounded that on that huge liner the bell took up less space than the average lampshade, maybe eighteen inches across. He studied the bell now on the monitor and calculated it could be no wider at the mouth than fourteen inches.

Moore landed the vehicle gingerly within three feet of the bell, directed the arm outward, tipped the bell upright, and slipped the jaws up under it like a fist in a slow-motion uppercut. As white silt trailed away, he guided the arm upward and held the bell up close in the lights, rotating it slowly in front of the cameras. It had a greenish cast and was mottled in orange and brown. An inscription encircled the upper portion, and another wound around the flange at the bottom. Out of both inscriptions, they could read only a few random letters, maybe an "e," maybe a "w," maybe an "a," maybe an "r," and one word, "York." They still couldn't tell its size.

After shaking the bell to check its stability atop the arm, Moore lifted off, flew the vehicle away from the core, and landed next to the basket. There he laid the bell on its side, partially buried in silt, and they photographed it from various angles another dozen times. Then Moore lifted the bell again and lowered it on its side in the basket. A small cloud rose, the mechanical arm pulled slowly away, and the camera came in closer. The basket was eighteen inches deep, and as the cloud cleared, they could see that the bell poked up at least another six inches. Doering was amazed. "I can't envision why anyone would have a bell that big."

After they had photographed the bell from several angles and every-one had wondered at its size, it was three-thirty in the morning. If they retrieved the vehicle now, by the time it arrived at the surface, the sky would be light, and Tommy wanted the bell to come aboard in dark-ness; he wanted no one but the tech crew to see it. He told Moore to leave the basket and the bell in the silt outside the ship, and they would re-trieve it the following evening.

THE TECHS RECOVERED the vehicle at dawn and slept till noon. When they awakened, Bob, Doering, and Milt went to the lab or returned to the darkroom to develop more film and continue scrutinizing earlier photographs of the site. Understanding what they had seen in previous dives would help them see more clearly what appeared on the monitors in the next dive. Tommy called it the learning curve, and having fol-lowed this procedure for one extended season and now part of another, they had become more skilled, yet ever more cautious with their under-water vision. They knew that colors and shapes and textures could all combine to turn lumps of coal and bits of wood into what appeared to be a bottle of wine, or a pile of coins, or a mottled bar of gold. On one dive they had landed close to the core, and as they were studying the near landscape, Bob and Milt both thought they saw a brick lying in the timbers. Moore closed in with a camera, and the bluish scene evolved with more color and detail, and now everyone could see the sharply delineated corners dusted with sediment.

Moore reached out with the manipulator and caught the brick from the side with the tip of the jaws, nudging it upward until it flipped over, raising a tiny cloud of sediment. Bricks made of clay weighed two and a half times as much as water, but bricks made of gold weighed seven-teen times as much and would not flip over so easily. They now knew that this brick was made of clay, that it was probably a fire brick, and that hundreds or even thousands of these fire bricks might litter the site: sharp angles, perhaps in piles, hidden beneath a fine coating of white sediment, and masquerading as a shipment of gold bars.

Another time, they saw an octagonally shaped coin standing on end, and just to the right of it lay a small gold bar. As Moore brought the camera in closer, the two artifacts looked even more like a coin and a bar, and then the camera got within two or three feet and the coin magically transformed into a dark blotch smeared with orange color on the surface of a decaying timber, and the small bar became the bloated grain of the wood.

"It's good to maintain your enthusiasm," said Bob, "but it's frustrating when what you thought was gold turns out not to be gold." It had happened so many times that now when anyone thought he saw gold, his second thought was, No, it must be a fire brick, wood grain, a knot, a lump of coal, a tubeworm.

The most confusing of all the misleading signs of gold at the site were the casings left behind by the tubeworms. As the worms bored into a beam, they created curlicue shavings of calcium carbonate, like shells, six to ten inches long. When the worms burrowed into the timbers and the timbers finally broke apart, the white casings dropped to litter other timbers in the wreckage. Then the iron oxide bleeding from hundreds of tons of corroding steam engines drifted through the water, staining the shavings a yellowish orange. At a distance of only several feet, the glare of bright lights shining on those stained shavings made them glint like collapsed piles of gold coins. The crew had seen similar trailings on Galaxy the year before and been fooled time and again into thinking they were seeing gold. The site at Galaxy II lay covered with them.

Milt had developed the film from the long dive on the 23rd, when they had shot for hours among the timbers before plucking the bell from the foredeck. As he studied the pictures, he saw something in one he had seen in none of the hundreds of other stills they had shot: a series of tiny bright spots with a faint, yellowish cast. The lights seemed to have hit something metallic at the right angle to ricochet back at the cameras just as the vehicle passed over a riddled beam. The beam ran way back into the timbers, under collapsed parts of the decking. Other beams ran parallel to it out of the debris, but this beam angled upward by itself another four or five feet. It was worn, jagged, its original ten-inch heft eaten down, and it tapered toward the end, where Milt had seen the glimmer. Whatever had caused the glimmer, the scene looked as if that worn piece of timber had reached out like a catcher's mitt reaches out to cup a foul ball. Milt told Doering about the slide.

"He was always doing that to me," said Doering, "And it was a joke. 'Oh, Milt, what'd you see now?'"

But Milt had seen the bell first, Doering couldn't deny that, and he had seen it in a complex area of debris, at an odd angle, with poor lighting, and other shadows and shapes around it to confuse what he was seeing. So maybe Milt was developing an eye for discerning three-dimensional objects in a distorted, two-dimensional world. Then again, maybe he got lucky.

With the two of them in the lab and the door closed, Milt said, "I just developed these pictures. Take a look at this one."

Doering leaned over the light table, squeezed one eye closed, and peered through the loupe. He studied the picture for a long while, saying nothing. Then he looked up with a by-golly grin and said, "You know, Milt, that does look good. It looks damn good, like gold coins laying there."

When the vehicle flew over that part of the ship two or three meters up, everything around the beam had appeared on the monitors in varying shades of blue. No one had noticed anything unusual. And they still couldn't be sure. "Either the camera wasn't in focus or we were still too far away," said Doering, "but it was really the first time we both agreed that that definitely had a possibility of being gold."

Tommy planned each dive for efficiency. He had little time left, and much to do, and he could not land until he understood the site. He was always prepared to deviate from a plan but not abruptly and not completely, even to look for gold. When Milt and Doering showed him the transparency of the beam, he had already written the dive plan for that night, a short dive only to recover the bell. That would remain the single objective for the dive, but in the plan for the following dive, he scheduled time for an additional objective: to take another, lower photo run at that beam sticking up and analyze it for a way to land nearby without disturbing the site.

THAT NIGHT, MOORE had the vehicle on the bottom next to the bell at nine-thirty, and in forty-five minutes he had snagged the basket with the grappling hook, winched it up snug beneath the vehicle, and the vehicle was on its way to the surface with the bell wedged into the basket. When the vehicle swung aboard on the crane, they slid the basket from underneath, covered it with a tarp, and Tommy had everyone leave and stay away from the area.

Late that night, alone on the deck, Tommy, Bob, Milt, and Scotty slid the basket into a flood of light underneath the awning, rinsed the bell with fresh water, and brushed it lightly. Patches of pale orange and verdigris mottled the dark gray-green of the bell, and a blotch of cobalt blue appeared in the middle of the word "Morgan." In the band around the top of the bell, they now could read the inscription "Morgan Iron Works New York," the same foundry that had cast the fittings and the huge steam engines of the *Central America*. Although the first two numbers of the date were lost in orange rust, they could read half of the number "5" and right next to it the number "3." The *Central America* had set to sea under her christening name the SS *George Law* in 1853.

In Tommy's mind, the odds that this site was the *Central America* had jumped dramatically the moment they saw the sidewheels and the engine works almost two weeks earlier, and with each successive dive, studying the configuration of the ship and seeing Bob's tiny flakes of gold, he had become more convinced. But it was infinitely more difficult to excite his partners with another chain of circumstantial evidence than it was to say to them, "Here's the bell from the SS *Central America*." It wasn't gold, but it was the next best thing. "That really confirmed it," said Tommy. "You start thinking about the odds now: We got sidewheels, we got the right engines, we got the bell, this is really it!"

From the loop at its top down to the flare of its skirt, the bell was cast from solid bronze two feet wide and two and a half feet tall. It weighed about the same as two average-size men. To move it, they had to run heavy deck lines through the loop and knot the lines over a thick pipe. Then, with a tarp draped over the pipe, Scotty squatted at one end and Milt at the other, and the two stood together and walked with quick steps across the deck to the starboard side of the housing like hunters shouldering a carcass.

The bell was so wide, they couldn't get it through the doorway until they took the door off its hinges, and even then they had to tilt it and twist it and jam it through in one long, continuous push with help from

Tommy and Bob. Once inside, they removed the pipe and the ropes, then Milt took two steps down the narrow stairs leading to the lower deck, the others rested the bell against his back, and he slowly descended the steps to Bob's small laboratory, where they took the lab door off its hinges and worked the bell through the doorway as they had topside, and carefully set it on the floor. Then they put the door back on its hinges, and Bob locked the door, leaving among his specimens and artifacts and books on biology, oceanography, history, and life at the bottom of the sea, the same bell that Captain Herndon had used to sound his departure the morning the SS *Central America* had steamed from Havana.

THE DAY AFTER they recovered the bell, the weather crept back up to fifteen-knot winds and six-foot seas, still low enough to launch and recover the vehicle, but that day they did not dive; they waited for parts from the beach. With the vehicle cinched down and covered on deck, Tommy, Milt, Doering, Barry, and Bob continued to study and talk about the beam. Three of them had persuaded Tommy to increase the time he had allotted in the next dive plan for shooting closer stills of the curious yellow-orange glint. Barry thought that the glint looked like mushroom caps growing on the beam, but Tommy now agreed that the beam deserved a closer look.

The following day the winds edged up another ten knots and the seas another two feet, and again the vehicle remained on deck. On September 27, despite the seas' hitting five to eight feet, they launched the vehicle at nine-thirty in the morning. Tommy's dive plan was to run still-camera flyovers from the rudder up the keel, then work over to the beam Milt had spotted and shoot it as close as they could. First, they dragged forty-meter track lines up and down the stern deck; although they shot 172 stills, they didn't know how many would turn out, because the camera shutters seemed to be sticking and the strobe often was not in sync.

Moore then flew the vehicle at Tommy's direction and came in above the area Tommy wanted to shoot. "I remember that day quite clearly," said Moore, "because they were very interested in this one spot, and we could see some bright stuff."

The timbers there were sharp and jagged and riddled with holes. Time and the sea and the animals of the sea had conspired to reduce the once stout beams to a forest of decayed wood with pockets and crevices and holes all filled with white pools of sediment.

As Moore flew the vehicle slowly toward the beam, Tommy said, "This is it, right here. It's one of those."

Moore eased down closer.

"Do not knock it off," said Tommy. "You close enough to get a shot of that?"

"No," said Moore.

"How close are we?" asked Tommy.

"Twelve feet," said Moore.

Moore dropped closer. The timbers in this whole area looked so thin and fragile that if he were to touch anything, even being underwater, it would snap with a dry crack. The vehicle inched closer, now within seven feet of the beam, now within six. As it crept in, shadows moved in the background, giving Moore a better sense of the third dimension.

"It's that thing sticking right out on the end there," said Doering.

Down closer, the beam looked almost like the head of a mythic animal, its mouth open, a gargoyle jutting out from the front of an old library. With the vehicle five feet up, Milt fired the still cameras, the whole scene brightening in the flash. Light bubbles of particulate matter floated by. Milt fired the cameras again, and a couple of shiny spots reflected the vehicle's lights.

Everyone watched the monitors as Moore eased the vehicle closer and closer, and Milt snapped pictures of decayed wood covered with silt and the occasional spindly-legged sea star.

"It's these bright things, right here," said Tommy, pointing at the monitor. He agreed with Milt and Doering—they could be looking at coins. "Those things here are what we're . . ." His voice trailed off. "Boy, look at those. What do you think, Bob?"

"In this area," said Bob, "it's deceiving."

The vehicle now was directly above the beam, only two feet away. Except for the intermittent moan of the thrusters, the control room was silent as the techs studied the monitors and strained to see beyond what the resolution of the video cameras would allow them to see. "Those are the droppings of the infamous tubeworms," said Moore.

"I don't think so," said Tommy.

"Yep," said Moore. "We have lots of tubeworms that are exactly that color in the photographs."

"I don't think so," Tommy said again.

"Yep," Moore said again, "they're tubeworms."

"Okay," said Tommy, "down on that . . ."

". . . there are tubeworms," Moore finished Tommy's sentence. "Can show you in other photographs why. It's that color they turn out."

The vehicle was rotating slowly to the right, the cameras looking into a depression where they saw many short, straight, gray lines, like small bricks covered with dust.

"What I find interesting," said Moore, "are those blocks right there." Milt fired the cameras at the depression several times, and Moore got increasingly excited about what he was seeing. "I don't give a shit about your tubeworms," he said, "but those are goddamn bricks! I don't know if they're bricks out of an oven or if they're bricks out of a boiler, but those are goddamn bricks right there, and that's where I'd go down and take a look!"

Milt shot another forty-three photos around the gargoyle, until just shy of five o'clock in the afternoon, when the power blew in the subsea computer and they had to end the dive.

THAT NIGHT THEY worked on the vehicle till late, and the next morning they were on deck again at seven o'clock, continuing their work. Although the seas were at eight feet, they would try to launch. At midmorning Milt began preparing the cameras for the dive and Doering went to the darkroom to develop the film of the gargoyle from the previous day.

Between the two cameras, they had shot 215 stills during the flyovers and the hang-and-shoot above the beam. Doering developed the 164 frames from the port camera first. When he pulled the film from the chemical bath, cut it into strips, and hung them up to dry, he noticed that the first strip was black. So was the next, and so was the next. The entire cassette had somehow been overexposed.

Doering knew they had had problems with the camera during the dive, but he thought at least some of the pictures would turn out. Now he wondered if he was doing something wrong in the developing. Without the stills, they couldn't tell what they were looking at, and without that finer vision, they could be on top of gold and not even know it. Weather and technical problems already limited the number of dives, so each opportunity to shoot stills was becoming more and more precious, and they couldn't afford to waste these opportunities. Any dive could be the last, and then they would have to head back to the beach with only a bell and hundreds of interesting photographs to present to their partners.

When Doering saw the black strips of film, he was so disgusted with himself and the cameras, he left the other film cassette in the darkroom, locked the door, and went to the galley for lunch. At lunch, one of the other techs asked him how the film looked from the last dive, and Doering drooped his shoulders and widened his eyes without looking the tech in the face. "Well," he said, "they didn't turn out at all."

"It was fairly demoralizing," Doering said later. "I was feeling kinda bummed out about it."

After lunch, he returned to the photo lab to process the film from the starboard camera. According to his records, they had shot that camera fifty-one times. When he pulled the film from the chemical bath, he cut it into five-foot strips, hung them from clips, and screened off the chemicals. As he screened them off, he glanced at the strips to see if he had images; he did, and that made him happier. After he had all of the strips up and drying, he cleaned up the chemicals and put everything away, and when he returned to the film, he laid the strips on the light table to check the exposure. At those depths using artificial light, they constantly had to guess at the aperture setting and how fast they were firing the shutter; that was usually the first thing he checked when he developed the film. Then he would cut them into six-slide strips, coax them into plastic sheaths, and hook them into a three-ring binder.

For over fifteen years, Doering had been looking at things underwater. He had seen artifacts lying on the bottom of the ocean that dated back three or four centuries, back to when explorers and conquerors still plied the Caribbean in ships stuffed with booty from the New World and an abundance of their own tools of daily existence that would become archaeological treasures after aging at the bottom of the sea. From societies long past, he had seen jugs and dishes and bowls and canisters and vials and cut glass. He had seen armor and swords and arquebuses and cannons. He had seen collections of jewels and bars of silver. He had seen gold. But nothing prepared him for what now lay on the light table waiting for his eye.

With the first strip laid out, he placed the loupe over one of the pictures, and in a twinkling appeared the most spectacular sight he had ever seen.

"It was just . . . it was just . . . *covered* with gold! I couldn't believe it! I *couldn't* believe it! That was the most thrilling. . . . We had hit right on a pile, nice low pictures, nice and clear. I mean everything was perfect, man. It was incredible! But I looked at it, and I looked up, and, Naaaah, this can't be. I thought, That's gotta be a bunch of brass laying there. So I looked again! Holy! And I just started looking at the other shots, and I . . . mean . . . it . . . was . . . PILES! I'm not kidding you, it is awesome! It is absolutely awesome! Stacks of coins and bars of gold of every size and shape are just sitting there!"

Doering picked up that five-foot strip, grabbed the loupe, and took the steps two at a time up to the next deck, where he found Tommy and Barry in the communications room. The door was open, so he walked in and ceremoniously closed the door behind him.

"I'm telling you my heart was pounding," he remembered. "I mumbled something incoherent like, 'Wah, wah, wah.'"

He handed the strip to Tommy, who held it up to the light.

Doering heard Tommy say something like, "Boy, I never thought . . . ," and then he heard a yahoo.

"God, he was happy," said Doering. "We were just on top of the world! We had done it! We had found it!"

But Tommy warned Doering to say nothing about the pictures and do nothing that would make the others suspicious. He was excited, he felt great, but they couldn't let on to anyone what they had just seen. Tommy needed to deal with the information responsibly.

Bob was on deck with Burlingham, Tod, Bryan, and a few others, preparing the vehicle for the day's dive. Doering had to get down on deck to run the crane. He hurried back to the photo lab, laid the strip out on the light table, locked the door, and went out on deck. Bob was in his life vest standing just behind the crane, ready to handle one of the lines on the vehicle during launch. Doering knew that Bob had a key to the darkroom; he also knew that when they had the vehicle ready to launch, Moore or somebody would spend another twenty minutes checking the switches and doing other little deck checks. He perched himself on the crane pedestal with a sassy grin on his face and aimed it at Bob. Bob saw the grin and hoisted himself up onto the rail next to the crane; Doering leaned over.

"I think there's something you ought to see," whispered Doering. "When we get a break here, there's some film laying on the light table. Take a look at it, and let me know what you think."

After listening for a second, Bob started hearing between the lines, and then a grin sprouted on his face, too. He set his line down, but just as he turned to walk down the deck, Doering stopped him.

"Oh, Bob," he said, "better take along a change of underwear."

BEFORE THEY COULD launch the vehicle, the weather dropped further, and they could not dive that day. Nor could they dive the next day or the next. For three days the seas rolled in at eight to ten feet, and the wind peaked at thirty knots. And during those three days, the vehicle sat on deck and Tommy and Bob studied the pictures of the site, those taken earlier at higher altitudes and the later tight shots, looking for a place they could work the vehicle in close enough to photograph the area and then reach out with the manipulator and recover coins and bars. Because the vehicle left a footprint four feet by twelve feet, Moore needed a flat patch of sediment that size plus a margin of a few feet on each side. But the gold lay in a fragile area near the edge of the core, where remnants of timbers crisscrossed and jutted out. "It looked like it could all just cave in," said Tommy.

Now into October, the seas remained at five to eight feet, and the wind came in at a stiff twenty to twenty-five knots. They launched on the morning of the 1st anyhow, and the vehicle arrived on the bottom at 9:19, suspended just above the area they now called the Bank of California. Tommy had one objective on this dive: Shoot magazine-quality photos of the gold.

After they had surveyed the scene for ten minutes, Moore eased the vehicle lower and settled in next to an area Tommy had selected to begin their photography. It was safe and at the edge of what looked like a dense concentration of gold. With the vehicle resting on the bottom, they all stared at the monitors for a long while as Moore panned the cameras back and forth across the site. The piles of coins and bars that had shocked Doering still lay mostly covered with sediment. Doering had seen gold peeking from beneath the soft whiteness, hard angles of yellow bars askew and large spills of small, round, flat objects with hints of orange and brown and yellow showing through the silt, and piles obviously of coins and bars coated in white. Along the gargoyle, the hundred or more coins that had caught Milt's eye glinted because the beam stuck up a few feet from the debris, and the slow current constantly washed them clean.

Moore aimed the forward thruster downward and dispatched a gentle wash of water to blow away the cover material. The sediment was thin, but when the wake of the thruster hit, it exploded upward, swirling into clouds, blotting out the rotted timbers, turning the monitors white. For several minutes the techs could see nothing but the roiling sediment. Then the clouds began to drift with the light current; the picture on the monitors began to clear and slowly revealed a scene few people could imagine.

"The bottom was carpeted with gold," said Tommy. "Gold everywhere, like a garden. The more you looked, the more you saw gold growing out of everything, embedded in all the wood and beams. It was amazing, clear back in the far distance bars stacked on the bottom like brownies, bars stacked like loaves of bread, bars that appear to have slid into the corner of a room. Some of the bars formed a bridge, all gold bars spanning one area of treasure over here and another area over here, water underneath, and the decks collapsed through on both sides. Then there was a beam with coins stacked on it, just covered, couldn't see the top of the beam it had so many coins on it."

Like deep-ocean sentries, sea creatures guarded the treasure: gorgonian corals, feathery and white, stood erect above the gold; brisingid sea stars, a brilliant pink-orange, sprawled across piles of yellow bricks or perched atop a single bar, their arms drooped possessively; red anemones, their tentacles splayed, stuck to ledges and inside crevices spilling with coins and bars. The scene was live, yet seemed forever like a photograph: piles of gold, much of it as yellow as the night it went down, surrounded by the neighbors it had known since that night in 1857.

"There's so much of it," said Doering, "and it's shiny, and some of it's real bright, and some of it sparkles at you, and some of it has this reddish cast to it, beautiful hues, all unmistakable. I wanna pick it up! I wanna bring it home!"

So many bricks lay tumbled upon one another at myriad angles that the thirty-foot pile appeared to be the remnants of an old building just demolished. Except these were bricks of gold: bricks flat, bricks stacked, bricks upright, bricks cocked on top of other bricks. And coins single, coins stacked, coins once in stacks now collapsed into spreading piles, some coins mottled in the ferrous oxide orange and brown from the rusting engines, others with their original mint luster. Besides a tiny squat lobster carefully picking its way across piles of coins, the scene lay perfectly still.

"Look at those damn fire bricks," laughed Moore.

Milt saw a collection of bars and coins watched over by a pink-orange anemone softly fluttering in the current. "Boy, that's nice," he said.

Sticking up out of another area was a coin tower, eight stacks of gold coins, twenty-five coins to a stack, all of the stacks abutting one another like poker chips still in the rack, the whole thing frozen together and angled upward at sixty degrees.

"Isn't that amazing," said Bob.

"That is amazing," said Doering.

In a couple of hard days, they could have grabbed every brick and most of the coins, but Tommy wanted the scene fully documented before they touched anything. Then he wanted to document the recovery of each bar and each coin carefully, one at a time.

Directly in front of the vehicle, only a few feet apart, were the first three scenes Tommy wanted to photograph and film, and the new rotating base allowed them to swivel from one to another without moving the vehicle: from the coin tower, to a pile of collapsed coins, to a mound of gold dust frozen ten inches high, dotted with nuggets, and capped by two small gold bars. As Tommy orchestrated the shoot, Moore rotated the vehicle, extended and retracted the camera trolley, and operated the light booms; Milt combined aperture settings with shutter speeds and tried as many lighting angles as he could think to use. Sometimes they fired the cameras every ten to fifteen seconds, sometimes they rested the cameras to think through the next sequence. With the cameras sweeping the piles, everyone watched the monitors as gold eased into view, passed by, and was replaced by even more gold, the scenes continuously brightening with the pop of the strobes. After two hours, they had taken dozens of pictures, but they had touched none of the gold.

When they resumed the dive in the early afternoon, Tommy had Moore shoot another ten pictures before he directed him to recover the first artifact. Eight minutes later, he told Moore to pick up another artifact; five minutes later another artifact; five minutes later another; six minutes later another; twelve minutes later another. Each recovery came only after shooting several more stills of the scene and ended with Moore setting each piece of gold carefully in its own numbered compartment in a plastic tray resting on the seafloor. They now had six artifacts, each a small bar or coin collected at the edge of the scene. But Tommy would allow Moore to touch nothing else.

After they had completed the final shots of the three scenes, Moore rotated the vehicle beyond the mound of gold dust, and Tommy directed him where to aim the cameras for the pictures he wanted of a pile of fifteen to twenty gold bars, each about the size of a house brick. Moore swiveled the vehicle slightly, and into view came what looked like the head of an animal with scales: the coin-encrusted beam they now called the gargoyle.

"There's the lure," said Bob. "That's what brought us here." The gold coins stretching the length of the rotted beam, the coins that had glinted with just enough light to catch Milt's eye, now looked like gold coins: shiny, yellow, round, engraved. As the cameras focused on the beam, more particulate matter drifted slowly by in the water, little white reminders that the otherwise still scene was not a photograph. Moore shot a dozen new stills of the beam. Then they focused on a pyramid of at least a hundred more coins, slowly moving the cameras in closer and closer. In the background, slightly blurred, a light pink-orange anemone opened its frondlike tentacles, which waved hypnotically in the current. Moore slid the camera so close to the pile, only inches away, that the crew could see the precise grooves etched around the edges of the coins, which now appeared huge on the monitors. One had its back turned, and when Moore adjusted the camera, across the top they could read, "United States of America." At the center was an eagle shield and the rays of the sun, and they could count thirteen stars in a tight oval above the eagle. Curving upward like a smile along the bottom appeared "Twenty D.," and right above the "n" in "Twenty" was a tiny "s," the mark of the San Francisco Mint.

"Look at that eagle shield on the back," said Bob, "the luster on it!" Bob wanted an even closer shot of the mint stamp. "Just to the left and down," he said to Moore.

Moore tapped the camera to the left.

"Look it!" yelled Milt. "Eighteen . . . you can read it!"

In adjusting the camera for Bob's shot, Moore had passed another coin with the front side facing the camera, but he had swung about two inches too far to the left.

"Wow," said Tommy. "Read that date, John."

"We're working on it," said Moore.

Everyone laughed as Moore tapped the camera back again. There stood a coin upright, face front, just as pure and lustrous as the day it left the San Francisco Mint. It was emblazoned with the bust of Lady Liberty, lovely in profile, her hair crossed with a tiara and cascading in ringlets down her neck, thirteen stars surrounding her, and her ringlets stopping just short of the date "1857." In a pocket thirty feet across, the ocean floor lay covered with these coins.

Doering figured he had now seen more gold in one place at one time than any other treasure hunter in history, and that included Cortés and Pizarro. He was ready to pluck some of that gold from the ocean floor, drop it into the artifact drawer, and bring it to the surface, so he could feel it right there in the palm of his hand. "Harv's rationale," said Doering, "was that this is the last time this gold will ever be in this untampered state, that you can't have too many pictures of it 'cause you'll never have that opportunity again. Couldn't agree with you more. So we started taking photographs. And we took photographs. And we took photographs. And we kept on taking. . . . 'Okay, Harv, we've got thousands of photographs from every conceivable angle, every conceivable lighting situation. Let's start picking this stuff up!' He wouldn't do it. We just tried everything, and of course, he's still not satisfied. It drives me up a wall!"

For another hour they filmed and photographed the scene, panoramic sweeps of piles of gold, and tighter shots of crevices choked with bars and strewn with coins, much of the treasure draped with orange sea stars and sprouting stalks of gorgonian coral. Since recovering the last artifact, they had shot over a hundred more photographs before Tommy directed Moore to recover a coin. Then they shot several more photographs, after which Moore plucked a large brick from the edge of the pile. They continued to shoot and retrieve until bottomside computer problems forced them to end the dive. Before the vehicle ascended, Moore recovered a cluster of eight coins directly in front of the vehicle. In plastic trays, each separated from the other, they now had twenty-seven gold artifacts, and for every artifact they had collected, they had shot nearly ten photographs.

When the vehicle came to the surface at six o'clock that evening, Bryan and Tod put it on the tugger and pulled it forward under the awning with the front of the vehicle facing away from the bridge. Then Tommy ordered the tech crew to leave the deck and the men on the bridge to step back from the windows while Bob removed the plastic trays from the artifact drawer and took them into his laboratory, which he then locked; no one else was allowed to touch the few pieces of gold, no one else was allowed to look at them; no one else was even permitted into the room.

"The whole notion of not letting anybody handle the gold but Bob Evans just drove people crazy," recalled Tommy, "absolutely drove them crazy. But that way there's one guy entirely responsible for it, to make sure we've got a solid pedigree for the gold."

EARLY THE FOLLOWING morning, the tech crew was on deck, dismantling the subsea computer and rebuilding it with parts delivered the night before. Milt had returned to the beach on the delivery boat to resume his teaching responsibilities at the University of South Carolina. All day the rest of the crew worked in seas rocking four to eight feet and winds gusting to twenty knots. That evening, they prepared to launch again, but Tommy himself canceled the dive when the weather appeared to be dropping. At sunup the next day, Burlingham had the thrusters down and the *Discoverer* on station. An hour later, the crew stood on deck, ready to launch. But as Doering raised the vehicle from the deck and rotated the crane slowly over the water, a gear at the base of the crane cracked under the weight and a moment later blew apart. Doering had no way to control the crane. It swung out over the water with the vehicle rocking at the top of the boom, and then the ship began its roll back to starboard, and the vehicle skimmed the waves and crashed into the bulwark. The ship's roll then shifted to port, and the vehicle again flew out across the water, and again it came crashing back against the side of the ship. Before they could get tag lines onto the crane and harness its swing, the vehicle had collided with the bulwark several times.

Tommy examined the damage and decided to lower the vehicle a hundred feet to see if the systems still operated. The weather wasn't perfect, but it was decent, and it could be the best they would see for a long spell. Using the tag lines to control the crane's swing, the crew paid out on the lines, the boom rotated slowly outboard, and Doering lowered the vehicle into the water. From the control room, Moore tested the systems, and except for reduced power with a lost battery pack, everything seemed to work. By noon they had contact with the bottom, but no one knew how long they would be able to stay. Although the vehicle was big and heavy, it was no match for the side of a ship and the roll of the sea. It had hit hard, and a system testing out at the surface could still blow suddenly.

On the bottom they fired the cameras every thirty seconds, but within a half hour, they started losing hydraulic pressure; then part of their subsea navigation blew. The cameras, strobes, and lights still worked, but after two hours on the bottom and 187 stills, the topside software died; and then the power began to drop, and they lost more of their navigation. The dive ended at three in the afternoon.

While the tech crew worked on the bottom, Burlingham, Tod, and Bryan had repaired the base of the crane, and it held when they recovered the vehicle that evening and swung it on board. For the next two days, with the weather just shy of marginal, everyone worked on the vehicle. The collisions against the side of the ship had destroyed one battery pack and blown fuses in the other two. Other parts of the vehicle were dented and hoses were ripped, and the rotating base had to be removed and welded again. Tommy still wanted hundreds more photos of the Bank of California area, but one of his objectives on the next dive was to recover gold.

EARLY THE MORNING of the 6th, a squall of twenty-knot winds passed over them in the dark, but by daybreak the sea again had dropped, and the ship now rocked gently in three-foot blue swells. The weather was as close to calm as they had seen in almost two weeks. At 11:35, they launched the repaired vehicle, watching the crane closely. Once the vehicle was in the water, Doering lowered it a hundred feet and held it while Moore checked the cameras and tested the systems. Then they sent the vehicle to the bottom. Within an hour of launching, the weather started to change again, the gentle blue swells rising higher and darkening.

In the southwest quadrant of the gold spill, adjacent to the gargoyle, sat a perfectly rectangular arrangement of gold bars. Tommy wanted complete photo and video coverage of this composition and the immediate surroundings, including more studies of the gargoyle. Scotty recorded every move in the dive log: "lead in photo, pan left, photo #12 end pan left, back to center . . . pan right, end pan #41, start dead zone, end dead zone, trolley outside of box, start blow, photos while blowing. . . ." By locking the camera angle, extending the camera boom, and very slowly rotating the vehicle a short distance, the photos later would appear as though the camera had traveled in a straight line.

As the dive continued throughout the afternoon, the swells topside gradually sharpened and turned gray, but Burlingham couldn't tell how far they would rise. When the weather shifted at this time of year, the sea typically ran up to six or eight feet and the winds peaked at twenty to twenty-five knots, as they had for the past few weeks. He could still recover in those conditions, but he never knew. Sometimes when the weather turned, it looked like a storm front sweeping across the sea but was really a localized squall that would rock the ship for only an hour or two and continue toward land. If it was a mere squall, they could leave the vehicle down and recover when the squall passed. But sometimes what looked like a simple squall was really a storm that would pummel the ship for five days.

Just before five o'clock, at Tommy's direction, Moore slowly opened the artifact drawer. Inside were the rectangular plastic trays with handles and large, numbered compartments, so Bob could catalog the recovery of each bar and coin before it was even touched by the manipulator. Moore removed the trays, placing them at the edge of the gold pile. Using a fingertip touch, like he was playing the flute, Moore moved the master arm on his console, and the manipulator on the vehicle reached out over a pile of coins, bent downward at the wrist, and paused, almost like it was contemplating the pile, then dropped carefully, spread its Teflon jaws, closed them, paused again, and lifted a single coin. When Moore opened the jaws again above the tray, the coin dropped and a tiny cloud of sediment rose. They shot each scene three or four times, then Moore plucked a coin or a bar from the scene, slowly cleaning up a discrete concentration at the edge of the gold spill.

At sunset, out beyond the far edge of the Gulf Stream, Burlingham saw clouds building: highs chasing lows across the sea, the dark clouds erupting between the two fronts, then arching upward and stretching along the horizon. Although Burlingham monitored these waters on a weather fax every four or five hours, localized disturbances often materialized so suddenly they never appeared on the fax; sometimes they did appear, but you couldn't tell what was behind them. Burlingham estimated this squall line was as close as twenty-five miles, and to him it looked thick, sweeping toward the ship. As the last light of day faded over Wilmington nearly two hundred miles to the west, the *Arctic Discoverer* began to rock over choppy, darkened waves, her bow lifting higher and higher. Then the temperature dropped, and the canvas awning on the foredeck began to undulate and snap in the wind.

Burlingham radioed the control room. "You'd better bring it up now."

Tommy heard the warning but continued to work. He needed to recover a significant amount of gold, yet do it carefully. With another 125 photographs of the site, they had finished the sweeping panoramic shots, a magazine-quality rendering of the treasure. By six o'clock, they had recovered seven artifacts. Then Moore trained the vehicle's thrusters on a portion of the pile, sending a stiff current downward, and they filmed and photographed the roiling sediment as it blew up off the gold. Thirty minutes later, they had recovered seven more artifacts. Moore dusted an adjacent area, and a half hour after that, they started recovering more artifacts, three within a minute. By 8:00 P.M., they had picked up seventeen more, thirty-two altogether. Now they were down to taking two, and sometimes only one, tight photographs of each setting before Moore reached out with the manipulator and grasped another piece of gold.

At 8:00 P.M., Burlingham recorded in his log that the wind had risen sharply to a steady twenty to twenty-five knots, the seas more consistent now at six feet. And according to the latest weather fax, the disturbance he had seen along the horizon at sunset was no squall but a frontal system that had formed suddenly and appeared huge: The sea and the wind could still double in size and intensity. When the storm hit, Burlingham wanted the vehicle on deck, not hanging off the side of the ship.

He radioed the control room again. "Harvey," he said, "this front's going to be on us in two hours. We should start recovery operations immediately."

"We're onto something important," said Tommy.

"Fine," said Burlingham. "We're going to be running ten- to fifteen-foot seas."

Burlingham didn't need to tell Tommy what that meant. Ten-foot seas made recovery almost impossible. And once the storm hit, it would be too late to begin. The winch took an hour and a half to wind eight thousand feet of cable to the surface. Then they had to raise the vehicle through the air-sea interface with the crane and swing it onto the deck, and that would take another half hour, especially in rough seas. But Tommy decided to stay down just a little longer, take a few more pictures, recover a little more gold, hope that the storm was less intense than it appeared or perhaps farther away. "Those were tough decisions," he said later. "Burlingham didn't know what we were trying to get, and we all knew that this might be the last dive of the season."

At the center of the ship, the heavy insulation that kept the control room cool and dry also deadened the sound and feel of the sea. The tech crew could hear only the dull, intermittent hum of the thrusters and feel only the slightest rise as the ship lifted over the incoming waves. Other than the slight rocking, they had no sense of the approaching storm.

In the next twelve minutes, they photographed and recovered seven more artifacts, three of them coins cemented together, and placed them safely in a compartment in the plastic tray. As they worked, a violet fish two feet long with black buttons for eyes hovered above two shiny coins, sometimes nibbling between them. A long upright spike rose from the back of the fish's head, and the last two-thirds of its body flowed like fine ribbon before a fan. The eerie moan of the thrusters changing azimuth seemed to accompany the silent rippling of the fish.

For another hour, they remained on the bottom. When they had secured the seven artifacts, Tommy had Moore rotate the vehicle and aim the forward thruster at a rippling in the sediment. Moore ran the thruster for a quick dusting, and from under the sediment appeared another pile of coins and bars. They photographed and filmed this pile, recovered a bar from the edge, then pulled the cameras back for long shots of the area and more angles on the gargoyle. After a while the fish departed, and only the deep-ocean sentries watched in silence. In the dark blue background, a white crinoid sometimes appeared, swimming with tentacles like a wispy tangle of pipe cleaners, each waving up and down as though to the sound of a harp.

THE STORM SPUN across the water toward the ship, the winds bucking the north flow of the Gulf Stream, kicking up high, jagged waves that collided with the cross swells still left from the early-morning squall. In the control room, six monitors cast a bluish light into the darkness, and the tech team watched Moore continue to probe with the vehicle in the quiet of the deep Atlantic, until they had photographed the small area 250 times and filled the artifact trays with forty gold coins and bars.

By nine o'clock that night, the seas had reached eight feet and the wind blew at near gale, foam from the cresting waves beginning to streak the surface of the sea. Thirteen minutes later, Tommy ordered the vehicle shut down and the team to begin recovery.

"They stayed down about an hour and a half, two hours after we told them we saw it coming," remembered Tod. "But what was amazing was the strength of the squall. When we finally did start to come up, the waves had already built from three feet to eight feet, and it was dark. By the time we got it up to the surface, we were in about twelve-foot seas and forty-knot winds and it was raining."

At ten-thirty the vehicle neared the surface. By then the wind had intensified to full gale, and the waves had grown by over half since they ended the dive and started winding in the vehicle. When the recovery team ventured onto the deck, they saw waves, some approaching fifteen feet, breaking over the side of the ship, and the rain came out of the darkness sideways across the deck, suddenly sparkling as pellets in the white lights.

They stopped the vehicle thirty feet beneath the surface on the port side, its lights intermittently visible beneath the waves. Crouching on his hands and knees, Bryan lifted himself over the bulwark onto the hero boards just as the ship dropped away and a twelve-foot sea hit broadside, ripping his hand from the platform. The wave exploded off the ship, jerking him up the cable, but he held tight to the cable, and the wave fell away as Tod dropped onto the boards next to him. Some waves were so big, the ship could not rise above them; it plunged, and the waves rolled along the sides, taking them under. Burlingham was the last of the three onto the boards, poised between Bryan and Tod to run the recovery. By then, Doering had fought his way across the deck to the crane, and everyone else on the tech crew was holding tight to the bulwark with one hand and hanging on to Tod, Bryan, and Burlingham with the other.

The waves swelled, and white patches of foam clung to them as they broke into the light and slammed the hull, the foam and the waves blasting into spray high above the ship. With the vehicle just below the surface, Burlingham signaled to Doering to begin swinging the crane toward him. As the crane drew near, the three men on the hero boards moved quickly, now high above the water blasted by the wind, now holding on, water up to their chests. They opened shackles, cut through tape, and sorted out cable lines. Spray stung their eyes, and wire leads blew stiff in the wind just out of reach, but in fifteen minutes, they had shut down the power, disconnected the cable, and hooked the vehicle to a wire coming off the crane. Then Burlingham signaled to Doering to start winding in the wire.

The vehicle broke the surface in a trough, then disappeared in the swell of the following wave. Doering had the crane lowered over the water, winching the vehicle higher, until it was snug against the tractor tires at the top of the boom. With the elbow of the crane straightening, raising the vehicle from the water, the ship rolled far to port. As soon as Burlingham felt the ship coming back on its roll to starboard, he shouted to Doering to start swinging in with the crane.

The ship began to roll back to starboard, and the crane began to turn, and then Tod heard a loud pop and saw the vehicle hurtling at him. He threw himself over the rail and scrambled aft, as the vehicle skimmed the water and crashed into the deployment arm, the electronics sphere exploding and gold sparks showering the deck. The gear at the base of the crane had blown apart, and the crane was again at the will of the sea.

The ship rolled back to port, the vehicle twisting out over the water at the top of the crane, and then the ship rolled back to starboard and the vehicle crashed again into the deployment arm. Flashes lit up the port side and more sparks shot into the air.

The batteries inside the vehicle's battery packs ripped loose from their brackets, and the packs broke open, spraying oil across the deck. To try to control the swing, Bob reached up with a boat hook and snagged a tag line coming off the side of the crane. As the vehicle swung toward the ship, the line he had snagged slackened, and other techs grabbed hold and heaved together, stumbling and sliding on the oil-slick deck, trying to haul the line toward the starboard rail. But before they could pull it that far, the vehicle began its outward swing, dragging six of them back across the deck.

Doering was trying to raise the crane high enough so that on its next swing in, the vehicle would clear the rail and he could drop it on deck. But the crane's motor suddenly died, and the knuckle locked in that bent position with the swing gear gone and the crane flapping back and forth and the two-and-a-half-ton vehicle slamming against the side of the ship.

Doering could not control the crane. He jumped from the seat and grabbed on to the tag line with the others. The vehicle swung in again and crashed into the deployment arm, turned slightly on the swing outboard, then missed the arm on the next swing and crashed into the side of the ship.

Moore disappeared under the forecastle, found the circuit breaker for the crane, and restarted the motor. When he rushed back into the storm, he saw people running back and forth in the unnatural brightness of the night deck.

"Harvey was in the thick of it, running back and forth," said Moore, "and that's what I was really worried about, was them being there instead of out of the way. They were sucking up and running across the deck, and then when it went to swing outboard, they were running back across the deck. They never had time to take one of the tag lines and get it made around something so they could control it. You cannot reach out and grab something that weighs five thousand pounds and stop it from whipping around."

Moore jumped onto the crane seat. The rain drove in sideways. The ship heeled to port, and with Moore in the seat, the crane swung out, flinging the vehicle at the height of its arc over the water, and then the ship suddenly righted and the vehicle flew back toward the ship and again slammed into the deployment arm. Moore couldn't control the swing, but if he could straighten the knuckle, on the next swing the vehicle might be just high enough to clear the bulwark; and once it was over the deck, he could let cable out and try to drop it, trapping it inside the rails. But Moore had never operated the crane, and when he looked at the panel of levers with black knobs he saw that the labels had rubbed off: He couldn't tell which lever straightened the knuckle.

On the next roll to port, Moore waited until the crane reached its zenith, jammed the lever he thought might control the knuckle, and the knuckle began to straighten, raising the vehicle as it swung back toward the ship.

Tommy and Bob stood five feet from the starboard rail, still tugging at the tag line, still trying to coax it around a cleat on the starboard side. Bob heard a loud thud and then voices shouting, and when he looked up, he saw that the vehicle, twisting wildly on the boom, wasn't going to crash into the deployment arm this time but had skipped over the port gunwale and was flying across the deck.

Tommy ducked and stumbled forward. Bob dropped the tag line and dived inside an empty aluminum cube pinned against the starboard rail. The vehicle swept the space where they had stood, then twisted at the end of the boom and began its swing back across the deck, over the gunwale, out across the water.

Moore had tried to drop the crane as the vehicle came over the deck, but the crane moved too slowly. By the time it was low enough to trap the vehicle, the vehicle had cleared the rail, was outside the ship, and had swung back into the deployment arm. Moore raised it again, so it would clear the rail coming in and he could try once more to drop it on the deck.

The ship rolled to port and the crane swung out. As the vehicle stopped at the end of its swing, Moore began to lower the crane, and as the vehicle came in this time, it bounced off the gunwale, continued across the deck, and had begun its outboard swing when Moore slammed a lever, dropping the whole boom down on top of the vehicle.

The vehicle slammed into the deck, skidded toward the port rail, hit, skidded back toward the center of the deck, then headed again toward the port rail and crashed into the base of the deployment arm, where it stopped. Suddenly, the tag line went limp.

Despite the wind and the waves and the falling rain, the night seemed still. No one spoke. Exhausted and shaking, they stood in the rain, not moving. Bob's left lens was covered in blood from a gash in his forehead, but he was coherent and otherwise uninjured. Everyone else was unharmed: Burlingham, Moore, Scotty, Doering, Bryan, Tod, and Tommy.

"It was the strangest feeling," said Tod, "just a rush. We all looked around and everybody made it alive, and we started laughing. Maybe we were all delirious, I don't know. Then for some reason, I remember looking over, and I saw Harvey, who was shaking from the cold. I'll never forget that, because he had the biggest smile on his face."

Although much of the vehicle had been destroyed and the artifact drawer was smashed closed, they had the vehicle safely aboard, and no one left the deck until they had it tied down. The ship was rolling so hard that that took another three hours, one foot at a time, removing and reattaching lines as they went, until the vehicle was snug in its parking place and anchored to the deck with half a dozen lines. Then

Burlingham took the bridge in seas running twenty feet and blue water coming over the forecastle. With the vehicle strapped tight to the deck, the *Arctic Discoverer* traveled through the storm at eight knots, headed due west for Wilmington.

TOMMY HAD A large cache of antique gold strapped to the deck of the *Discoverer* and photographs and film of a lot more gold lying in piles back at the site. He had solved the problems of finding and recovering the treasure of the *Central America;* now ahead were the problems of trying to keep it. In his failure-mode thinking he could conjure infinite ways that the treasure still could be lost, destroyed, or defaced or otherwise diminished in value. Two groups already had tried to run him off what everyone had thought was the *Central America* site, and a federal judge had warned them away. But others still might try, and of course there were far more subtle ways to lose the treasure than having someone try to drag a sonar fish across your work site or broadside your ship. This was the one that concerned Tommy most: Now that he had found the treasure, someone might try to take it away.

Weeks earlier, after they saw the sidewheels at Galaxy II and found the bell and then the gold and knew that this was the *Central America,* Tommy, Barry, and Bob had begun planning what they would do with the treasure once they brought it to the surface. "There's no precedent," said Bob, "for taking several tons of gold in small parcels off the ocean floor, moving it to shore, and securing it." That was the immediate problem: where to secrete the gold on the ship, when to remove it, how to carry it off with no one seeing, and where to take it for safekeeping. On the way into Wilmington, they refined the plans they had begun earlier, but those procedures and times they would divulge not even to the rest of the tech crew.

The next problem was less immediate and more subtle. The Columbus-America Discovery Group was now a partnership of 160 people, most of whom did not know one another. They had supported Tommy even when competitors appeared on the horizon and costs went up, their shares were diluted, and Tommy returned from the site with little more than stoneware and coal; the partners had believed in him and his ideas and had sent him back out to try again. Benjamin Franklin once said that three could keep a secret if two of them were dead. The partners deserved to know and needed to share in the stories about the gold. But how did Tommy keep 160 people quiet until he could return for the rest of the treasure?

In Wilmington after the storm, Tommy and Barry wrote a series of letters to the partners, one reconstructing the recovery of the bell, one describing a bar of gold they had brought to the surface, and another telling the story of two gold coins they had found. "As for the bar of gold," they wrote, "our preliminary information indicates that only a few like it are still known to exist, because the government melted down most of them during and shortly after the Civil War." They told the partners that the coins were "uncorroded as expected, but portions of them are stained with reddish orange rust."

In each letter, Tommy closed with an admonition. "It may seem impossible to keep a lid on our discoveries until next summer," began one final paragraph. "Keep in mind that our Partnership has done other things that are still considered impossible. The next few months will really test our resolve to contain the information about our gold recoveries in order that we may continue to operate efficiently when it really counts. We all must be patient until the rest of the gold is in hand. Once this critical-path goal is met, then we can safely go public with our discoveries."

Tommy had decided that revealing the recovery of only one bar and two coins offered the perfect compromise: It provided the partners with a sense of wonder and excitement and allowed them to enjoy their role in the accomplishment; it also gave Tommy the ideal cover if the information that they had found gold did leak out. Anyone in the deep-ocean community who heard third- or fourthhand that Tommy claimed to have recovered one gold bar and two gold coins from the *Central America* would immediately assume that he had "salted" the site: either placed a tiny amount of treasure on some ship, then recovered what he had placed, or merely said that the treasure came from the site and used it to show to potential investors to raise more cash to continue the search. To the deep-ocean community, the logic was simple: The ploy might dazzle potential investors, but if the treasure really was there, no one would leave it to come back to later.

So if Tommy's story ever leaked out to competitors, they wouldn't believe it anyhow.

WHEN HE RETURNED to Columbus in late October, Tommy scheduled a meeting of the partnership for Saturday, November 26, at the Columbus Athletic Club. About one hundred people attended, partners and their spouses. Wayne Ashby described the gathering as "a very excited and happy group of investors."

Tommy stood at the podium, his black beard thick from months at sea. The lawyer Bill Arthur thought he had lost a lot of weight since the last time they had seen each other. "He was all blue twisted steel," remembered Arthur. "I don't think there was *any* fat on him." A few feet from Tommy sat the bell from the *Central America*. Bob and Tod had wrestled the three-hundred-pound artifact out of its wet storage, covered it with damp towels, and with some help had set it on top of its storage box and carted it into the room to display for the partners. On a table in front of Tommy sat a two-and-a-half-gallon aquarium filled with water, and in the water sat a bar of gold weighing twenty-five pounds.

Tommy talked for over an hour about their accomplishments on the site that summer, the dramatic viewing of the sidewheels on the first dive, and the progression of finds at the site: the bell, Bob's tiny gold filings in the sediment, the false sightings of rust-stained tubeworm shavings, then the discovery of a gold bar, and finally the storm that almost wrecked the vehicle as they brought the bar to the surface. He talked of his plans for the following season and how he intended to upgrade the vehicle to the next level of capability and find and recover the rest of the gold. He cautioned them again about keeping in confidence everything they had read in his letters and everything they were hearing and seeing that morning.

"Tommy," asked someone from the audience, "what about the coins?"

"Oh," said Tommy, as though he had forgotten. "Just a minute." Then he reached into his pocket and pulled out two 1857 double-eagle twenty-dollar gold pieces and held them up. "I don't know if you can see them," he said, "so I'll pass them around." As he walked over to hand them to the first partner to examine, he looked up at the audience and grinned. "Don't anybody pocket 'em," he said, which brought loud laughter from the group.

Bill Arthur loved the whole setting. "This gold bar, the double eagles," said Arthur, "I mean, you couldn't have written a better story. That was absolutely a coup. These guys all brought their wives so their wives could go *ohhhhhhhh!* and of course Tommy was so blasé about it."

Tommy could not reveal even to the partners the fairy tale scenes of gold he and the tech crew had witnessed at the site. Many suspected that he had already found more than one bar and two coins anyhow, but they understood his decision not to reveal everything the crew had seen. "There was absolutely no reason," said Jim Turner, "for Tommy to share with me, or the other partners, or a reporter from the *Columbus Dispatch,* the fact that they had recovered twenty bars, or fifty, or three. It was important for a lot of reasons that they could show that they recovered one bar and the coins. But in one of those letters they made it very clear that they weren't going to talk about the amount of gold until they had recovered a significant enough volume."

When Tommy had finished his speech, a partner named Donald Dunn stood up and said that this was an incredible moment and he wanted to reflect on it. "I don't know how the rest of the people in this room feel," he said, "and I don't know whether you're really going to find any more gold, Tommy, but whether you do or not, I want you to know it's been a privilege and a pleasure to be associated with you and to have been on this journey with you." The room erupted in applause.

"Naturally, the investors were really feeling good about being a part of this success," remembered Wayne Ashby. "People like to be a part of anything that's successful, but being a part of something so unusual and spectacular and adventurous, there isn't any question they were feeling good about that."

Some of the partners, "the more naturally worrisome members of the investor group," as one put it, were still concerned about pirates, protection from the court, and how much of the treasure the court would award to the group. But for now they enjoyed the feeling and secretly applauded themselves for having the perception to invest in the venture. Tommy continued to worry far beyond the partners' worries about technology, the court, science and archaeology, the further recovery and eventual disposition of the gold, but for a few moments, in the presence of these people who had believed in him, he enjoyed the warm satisfaction of success.

Arthur was in line to look at the gold bar just ahead of Jim Turner and his wife. Arthur, who is a big man, held the bar in both hands, and as he turned slowly to face Turner's wife, he said, "Now before I hand this to you, let me tell you, IT'S REAL HEAVY!" Arthur was afraid she would drop it on her foot, and after Turner held it he understood why. The density of gold always played tricks on the eyes. "It's one of those wonderful problems to have," said Turner.

Within one week of the partnership meeting, they had oversubscribed the $4 million remaining from the stock offering the previous spring. "It was a very exhilarating week compared to the other four offerings," said Ashby. "I mean, they were very responsive."

One partner had told Art Cullman he'd never go near Tommy again. "Tommy hasn't done a damn thing all summer," the partner had said. "I'm not gonna do a third round. I'll stay where I am. Why don't you get somebody to run that damn place?"

"So he got down to this meeting," remembered Cullman, "and saw the gold bar and the coins come out of the pocket, and two days later, I heard he'd put a hundred and fifty thousand dollars into that last round."

One of the partners closest to Tommy, and the one who had waxed most eloquent for that million dollar "insurance policy" in 1987, was Buck Patton. Patton had been out of town the weekend of the partnership meeting, but he had read the letters, and he met with Tommy the following Monday. When Tommy showed him the bar, Patton said, "Tom, if this is the only thing that's down there, we will have more than paid for the project in a number of ways." Patton called Tommy a "rascal" for the way he held out that gold bar and grinned. "That's about the best closing tool I've ever seen," said Patton.

Patton thought that the majority of big risk was behind them, that what remained were the need to recover and the need to keep quiet. Out of the remaining eighty units, he took ten and gave Tommy a commitment of $500,000.

"A lot of things depend on how you feel about them," Patton said later. "Sometimes, financially, you do okay, or real well, or extremely well. Occasionally, you hit the Babe Ruth home run, and other times you might just lose a whole bunch of money but everybody worked hard and tried, so it's okay, because everybody put their heart into it. It's when you get into deals where you've got scabs running a partnership and you know the money never gets to whom it's supposed to and you say, 'I'm gonna file suit, here's my attorney, and I just took his leash off.' That's when you feel bad about things. But you couldn't help but feel great about this. I told Tommy, 'I don't care how much gold is down there, you're a hero. I think every investor feels the same way. I can't speak for them, but they've got to feel like you just went to the edge of the world in terms of commitment and innovation and creativity and just sheer love, guts, and determination, and homework, and you did it. You put the points on the board! This may be the only gold bar there is, but it's real, and you did it.'"

Now Tommy had the money to build an even bigger, more complex, more capable vehicle to explore the site further and recover the rest of the gold. And next season there would be no competitors, no rush to get to sea, no wondering if he was on the right ship.

ONE OF THE more remarkable accomplishments of the Columbus-America Discovery Group is that from October 1988, when Tommy announced to the partners in a letter that they had found gold eight thousand feet beneath the sea, until they finally announced to the public that they had recovered the treasure of the *Central America,* that news never traveled outside the partnership. "We don't discuss it," said Mike Ford. "We just don't discuss it. It's almost like the Manhattan Project."

The partners' commitment to secrecy allowed Tommy and his tech crew to concentrate on refining their technology and returning to the site for further exploration and recovery. After much thought and conversation with the lawyers, Tommy decided not even to file a claim on the new site, but to let the matter rest until he was ready to return the following summer. Because of the hearing in federal court in July 1987, everyone already assumed that the Columbus-America group was on the *Central America* at the old site; why announce to the world that the treasure really lay somewhere else and give them the coordinates to pinpoint it? "It was a pretty bold move not to file on the new site," said Tommy. "If you want protection from the court, then you file; but you're not obligated to file anything. We were still somewhat convinced that Lamont-Doherty and Kreisle had gotten our coordinates from the court on the first site. So from a business point of view, we had to keep things in perspective. I thought about that a lot, and I just kept running scenarios on it. What if we go out to sea and our equipment doesn't work? This stuff we're doing was almost impossible to do. We ought to get some sense that we're going to be able to operate all season before we file. So that's what we ended up doing. We took the chance of not being protected by the court."

Robol filed a substitute custodian report, informing the court that gold had been recovered, and that it appeared as though the gold was from the *Central America*. "But I felt kind of naked out there," said Robol, "without an injunction." Robol thought that Judge Kellam should view some of what they had found. Late that fall he called the judge and told him he had a matter of some urgency to discuss and that he preferred the meeting take place in a secure and confidential environment. He also mentioned that some members of the Columbus-America Discovery Group would be in town over the upcoming weekend, and the judge invited the group to his home. On a Saturday morning, Robol, Tommy, Barry, and Bob drove a van to the judge's house in Virginia Beach, the three-hundred-pound bell preserved in back in a tank of water. When they arrived, they found the eighty-year-old judge wearing an old hat and khaki pants, mowing his lawn with a push mower.

The judge offered them some refreshments, then led them the back way into his home to a study. Bob carried a cosmetics case, and when they reached the judge's study, he placed it on the judge's desk. Robol thanked the judge for agreeing to meet with them on the weekend and spoke for a minute about why he had requested the meeting. Then Bob opened the cosmetics case and pulled out a towel, inside of which was a layer of felt, and inside that was the gold bar they had showed to the investors. Bob peeled back the felt, and the judge examined the bar and its markings. "Lordy, lordy," said the judge, and everyone, including the judge, laughed. The meeting lasted ten minutes.

DURING THE OFF-SEASON, Tod had a lot of long talks with Tommy. Now that they had found the *Central America* and recovered some of the gold, Tommy seemed more relaxed and approachable. "He realized what everybody had been through," said Tod. "He made a point of getting back in touch with everybody, and he basically conceded that what happened last year was wrong, because everybody was always upset about this need-to-know thing. But he said it was a business decision that he needed to make, and I said I understand that."

Throughout the winter and into the spring of 1989, Tod and Bryan worked in the back of the warehouse with a welder, cutting channel aluminum into four-foot lengths and welding the cubic frames together for the skeleton of the new vehicle. Tommy directed Hackman, Scotty, John Moore, and other engineers from Battelle and Ohio State in designing the new systems. That year, they would have more cameras and more reach, finer controls, better storage for the ascent, and more power on the bottom.

The office space at the front of the warehouse served as a mock control room. Cables ran from the growing vehicle back in the warehouse forward into the office, connecting the vehicle to the master arm control that John Moore would use at sea. As Moore stood in the warehouse next to the vehicle, he gave commands into a microphone, like, "Extend camera boom number three," and another engineer in the office moved the controls so Moore could see if the vehicle responded appropriately. "Rotate camera boom clockwise." The camera boom moved clockwise. "Counterclockwise." It moved counterclockwise. As the parts moved, Bryan watched for leaks in the hydraulic system and tightened them with a wrench. This year they had the luxury of running the systems and searching for problems before they went to sea.

The first week in June, Tommy drove his mentor Dean Glower out to the warehouse to show him the vehicle. Almost ready for shipment to Wilmington, it now stood over seven feet tall, five feet wide, and fifteen feet long—a mass of aluminum framing, booms, cameras, batteries, junction boxes, and electronic spheres all crisscrossed and interlaced with twelve hundred feet of orange hydraulic hose. The vehicle had over a hundred electrical functions and ninety hydraulic functions. One manipulator alone had seven functions, at the shoulder, the elbow, the wrist, the fingers, and each function required two hydraulic hoses, fourteen hydraulic lines all massed together, each about the width of a little finger. Drooping off the front of the vehicle like praying mantises was a thick assortment of camera and light booms.

"When I first looked at it," said Glower, "I was in a state of shock. It's an enormous design." Glower was struck by the vehicle's complexity and its place in history, "almost like designing and building an automobile," said Glower. "He was building a whole new apparatus, and it looked complicated as hell."

His question to Tommy in 1973—How are we going to work in the deep ocean?—had led to a two-and-a-half-ton vehicle that eventually would grow to six tons, with nine mechanical arms, some having as many as eleven segments, with pan and tilt capability and telescoping booms, 3-D video, and seven other broadcast-quality cameras, two still cameras, five strobes, and twelve spotlights at all angles for backlighting, two mechanical arms to collect, a suction picker, a vacuum, blower nozzles, a silicone injection machine weighing over one thousand pounds, thrusters, dusters, a retractable drawer, collection trays, and excavation tools, many of which could be added at will as they worked below. It was similar to the design Tommy had envisioned five years earlier but, because of competition and then thruster problems and a closing weather window, had never been able to build.

BEFORE HE LEFT Columbus to return to sea that summer, Tommy consulted with an international coin expert named James Lamb, head of the Coin Department at the world-renowned auction house, Christie's. When Lamb examined one of Columbus-America's 1857 double-eagle twenty-dollar gold pieces, he pronounced it "gem" quality. "A coin expert can go through an entire career and see about two of these coins in this condition," said Lamb.

One thing concerned Lamb: A small nick or scratch on a gem coin could reduce its value to one-third, and two blemishes could drop it another third. That was part of the reason Tommy refused to recover more than a small amount of gold in 1988: He wanted to make sure they were doing it right. The problem was that so many coins lay piled at the site that meticulously plucking one at a time would require too much time and expense, yet to recover two or more at once might mar the finish. Tommy had to create ways to speed up plucking the coins from the site without one coin touching another.

THEY LEFT JACKSONVILLE on the *Arctic Discoverer* July 19 and arrived at the Galaxy 11 site the following evening. Tommy held a meeting that night on the foredeck, explaining to everyone, new and old, his policy on security dives: who would be allowed in the control room, who would not be allowed in the control room, and that everyone but Burlingham and the tech crew would have to leave the deck during launch and recovery.

For the next two weeks at sea, they tested the crane and the new thrusters and Scotty's new logging system. Scotty had created software that allowed Bob to document and catalog all video, stills, and crew comments on the recovery of each artifact as it happened. As soon as they took a photograph below, they could print it out topside in the control room; they could freeze a video image, the computer would digitize it, and the printer would give them a hard copy of the scene. Later, for an investor considering the purchase of a coin, they could show the actual recovery of that coin; for a scientist wanting to see footage of a unique sponge, they could locate the video segment immediately or return to the sponge at the site for additional observation.

With all of the new capability and systems to be tested during those first two weeks, there was so much to do that often the tech crew operated on no more than four hours' sleep a day, and sometimes that was composed of two two-hour naps. Just the day-to-day routine of keeping the vehicle maintained kept them up late and got them up early. "Then Harvey comes up with something new he wants to do," said Battelle engineer Mike Milosh, "something a little off the wall, and the time to do this extra work comes out of your sleep."

Over the winter and spring back in Columbus, Tommy had met frequently with Milosh and other engineers to discuss how they might recover the coins and bars faster than having John Moore pluck them from the piles one at a time. But whatever method they came up with could not nick the treasure. "Our ground rules," remembered Milosh, "were that we couldn't scratch any of them, period. Harvey was very intent about this." In 1857, stacking all of that gold on top of itself to be transported would not have diminished its value. But 131 years later, the value of that gold as historical artifacts had risen so dramatically that each piece had to be carefully handled.

Doering thought it would be impossible to recover all of the gold without one coin or one bar touching another; there was just too much gold. The coins far outnumbered the bars, and Doering couldn't even estimate how many bars there were. Many times he had sat in front of his monitor and tried to count them, but he got so lost in the interrupting angles and the overlapping curves, in the sheer quantity of geometric shapes in orange and yellow, that he quit. When he had seen that first film strip of the gargoyle area lying on the light table, he couldn't imagine gold being found in piles that enormous. "And there's probably six, seven more piles like that," said Doering. "There's no way you can clean up this area in two months. You couldn't do it with a bulldozer."

Doering's solution was to pick it all up with "big scoops" and be done with it. "Harvey wants to pick each coin up individually, photograph it, take a videotape of that coin, and then place that coin in a special holder. We'll be here for a hundred years, for Christ's sake, doing it like that."

Milosh had been pondering another approach, simple and seemingly far-fetched, but just the sort of ingenuity Tommy prized: Place a mold over a pile of coins, inject silicone into the mold, let it harden, and raise the whole mass in a rubber block. You could pick up a hundred coins at once, and the silicone would envelop the coins and protect them. When Hackman heard the idea, he volunteered to take all of those silicone blocks to his house over the winter and peel the silicone away from the treasure while he sat in front of his fireplace. He'd do it for free. All they had to do was find a silicone that was heavier than water, figure out a way to mix the silicone with the catalyst, clean the nozzle regularly, and get the solution to flow and harden despite a temperature of thirty-eight degrees and pressure that would crush the steel hull of a nuclear submarine.

In the dining room of the old Victorian that served as Tommy's office, they set up a thirty-gallon fish tank filled with salt, water, and ice and began experimenting with various silicones. The pressure wasn't there, but they worked with the salinity and the cold until they got the silicone to flow and set up in a reasonable amount of time. To test the method, they coated Krugerrands with bluing, then placed them in a pile at the bottom of the tank, and covered them with the silicone. "We did quite a bit of testing," said Milosh. "Our criterion was that we couldn't even mark the bluing. So we were being pretty strict." The first time they tried it at sea, the catalyst was too weak, and the mixture ran out the bottom of the mold. But when they doubled the amount of catalyst, and then doubled it twice more, it oozed among the coins and solidified, encasing and protecting and uniting the coins into a block.

To prepare a small pile of gold for recovery, they first dusted the area lightly with the wash from the thrusters, then used a tiny suction picker to remove individual coins scattered at the edge of a small pile, or coins that might prevent the silicone mold from lying flat over the pile, and dropped them into a tray with foam-lined compartments. The new logging system allowed Bob to assign each tray a number and identify each compartment with a coordinate system that electronically recorded the time of collection and its location on the site.

When the pile was prepared, they placed the mold over it and injected the silicone. As that was setting up, they moved to another pile, dusted, picked around the edges, and placed another mold for injection. Later, they returned to recover each of the gray blobs—roughly a foot square and about eight inches thick with small gold bars and gold coins embedded in them, some of the gold at odd angles, some in neat stacks. Often they were filled with hundreds of coins and sometimes several small bars.

"The coins we bring up in that block of silicone are in beautiful condition," said Milosh. "We've got ones that are untouched, just perfect."

The gold bars that came up in the silicone ranged in size from little squares of about five ounces, what Bob called chocolates, to larger squares and rectangles of a few pounds that he called brownies.

In one block of silicone, they recovered the "coin tower," stacks of gold coins each thirty coins high, the stacks three wide and five long, or about 450 perfect double eagles, all naturally cemented together.

The first dive after they had completed all of the tests was August 5, but they did not begin bringing gold to the surface until the first security dive on the 18th. "This thing went unbelievably well," said Burlingham. They launched in the early morning and recovered in the early evening. Tommy planned the security dives that way, so that by the time they brought the vehicle back to the surface, night had come. The tech crew would rinse the vehicle with freshwater, pull the cameras off, hook up the battery charger, and remove pieces of the vehicle that needed to be worked on. Then they would clear the deck, turn out the deck lights, and in the darkness Bob would duck under the tarp, remove the trays filled with gold, and place them in a black trunk. Tommy and John Moore would help take the trunk below to Bob's lab, where Bob removed each artifact, tagged it with a computer-generated number, cataloged it, and stored it, so it could be correlated later with the recovery footage.

Night after night the vehicle came up filled with gold bars and gold coins, nuggets, gold-veined quartz, even piles of sparkling dust. And then one evening, after a week and a half of security dives, they loaded the artifact drawer on the bottom, and before they could close it, the cable that carried all of the topside commands down to the vehicle suddenly shorted out, and they lost control. They couldn't close the drawer, they couldn't empty it on the bottom, and they couldn't fix the problem without recovering the vehicle and having it on deck. Tommy had no choice but to recover with the drawer open. He decided to announce what had happened and allow everyone to view the treasure when it came on deck.

"They told us beforehand that there was stuff in the drawer and we had to be careful," remembered Burlingham. "The drawer was out, full of coins and some bars—everybody saw it. It was a pretty good haul, well over a hundred coins and a couple of bars, all in their little cases and boxes." Burlingham saw one coin much larger than the others and over an eighth of an inch thick, a fifty-dollar gold piece cast in 1851 and shaped octagonally. "This thing looked awesome," remembered. Burlingham. After all of his research on pioneer coinage in California, Bob recognized immediately the uniqueness of the piece. "This coin paid for the dive," he said. Besides the coins, brownies also sat in the tray and one bar five inches long and weighing about 150 ounces. "We were all there," said Burlingham. "Harvey made the best of what he probably felt was a terrible situation. But everybody got a chance to see the stuff."

THE INSIDE OF Bob's tiny lab was painted lime-green. The basin was stainless, the microscope beige, and the lime-green bookshelves above the basin lined with books like *Conquering the Deep Frontier, Ocean Salvage, Exploring Ocean Frontiers, American Seashells*. A refrigerator and a computer sat opposite the sink. Around the sink, carefully placed on terry-cloth towels, Bob had samples of coins, bars, and quartz-lined nuggets, parts of the treasure he was examining and cataloging. Most of the gold remained in a storage locker just forward of the lab, where Bob left it in the trays and silicone blocks until he had time to work with it. "The bottom is just covered with this stuff," he said. "It's astonishing."

Bob spent most of his time now in the lab, studying the gold, noting peculiarities, searching for understanding and connection. As a geologist, he was particularly fascinated by the chunks of quartz veined with gold. "There's something almost spiritual about reperforming the type of work that these same people did over 130 years ago," he said. "I'm gold-panning on a ship in the Atlantic Ocean, whereas before it was in a gold pan with a guy wading in a stream along Grizzly Flats or Mormon Island. You find these nuggets and you reexperience the thrill of the original discoverer."

About the fourth time he examined the Harris Marchand gold bar they had displayed for the partners, he noticed something peculiar: The bar was thicker on one end than it was on the other, which meant that the ingot mold back in 1857 had sat on a lopsided table. "I could imagine the Harris Marchand office in Sacramento, where this bar was stamped," recalled Bob. "He had his ingot molds on this rough-hewn table. The whole picture came flashing back to me, this snapshot, and it gave me such a sense of connection with the historical story. It's like the scales fall from your eyes and you go 'Wow.' Finding the gold on the bottom was not the great historical connection moment for me. Some of the things that have happened to me in the lab have been more along those lines."

In September, James Lamb, the coin expert from Christie's, visited the *Arctic Discoverer* dockside in Wilmington to see more of the treasure. Back in Columbus, Lamb had seen only two double-eagle twenty-dollar gold pieces and had been told there were others like them at the site. He knew nothing about the quantity or the diversity of the coinage, but he remained concerned. "I didn't know what state of preservation the vast majority of the coins were going to be in."

Lamb arrived at the ship wearing a gray suit and a gray tie, but within minutes of entering Bob's laboratory, he had his coat off and his sleeves rolled up and an 1857 double-eagle twenty-dollar gold piece between his thumb and forefinger, holding it at an angle to catch the light. "That's how it looked when it came out of the mint," he murmured, "and the more the coin looks like the minute it was made, the more valuable it is, and this is just about perfect. This coin has a wonderful and complete mint luster and no wear whatsoever."

Not all of the coins from the San Francisco Mint were twenty-dollar double eagles. Some were ten-dollar eagles or five-dollar half eagles or two-and-a-half-dollar quarter eagles. According to Lamb, one 1857 ten-dollar eagle gold piece recovered from the *Central America* was "by a considerable margin the finest example known to exist."

Many of the coins coming up from the bottom had been struck by small, private mints that existed before the San Francisco Mint opened in 1854, and the coinage produced by those mints was extremely rare. From 1848 to 1854, miners coming out of the mountains had only nuggets and dust, neither of which made a good medium of exchange when purchasing food staples and supplies. Business often was conducted with a "pinch"—two pinches of dust for a sack of flour, one pinch for a shot of whiskey. Since miners in from the hills often hit the saloons first and had no way to pay except with gold dust, saloon owners hired bartenders with big thumbs.

"There was no standard," said Bob. "It really caused quite a monetary crisis. The easiest way to establish the value of the gold was to have the miner go to an assay office and have it weighed and assayed, then he'd get back either a bar made out of it or he'd get back coins of equivalent value. This is the earliest step in this whole monetary process that we're seeing in the treasure on the *Central America,* 'cause we have both. We have the raw rocks that started it all. But we also have the bars and the coins made by the assayers out of such deposits." The coins were rare, because the pioneer mints operated only briefly, some driven out when caught underweighting, some ceasing operations when the San Francisco Mint opened.

Lamb saw coins in Bob's lab and more coins in storage in a forward hold. "My doubts," he said, "which very, very, very crucially affect the potential value of what they're finding, have now been completely allayed. The coins are mostly pretty much perfect and uncirculated."

Before Lamb arrived at the ship, Tommy, Bob, and Barry had not mentioned the gold bars. Lamb knew nothing about them. Bars of California gold are exceedingly rare because their intrinsic value as bullion was so great that they were melted down and stamped into coins, rather than kept as collector's items. By the end of the Civil War, few remained. Gold bars created by banks or government institutions are uniform in shape, size, and character. The largest bar of California gold known to exist weighed fifty ounces. The bars recovered from the *Central America* ranged from five ounces to over nine hundred ounces, and there were hundreds of them. And each one, from the chocolates to the brownies to the bricks, came adorned with a unique set of symbols and numbers: In one corner appeared a shiny cut, where the assayer had taken a small sample to determine the gold's purity and kept the sample as a fee. Then the assayer stamped the bar with his seal, recorded the "fineness" or purity in thousandths, e.g., "891 fine," or 89.1 percent pure gold, assigned the bar a serial or "identification" number, and gave its weight in ounces. Based on the fineness and the weight and a value of pure gold at $20.67 an ounce, the last number on the face was the value of the bar in 1857. Besides the unique markings on each bar, California gold contained silver rather than copper, which gave the bars unusual brilliance and luster.

One of the largest bars was no. 4051, Justh & Hunter, 754.95 oz., 900 fine, $14,045 value in 1857. Today, in bullion value alone, the bar could be worth almost $250,000.

James Lamb saw bars for the first time in Bob's small lab. As much of the treasure as he had now seen and heard about, he was not prepared for the sight. "This is the most extraordinary, most incredible, most exciting thing," he said. "Ever. It's impossible to describe the significance, the potential monetary value, the general excitement of this find. All of these hitherto extraordinarily rare, desirable artifacts in perfect condition, in huge piles is just . . . it's beyond my imagination. If I didn't have it sitting here in front of me, then I wouldn't believe it."

By late August, Judge Kellam had enlarged Columbus America's legally protected area to include the Galaxy II site, awarded them title to the artifacts they had already recovered, and made the injunction permanent. Nearing the end of summer, the Columbus-America group finally announced publicly that they had found the treasure of the *Central America.* The first article about their success appeared in the staid British journal, *The Economist,* and was followed by a lengthy piece in *The Washington Post.* Tommy appeared on the *Today Show* via satellite.

The crew continued to dive at the site into mid-September, when Hurricane Hugo chased them into port at Wilmington. After Hugo they returned to sea, but the weather remained the worst they had encountered. People calling from shore on the SAT COM could hear the crash of falling dishes and books. Storm after local storm spun across the water and hit the ship, until Tommy finally decided it wasn't going to get any better for a while. It was time, he thought, to bring the treasure home.

The night of October 4, the *Arctic Discoverer* hung offshore near the sea buoy. The next morning Burlingham was up at five to take over the bridge and head in to the Cape Henry Lighthouse at the mouth of the Chesapeake Bay. In 1857 Captain Johnsen on the *Ellen* and Captain Burt on the *Marine* had followed the same route past the same lighthouse, carrying survivors from the *Central America.* When Bryan and

Tod and the two deckhands rolled out, Burlingham set them to scrubbing streaks of rust from the superstructure until the upper decks of the *Arctic Discoverer* gleamed.

The day was Indian summer blue with a warm sun and just enough chill to freshen the air. Flags flapped in a gentle breeze, and along the river tugboats shouldered up against navy warships to maneuver them into the naval shipyard. As they crossed the Chesapeake and neared Norfolk, the crew began packing and storing their gear, readying the ship for family and friends to come aboard when they arrived at the docks. But they had no idea of the scene that would await them as they rounded the last bend and Burlingham steered the *Discoverer* slowly toward Norfolk's Otter Berth at the city docks.

When they picked up the harbor pilot, who would steer the ship through the inland waterways, John Moore knew something big was going on. "The pilot came out with a great-looking hat," said Moore, "and he was wearing a blazer and nice trousers." Then a news station helicopter flew low overhead, and a police escort boat arrived, and the customs agents boarded and sat down with Burlingham to review the manifest. Moore stood on the deck and watched all of this. "It was kind of exciting," he said, "kind of a dog-and-pony show when we got there. But you have to admit, it really is a big deal."

With five days' notice, Paula Steele had notified hundreds of people who had been involved with the project: family, friends, office workers, and partners; people from all over the country who had contributed to the success. On the dock now over two hundred people mingled in the VIP area: Wayne Ashby in an English driving cap with Fred Dauterman and Tom Jordan; Buck Patton with his wife, Jodi; clusters of younger partners like Don Garlikov, Victor Krupman, and Brad Kastan; Jon Jolly, a mentor of Tommy's from Seattle; Ken Ringle of the *Washington Post;* Tim Daniels, their insurance man from New Orleans; Larry Stone, who had created the optimal search maps; Don Craft and his wife, Evie; the Thompsons and the Butterworths all; Barry's mother, Suzanne, and his sister Sally and two of his old journalism buddies, Rick Ratliff and Dave Seanor; Bob's wife, Jane, and his mother and father, Darline and Larry. The list went on, and the area was filled with people wearing VIP name tags.

Between the VIP area and the water, forming a V on two perpendicular docks, 140 members of the Herndon High School Marching Band stood in their red, white, and black uniforms. The town of Herndon, Virginia, and its high school had been named after Captain William Lewis Herndon of the *Central America*. Coincidentally, the band had won forty Grand Championships in competition with the finest high school marching bands in the country. Near the band stood Norfolk police and security guards, some with M-16s pointed into the air, forming a path from the ship's berth to the three Brinks armored cars waiting near the dock with their doors open.

Behind the VIP group was a retaining fence, and behind that were several hundred more people and scaffolding supporting cameras and cameramen. The difference between the people in front of the fence and the people behind the fence was that the people behind the fence had come to witness the excitement. The people in front were part of the story, each contributing money, expertise, advice, love, or encouragement.

As the ship glided within two hundred feet of Otter Berth, a cannon fired from the docks and the band burst into John Philip Sousa's "Stars and Stripes Forever." The crew stood watching from various places on the decks of the *Arctic Discoverer*; the crowd looking back at them, knowing what they had accomplished, feeling the warm sun of a crisp fall day, and hearing the rousing notes from John Philip Sousa broke into broad grins. It was almost too perfect.

Fred Dauterman described it as "a happening." "During a lifetime, you don't have many days that are like a happening," said Dauterman, "and this was like a happening. Everything just came together perfectly."

Burlingham backed slowly into the berth, jockeying with his bow thruster to snug up to the dock, his face sometimes framed by one of the port windows on the bridge. The vehicle sat on the foredeck wrapped like a large package in bright blue plastic. The band continued playing a medley of Sousa marches and then songs from Disney, including, "When You Wish upon a Star," with many in the audience humming the tune and thinking the words. Dreams and dreamers.

The crew could not leave until customs had released them, so they continued to watch the spectacle from the deck, wondering if the people out there on the dock and on the lawn had any idea what had happened back in little rooms and a warehouse in Columbus, at sea for months, at night during the storms. One deckhand just wanted to see his girlfriend; Tod was anxious to say hi to his parents and his sister; his roommate was thinking about a hot bath and a soft bed that didn't rock. "The public's perception of it is a great deal different than what really happened," said Moore. "The people who were there at the pier, except for the family people, thought, you know, the big, glorious, treasure hunt. They don't realize the years of work that go into it and all the rest of the stuff that went on. I know certainly the media people there didn't have a clue what it was all about."

Still, many people, even some of those who did know much about what had happened, were touched by the scene. It seemed about as close as most of them would get to experiencing heroism. "Tommy's going to be a very, very, very wealthy man," said Buck Patton. "A very wealthy man. And he'll be in the limelight, and I wish it for him—he deserves it. That's a long time to be in the trenches."

In addition to being proud of Tommy and the group, a lot of the partners were proud of themselves and their hometown for deciding to back the venture. "I was very happy for Tommy and the group, proud of them," said Tom Jordan. "I was proud of Columbus, Ohio, and the fact that our community had, in its wisdom, wherever that came from, supported a very successful venture."

The wisdom came from Wayne Ashby, who had the foresight to appreciate Tommy's vision, the instinct to move on a hunch, and the realization that all that was needed was some credibility. Watching the festivities on the dock now brought sentimental feelings.

"When they docked," said Ashby, "that whole scene of the marshals and police lent a lot of realism to it. This is really a special and fabulous thing to have happen in anybody's life, but I don't think that this will be the climax of Tommy's career. I don't imagine him having as much impact on people's lives as Bell and Edison, but I can imagine him becoming as well known and famous as Cousteau."

Customs cleared the ship and the crew, and then on behalf of the federal court, and in keeping with maritime tradition, U.S. marshals boarded the ship and "arrested" the gold. With the gold arrested and safely packed in army ammunition cans, the crew began off-loading the treasure of the *Central America*. Moore, Scotty, Doering, Tod, Bryan, even the new cook, Mickey, in a tall white chef's hat and Bermuda shorts, carried the ammunition cans along the line of police and security guards to the Brinks armored cars by the dock.

Near the dock was a podium and a microphone. The mayor of Herndon, Virginia, spoke to the crowd and reminded them of the heroism of the man who was the namesake of his fair city. Then came the mayor of Norfolk to say a few words about his city's outpouring of kindness to the survivors of the sinking in 1857. The group's historian, Judy Conrad, had tracked down twenty-nine descendants and relatives of the passengers who had sailed aboard the *Central America* on its last voyage. One was Genevieve Gross, who told the crowd her family's stories of Alvin and Lynthia Ellis and their three children, how Alvin had bailed so Lynthia, little Alvin, Charles, and Lillie could be rescued, and how long she and others in her family had waited to learn the whole story of the *Central America*. Bob spoke; Barry spoke; the partners smiled a lot, and laughed, and shook their heads in disbelief as they recalled the early conversations with Ashby or Dauterman and then Tommy, and how far the ideas had progressed. Tommy had jotted down a few words the night before, people to remember and thank for contributing their time, resources, courage, and ingenuity. He thanked them all for helping to make the impossible possible.

The armored cars pulled away slowly, the crowd began to disperse, and the VIPs and the media repaired to white tents on the lawn, where Paula had organized a catered buffet for two hundred. That night, Tommy was to appear on *Prime Time Live,* and by late afternoon the front deck of the *Discoverer* was filled with two cameras and a dozen technicians hooking up cables, positioning lights, and checking sound. For their cameras, Bob had displayed some of the gold coins and bars on blue velvet next to an aquarium filled with water. In the aquarium sat a gray silicone block embedded with gold coins sticking out at odd angles, a deteriorated box containing gold coins in three neat stacks, a sixty-two-pound gold brick, and fifteen to twenty gold chocolates and brownies. Two police and a white-shirted security guard paced the deck.

While preparations and celebrations continued throughout the afternoon and into the evening, lawyers representing thirty-nine insurance companies filed a lawsuit at the federal courthouse a few blocks away. Tommy knew it was coming. The insurance companies claimed that they had covered the loss when the *Central America* sank in 1857, and that therefore all of the gold he had recovered from her timbers, and all of the gold he had yet to recover, belonged to them. They wanted it back. It was his greatest concern, that after all of the years he had spent achieving what many thought was impossible, someone would try to steal the rewards. Starting the following day, he fought the insurance companies for ownership of the gold, which would not be determined for another seven years.

EPILOGUE

In April 1997, William Broad, a science reporter for the *New York Times,* published a book called *The Universe Below: Discovering the Secrets of the Deep Sea.* An in-depth look at our endeavors to penetrate the deep ocean, *The Universe Below* offers perspective on two things: the enormity of the sea and the paltriness of our knowledge about it. Until the mid–nineteenth century, scientists assumed that beyond a depth of even a few hundred feet the ocean lay barren. In the latter part of that century, scientists on deep-sea explorations around the world deployed trawls and dredges to sample the bottom and found an abundance of deep-sea life. From these discoveries and the further sightings of Beebe and Piccard and those who followed, scientists realized that although vast deserts did lie at the bottom of the sea, the ocean beyond the shallows supported abundant life in many places at all levels all the way down to the seafloor. And with that understanding came the overwhelming realization that although the ocean may take up 71 percent of the earth's surface, its volume accounts for as much as 97 percent of the earth's biological habitat.

Broad reported that oceanographers in 1991 estimated that scientists had explored between one thousandth and one ten-thousandth of the ocean floor. But if you consider the volume of the ocean, wrote Broad, "humans have scrutinized perhaps a millionth or a billionth of the sea's darkness. Maybe less. Maybe much less."

"The main impediment to better understanding," opined Broad, "has been a shortage of instruments that can withstand the deep's crushing pressures and illuminate its inky darkness while advancing the exploratory job. . . . The entire history of deep exploration bristles with frustration, with jibes about having to blindly grope the sea's boundlessness 'by the square yard.' Like astronomers before the invention of the telescope, oceanographers often found the available tools inadequate for addressing the great questions at hand."

Submersibles like the *Alvin* had been to the bottom, but only with human life at risk, only covering tiny areas, rarely back to the same area for comparative observations, and normally for no more than three or four hours on a dive. Tommy's vehicle could reside on the bottom of the deep ocean and perform for days. And instead of having a pilot and one or two others in a submersible thousands of feet below, cold and risking their lives, ten people from different disciplines could sit in a control room topside, observing and advising.

When members of the deep-ocean community ask Don Hackman if he went down in the vehicle, he says, "No, I sat in an air-conditioned room and sipped on a cup of coffee." When they ask him, "Well, what can you do down there—look around, maybe pick something up?" Hackman tells them, "Sure. I can also drill a hole, tap a thread, and put a bolt in it. From two miles away."

No one Hackman has talked to can conceive of such intricacy at such depths. He showed pictures of the front end of the vehicle to ocean engineers in the navy, one of whom told Hackman he saw not one thing he recognized. Hackman showed videos of the vehicle working on the bottom to an admiral and his entourage of captains, and the admiral asked Hackman, "How do you protect the site when you're not there?" Before Hackman could answer, one of the captains interjected, "Well, Admiral, they don't have to worry about us." No one can figure out how the vehicle works.

"This vehicle," said Hackman, "is less sophisticated in most cases than anybody else's vehicle, but it has capability that no one else even has on the drawing board right now. No other company, no government, has excavation arms and telescoping manipulators and hydraulic drawers and rotating thrusters; no one else even has them in the concept stages. I'm talking about all of the communities—the secret, the unsecret, the industrial, the oil companies. I know, because I work for all of these guys."

BEFORE HE FOUND the gold, Tommy already had been planning for the scientific studies to be conducted at the site. Since 1989, he has used his new technology to provide an opportunity never before available to science: data, specimens, photographs, film, and on-site time at sea observing and experimenting in the deep ocean for over 150 scientists, researchers, and educators in the United States, Canada, Germany, Monaco, England, and New Zealand. They are corrosion experts, underwater archaeologists, marine biologists, marine geologists, ocean chemists, ocean physicists, material scientists, bacteriologists, fisheries scientists, and maritime historians. The scientists have been identifying life forms, determining life cycles, evaluating data, and providing insight.

Tommy asked professor emeritus at Ohio State University and his mentor from Put-in-Bay, Dr. Charles E. Herdendorf, to head the adjunct science program. Dr. Herdendorf calls it "a modern-day adventure which rivals the discovery of the New World explorers. When you're out there working, you're part of something really special and different. You always feel a little glow, because you realize this is something that's just not being done anywhere else."

Columbus-America now has several thousand hours of videotape and thousands of slides shot over a period of four years of bottom time, all at the same site. In 1991, one dive alone lasted for over a hundred hours. And by that year, Tommy and his engineers had incorporated fiber-optic cable to transmit broadcast-quality images from the ocean floor and display them on seventeen monitors in the control room, so the scientists could observe this living and growing environment with their own eyes, not through specimens and fossils. The scientists have now observed what they think are thirteen new life forms, or species. A new species of sponge has already undergone the rigor of identification, classification, and naming.

One afternoon, an octopus embraced the vehicle, and the crew captured the creature on film for an hour while it probed, its arms in constant motion. Most octopuses measure no more than one foot from the tip of one arm to the tip of the opposite arm. This one measured over six and a half feet, tip to tip, and the bulbous upper portion of its body was a foot and a half wide. After studying the film and the stills, four authorities on octopuses, two of them from the Smithsonian, agreed that the sucker alignment and the siphon design on this animal differed from those on all other species of octopus, that this was a species of animal no one had ever seen before.

Late one night, while Dr. Herdendorf and the tech crew were conducting experiments with fish carcasses and cornmeal placed at the site, a shadow rippled across the bottom, and then a creature passed within inches of the video camera lens. Later, the "shark lady," Dr. Eugenie Clark at the University of Maryland, identified it as a Greenland shark, one of the largest sharks ever captured on video at any depth, and by far the largest ever sighted in the deep ocean. The previous record was three and a half feet; this one was twenty-two. For scientists, the presence of so huge a fish, with its enormous need for food, provided more evidence that the food chain even at those depths was consistent and abundant.

Many of the scientists had assumed that at those depths, they would find little life except for fungi. They had assumed that microorganisms that die at the surface were consumed before they could reach the bottom. But these tiny organisms, the first step in the food chain, actually sank to the bottom and the *Central America* site acted as a mechanical break, collecting them into one large food source, creating what scientists call "a bloom and pulse of life." Some of the scientists were startled to find that in this barren and hostile land was a world of remarkable stability and great abundance, a world providing many interesting avenues for further study.

Scientists had thought that deep-sea borers, or shipworms, were mostly shallow-water life forms, but in experiments set up at the site, Dr. Ruth Turner at Harvard found a whole new species of deep-sea wood borer. Evidence of deep-ocean borers had been found in the past, but the borers were no larger than a pencil. At the site, Turner found this new species as fat as a cigar and up to half a meter long. Her experiments continue.

Other scientists have found a new species of gorgonian, or fan, coral. All species previously described had been found on a sandy seafloor in shallow water, and they had stringy rootlets that burrowed and grabbed in sediment, the way a plant holds soil. But this species had a different "holdfast" that enabled it to grasp, like a hand, so it could live on coal or iron or gold, an adaptation to survive in the environment.

In the heavy rusticles dripping from the ironworks at the site, scientists identified a strain of bacteria; later, they secured a rust sample

494

from the *Titanic,* collected from the luxury liner when a sub acciden-
tally collided with the steel hull. From that sample, they were able to
culture the same bacteria they had found at the *Central America* site.
With that knowledge, they can develop ways to prevent deep-ocean
corrosion, which will be one of the major problems to solve when someday we build deep-sea mining facilities or lay cable or establish monitoring stations on the floor of the deep ocean.

One of the most exciting outcomes of exploration for scientists is
simply a description of the diversity of new life forms available for
pharmaceutical testing. "We now have the location and the population,"
said Dr. Herdendorf, "for a ready, rich source of new life forms of the
type in which we have found medicinal drugs in shallower-water cous-
ins." These new life forms might provide anticancer drugs called "tumor
inhibitors." Scientists have observed such qualities in the Sydney Opera
House sponge, which was found at the site. Living within the sponge is
a tiny white gallathiad crab that scientists surmise is protected by the
sponge secreting a chemical retardant that repels predators. Immune
to the retardant, the crab in turn removes large food particles that clog
the pores of the sponge and could cause the sponge to starve. This same
symbiosis is found in shallow-water sponges that scientists associate with
having beneficial medicinal properties.

The scientists who have worked at the site or have evaluated data
collected at the site have come from the Smithsonian, Woods Hole
Oceanographic, Harbor Branch Oceanographic, Harvard, Yale, Columbia, Oak Ridge National Laboratory, Ohio State University, the U.S.
Navy Civil Engineering Laboratory, the University of North Carolina,
Scripps Institution of Oceanography, the Field Museum of Natural
History, the United States Geological Survey, Texas A&M University,
California Academy of Sciences, United States Naval Academy, and
other government, educational, and private institutions. Their efforts
continue and will intensify when Tommy takes the vehicle on to explore
new areas of the deep ocean.

In 1989, when the Columbus-America Discovery Group announced
publicly its finding of the *Central America,* historian Judy Conrad organized a media campaign to contact descendants and relatives of the

people aboard the *Central America*. First, she assembled a list of every city and town in which anyone on board the ship was said to have lived. Then she sent press kits to the newspapers in those towns, including the names and any other information she had on the passengers or crew who once had resided there.

Genevieve Gross, from Ohio, was the first relative to contact Judy. "I'll never forget that moment," said Judy, "a connection right back to those people I had been studying so long. It makes them come alive."

Judy's initial effort brought twenty-nine responses, including two direct descendants of Oliver Manlove, who provided her with Manlove's writings about his trek across the plains to California and his voyage aboard the *Central America*. She began corresponding with several descendants, sharing information with them, often enlightening them to details not known in the family lore, but more often learning things about the people she had not known. She has located four relatives of Captain Herndon, one named Herndon Oliver III, a graduate of Annapolis and himself a captain (ret.) in the U.S. Navy; twelve relatives of the Eastons, three of whom are direct descendants; three relatives of the Badgers; four great-granddaughters of the ship's purser.

Judy found one direct descendant of Mary Swan, whose husband died in the sinking, leaving her with a baby not yet two. Upon arriving in New York, Swan was the woman who told a reporter, "I have no friends in New York, nor in all the world, now that my husband is lost." The descendant, a great-great-grandson, told Judy that Mary Swan had later remarried, become a nurse during the Civil War, and returned to California, where she had seven more children.

Jane Renard of Seattle is the great-granddaughter of a steerage passenger named Samuel Look, who was traveling with his brother, Prince. From her, Judy learned that when the ship sank, the two brothers ended up floating on the same timber; and that during the night, Prince died, and Samuel held on to his brother's body until another survivor floated within reach and needed help. Samuel allowed his brother's body to slide into the sea, and took on the other man, Billings Hood Ridley. A few hours later, the crew of the *Ellen* rescued them both, and Samuel and Billings became the closest of friends. Billings's great-great-grandson lives in Maine.

Judy is now in touch with 106 relatives and descendants of passengers and crew who were aboard the *Central America* on its final voyage.

BOB EVANS BECAME so immersed in the stories of the passengers aboard the *Central America* he sometimes daydreamed of finding things he had read about in the stories. "Finding Herndon's sword would be great," he had once said. "Finding Oliver Manlove's poems would be great. Finding Badger's $16,500 worth of double eagles in a carpetbag would be great. Finding any of that stuff that we can really tie to the story would be incredible."

In 1989 Bob talked about two trunks they had found in the debris, which appeared to be made of leather. "They look fairly fragile," said Bob. "In fact the one is open, the top is off of it, and there are exotic life forms living on both of them. What you seek to do is create a permanent record of those objects in their original setting that shows proper reverence. This is the goal of the scientific work on the site, to create an immortal record of the site as we have found it. Inside the one trunk from which the lid is missing we can see the brim of a hat. Tucked over in one side are two what appear to be books. What if these things are some guy's journals? What if they're Oliver Manlove's book of poems? We've been in touch with his family. He had a sheaf of papers which contained a bunch of poems he had written. What if they're sitting there and what if we can get those back? Because if we can find these papers, how much richer does the story become? How much richer does the history become? It's possible that we could retrieve some of these things. Like this other trunk that still has the straps on it and it's still closed. It looks like the quintessential treasure chest on the bottom of the ocean."

In 1990, they recovered that trunk. It was lying in the debris field next to the white teacup. And indeed it contained treasure. Bob examined the contents in a freezer facility at Ohio State's Department of Textiles and Clothing. Supervised by a polymer chemist and anthropologist specializing in the preservation of submerged historical fibers and papers, Bob removed each item and immersed it in demineralized water. Wet garments they framed over a fiberglass screen and flash froze at $-28°$ C.

In the trunk they found a shirt wrapped in a steamer edition of the *New York News* dated July 20, 1857. The newsprint was still legible. They found thirteen linen shirts and many silk cravats, plus coats, waistcoats, trousers, collars, chemises, petticoats, bloomers, a dressing gown, a morning robe, and several pairs of stockings. They also found a brace of dueling pistols, a gunpowder flask embossed with an eagle, a gold watch fob cast in the head of a dog and set with ruby eyes, a quartz watch fob veined with gold, a small Oriental carving, a bottle of double-distilled "Bay Water," three bottles of cologne, a gold buckle, two quill pens, and ambrotype photographs. The trunk is an archaeological treasure, a time capsule sealed in the late summer of 1857, and every item in it belonged to the newlyweds Addie and Ansel Easton, much of it gifts from friends, wishing them a lifetime of happiness together.

The other trunk, the one with the top missing, belonged to John Dement, the last man rescued by the *Ellen*. He had floated upon a hatchway with Ansel Easton's friend Robert Brown and was picked up with Brown after Captain Johnsen promised Easton to tack one more time. The trunk contained sixty-four articles of clothing, including underwear, pants, coats, vests, shirts, collars, cravats, and shoes. It also held a leather shaving kit, three novels, one of which was *Lady Lee's Widowhood,* and a letter of introduction. This missive was dated July 30, 1857, and addressed to B. B. Lancaster, Esquire, of Baltimore:

Dear Sir:
    Allow me to introduce to your acquaintance Mr. John Dement of Oregon City. Mr. Dement is one of our largest merchants here and is well acquainted with Oregon and its resources. As he is visiting your city, I have given him this, hoping that you would introduce him to such of my friends as may be in Baltimore while he is there and extend to him your usual courtesy which will be highly appreciated by him and
    Your friend,
    J. A. Simms

The significance of the novels and the letter lies not in their contents, but in the enormous implications of finding printed matter at the bottom of the sea that is still legible 130 years later. John Woodhouse Audubon, who worked closely with his father, John James Audubon, and carried on the Audubon legacy of wildlife artists and naturalists, led an expedition through the southwestern United States and California in 1849 and 1850, sketching and painting the animals and birds of the region. When he returned east in 1850, he left behind the bulk of his work, two hundred drawings from the expedition, which he entrusted to a friend. The friend turned the portfolio over to another mutual friend, John Stevens, who had promised to deliver it to New York in the late summer of 1857. Stevens booked passage on the *Sonora* and the *Central America,* and realizing the value of the extensive body of original work in his care, he presumably protected and packed it well for the voyage. When the *Central America* went to the bottom, Stevens and the Audubon portfolio both were lost.

The Columbus-America Discovery Group has located many more trunks in the debris, and these trunks still lie waiting to be recovered and opened.

T RACING THEIR CORPORATE lineage back to 1857, the thirty-nine insurance companies who met Tommy at the dock with a lawsuit included Atlantic Mutual, ICNA, Great Western, and other American insurers, as well as consortia from Lloyd's of London and other foreign insurers. They contended that they had insured the gold shipments aboard the ill-fated steamer, that they had paid the loss, and that therefore they now, 132 years later, owned the gold. They wanted it back. Possessing little more than the minutes to a meeting at which the directors of one London company discussed the possibility of setting aside fifty thousand pounds to cover its share of the loss, the insurance companies relied mostly on newspaper articles of the day. They had no bill of lading, no receipt, no contract to offer as evidence. Before the trial, Judge Kellam in Norfolk dismissed the claims of twenty-one of the insurance companies.

In August 1990, Judge Kellam heard the claims of the remaining insurance companies, decided the case under the law of finds, held that over the 132 years the gold sat at the bottom of the sea the insurance companies had made no effort to recover the gold and therefore had "abandoned"—

the key word—their claim to the gold, and awarded the Columbus-America Discovery Group 100 percent of all artifacts at the *Central America* site. On appeal to the Fourth Circuit Court of Appeals, a three-judge panel, in a split decision handed down on August 26, 1992, vacated and remanded the case to Judge Kellam for another trial under the law of salvage. The court held that for abandonment to occur there has to be an "express" act of abandonment, and that the insurance companies had established at least some connection to the gold. The court also opined, "We are hazarding but little to say that Columbus-America should, and will, receive by far the largest share of the treasure."

Columbus-America appealed the decision to the United States Supreme Court. Several prominent groups, including the National Association of Academies of Science, the National Maritime Historical Society, the Marine Technology Society, and Ohio State University, filed *amicus curiae,* or "friend of the court," briefs supporting Columbus-America. But the Supreme Court declined to hear the case at this time, so the matter returned to Judge Kellam for a new trial.

For two weeks in July of 1993, Judge Kellam heard evidence of the difficulty of the salvage, the expense, the risk to life and to capital. It was the first time a federal court had asserted jurisdiction over a wreck site in international waters. It was also the first time a court sitting in admiralty had accepted the concepts of "tele-presence" and "tele-possession," recognizing that with technology taking us into the deep ocean, where we've never gone before, the law must go with it, and the law must protect these cultural treasures. The court found that Columbus-America's efforts "to preserve the site and the artifacts have not been equaled in any other case."

At the end of the trial, Judge Kellam awarded Columbus-America 90 percent of the insured gold and 100 percent of the rest of the treasure. He reserved judgment on the remaining 10 percent of the insured gold. Columbia University, Harry John, and a colleague of John's named Jack Grimm claimed that Tommy had used the sonar data they had compiled during their hasty February search for the *Central America* in 1984. Kellam found otherwise and dismissed the lawsuits. One man claimed that he was the first to tell Tommy about the *Central America* and that that entitled him to a share in the treasure, but Kellam dismissed that claim on summary judgment. Columbus-America, the insurance companies, and John and Grimm appealed Kellam's decisions, and again the case went to the Fourth Circuit Court of Appeals.

On June 14, 1995, the Fourth Circuit affirmed Judge Kellam in all respects. First, the claims by Grimm and John had been rightfully dismissed for lack of merit. Second, Columbus-America would receive 90 percent of the treasure. At the outset, the court admitted, "We are in no sense offended by the possibility that a salvor who has 'boldly gone where no one has gone before'. . . might be entitled to the lion's share of a long-lost treasure."

The court then reviewed the six factors to be considered when arriving at a just salvage award: 1) the labor expended; 2) the salvor's promptitude, skill, and energy; 3) the value of the property employed by the salvor; 4) the risk incurred by the salvor; 5) the value of the property saved; 6) the degree of danger to the salvaged property.

Under the first factor, the court found that from 1986 to 1992, Columbus-America had been at sea for 487 days, the crew working twelve-hour shifts. The personnel combined had logged over 400,000 hours at a cost of nearly $8.5 million. The court's research uncovered only two other cases involving the recovery of underwater property that had lasted longer than one month. Most of the operations lasted but a few days or even hours. Even if a discoverer had expended "but a fraction of the labor detailed in the above account," wrote the court, "that fraction alone would represent a huge investment. In the actual case, it is apparent that Columbus-America's effort was monumental."

Discussing the second factor, the court stepped outside its usually staid approach to rendering the law. "When we write an opinion," began the court, "our tendency to emphasize those discrete facts necessary to our legal conclusions sometimes serves to distort the broader perspective. Were we to allow that to occur in this case, it would be a disservice to Columbus-America and the significance of its accomplishments. Hence, a brief review is in order." The court then devoted a large part of its decision to praising Tommy and the Columbus-America Discovery Group and concluded, "We cannot imagine anyone demonstrating more diligence, skill, and energy than Columbus-America has shown here. Its efforts provide a standard against which all others should be judged."

Under factor number three, the court found that the vehicle, now called *Nemo,* the *Arctic Discoverer,* and the rest of the recovery equipment was worth over $6 million. That was for the pieces, not the technology. Columbus-America had also rented other equipment and spent money on related items that ran to more millions. The judges could find no cases where a recoverer had used property of a comparable value.

The fourth factor the court considered was the risk incurred. "The crew of the *Arctic Discoverer* constantly dealt with heavy equipment—not the least of which was the six-ton *Nemo*—and the vessel's 160-mile distance from shore meant that treatment for the most severe injuries was hours away. . . . As the district court noted, 'anytime a ship goes to sea, there is danger to the ship and the persons aboard.'"

Number five was the value of the property saved. The court wrote that, despite the testimony of trial experts varying substantially in their estimates, it "appears that the treasure's value will not approach one billion dollars, as earlier speculated. It is nevertheless clear that, when the salvage is completed and all the treasure is eventually sold, the haul will be one of the largest in history. One commentator has suggested that 'the combination of high values and highly meritorious service should result in a high award.' We entirely agree."

In assessing the degree of danger to the ship in distress, the sixth factor, the judges took a unique approach. "While it is true that the ocean itself presents no danger to the essential nature of gold and similar substances, it is also true that any value that our society attributes to gold depends entirely on the ability of someone to assert a property interest in it. Because property is far less certain of being recovered once it has sunk, especially when it has sunk in deep water, we perceive that its sinking sharply increases the degree of danger to its continued existence and utility as property." They called Columbus-America's recovery of the *Central America* "the ultimate rescue from the ultimate peril."

In earlier briefs, Columbus-America had suggested that when evaluating an award admiralty courts should consider a seventh factor: the discoverers' preservation of the historical, archaeological, and informational value of the wreck and cargo. Judge Kellam already had addressed the issue, and the Fourth Circuit deferred to him, noting that the district judge "was convinced that Columbus-America had taken extraordinary care in preserving the *Central America,* benefiting a range of sciences and disciplines." It quoted Kellam on how expert witnesses "'described the care and means used to preserve the artifacts and objects recovered . . . and how Columbus-America followed their advice and suggestions for handling and preserving the articles. They described . . . the discovery of new species, and the unparalleled opportunities while aboard the research vessel. . . . They demonstrated the particular care exercised in recovering and handling delicate items such as jewelry, china, cloth, papers, and so on. One of the items was a cigar, which appeared in perfect condition.'"

The court concluded, "It will not be often, if ever, that all of the principles governing salvage awards will so overwhelmingly militate in the salvor's favor. It will also be rare that a ship and cargo will have been imperiled for so long, and will be so difficult to objectively value, that a flat percentage award will be preferred to a sum certain. Nevertheless, this is that exceptional case, and though the district court's ninety percent award is a generous one, we cannot say that it is excessive."

But the court wasn't through. It then gave Columbus-America exclusive control over all of the marketing of the treasure. The insurance companies had wanted their part of the gold immediately, so they could dispose of it however they pleased, but the court held that "there is just too much gold involved to allow more than one party to market it. The loss of the gold aboard the *Central America* significantly affected the precious metal and financial markets in 1857, and there is evidence that the gold's recovery could, upon its re-entry into the market, have a significant effect today. Our review of the expert testimony introduced at trial convinces us that a unified plan is necessary to ensure the maximum return from the sale of the gold."

In all of the court proceedings since the fall of 1989, one major issue still had not been resolved: Could the insurance companies prove they had paid the loss? They may have made a prima facie case for their claim to the gold, argued Columbus-America, but before they shared in the recovery they still had to prove which parts they insured and that they had covered those losses in 1857. The court agreed. "That Columbus-America is entitled to a ninety percent salvage award does not automatically mean that the underwriters may, merely by virtue of bringing this suit, walk out the courthouse door with the remaining ten percent." The Fourth Circuit now sent the case back to Judge Kellam one more time to "weigh all of the evidence and decide whether each underwriter has proved all or any portion of its claim." The court noted that if the insurance companies' proof covered less than 100 percent of the gold recovered, "Columbus-America may retain the excess."

Then came the court's closing words. "What Thompson and Columbus-America have accomplished is, by any measure, extraordinary. We can say without hesitation that their story is a paradigm of American initiative, ingenuity, and determination."

The decision was a rarity in American jurisprudence: a stern and staid federal court of appeals effusive in its praise of one litigant. The appellate court's decision so brimmed with excitement and enthusiasm, even a casual reader could discern the judges' fascination with the story, both from 1857 and in the 1980s, a feeling that they, too, by having the good fortune of deciding this case, had entered at a select moment into one of the most extraordinary stories of the nineteenth and twentieth centuries to become themselves a part of history. The court even praised Judge Kellam, who had "intrepidly waded through the morass of records and filings," and who had "consistently evidenced good humor, notwithstanding the occasional contentiousness among the parties."

In the winter and early spring of 1996, both sides submitted final briefs to Judge Kellam, supporting their claims. After the briefs were filed, doctors diagnosed Judge Kellam with pancreatic cancer, giving him only two to three months to live. Kellam nevertheless wrote his opinion in the spring of 1996, then died in June. His Opinion and Order was filed by another judge on August 13, 1996.

Using mostly articles in contemporary newspapers for supporting evidence, Judge Kellam found that all but two of the insurance companies had covered at least some of the commercial loss. The other two had never been mentioned in any of the evidence. With those two companies no longer entitled to recovery, Kellam increased Columbus-America's share to 92.22 percent and held that the insurance companies would receive their 7.78 percent of the commercial shipment only after Columbus-America had deducted all of its expenses, including recovery, storage, and marketing. Because the insurance companies had produced no bills of lading, no insurance contracts, no receipts, no evidence that parts of the shipment were insured against loss or that the companies paid the loss when it occurred, many observers assumed that Kellam gave the insurance companies a small piece of the commercial shipment merely to discourage further appeals. Only three tons of gold comprised the commercial shipment out of a potential twenty-one tons aboard the *Central America*. In the final distribution, Columbus-America received 92.22 percent of the three-ton commercial shipment, 100 percent of the miners' unconsigned gold, which approximated the commercial shipment, and, if Columbus-America could recover the shipment guarded by the army, 100 percent of that shipment, or another fifteen tons. The court also appointed Columbus-America central authority over marketing and future recoveries at the site. It was by far the largest salvage award in the history of admiralty.

Tommy continues to explore the ultimate disposition of the treasure. The story of how it was lost and how it was found again is certain to widen the circle of people interested in owning a piece of it. The rare and beautiful gold aboard the *Central America* was not bought from a collector or handed down within the family; it came out of the California Gold Rush, a sidewheel steamer, a hurricane, a selfless group of spirited men, a rescue at sea, a couple named Easton, a courageous captain; it came from the dream of a young engineer from Ohio, and his story of risks and setbacks and breakthroughs; it came with the opening of the next frontier. More than gold, the treasure represents a bonding of two centuries, two pioneering spirits, the opening of two frontiers.

The value of the treasure will remain undetermined until Tommy decides what Columbus-America should do with it. As of February 1998, he continues to assess every avenue conceivable and some avenues no one has ever thought of. His percentage started at forty but has been diluted since 1985 to about one-third of the total treasure. Ultimately, he will receive a huge sum, but his friends, partners, and people at Columbus-America see him changing little. He will still wear the same mismatching clothes, ride a motor scooter in a T-shirt in January, and drive everyone around him crazy by pushing what everyone else perceives as limits. Except his new riches will allow him to dream on an even grander scale.

One perspective on Tommy's achievement: In 1961, President John F. Kennedy told a joint session of Congress that the United States should commit itself to landing a man on the moon within the decade. For the next eight years, over 400,000 people spent nearly $100 billion (current value) to put Neil Armstrong on the moon. Many scientists and engineers consider the challenges of exploring and working in the deep ocean equal to the challenges of exploring and working in outer space. In 1985, Tommy Thompson and about a dozen colleagues committed to developing a working presence on the bottom of the deep ocean. No one had done it; no one knowledgeable thought it could be done without the full force of the United States government and unlimited resources. Even then, some were skeptical, because the government had already spent hundreds of millions trying. Despite legal challenges, a time frame of only three years, the efforts of only thirty people, and a budget of no more than $12 million, Tommy and his group successfully searched the bottom of the Atlantic Ocean, perfected their technology, found and imaged a mid–nineteenth century wooden-hulled ship in eight thousand feet of water, and began recovering her many treasures.

They had proved the experts wrong. You could work on the bottom of the deep ocean, do intricate work carefully and heavy work delicately, and you didn't have to spend hundreds of millions of dollars to do it. You just had to shed old ways of thinking and reexamine old assumptions and do it smart from the beginning. You had to keep diverging, even beyond the point where it all became difficult and confusing. That's where Tommy lives, and he made those around him live there, too, some for far longer than is comfortable for most people. Yet just on the other side of that juncture is where impossibility sometimes vanishes and the world can be seen in a new way. Tommy, Barry, and Bob had talked about it before they ever went to sea: Historic ship recovery was really an adventure in thinking, a way of looking at the world. Finding the treasure of the *Central America* was a goal, but it wasn't the purpose. The purpose was to unveil the treasures of the deep ocean, to enhance our understanding of history, to advance marine archaeology, to further science, to form new corporate cultures, to develop technology. Going after the treasure was like plowing and planting a field. Tommy had broached the idea during the round tables back in 1985. "If you do that," he said, "all kinds of things can blossom."

Index

Johns-Manville, 80, 83
Johnson, Jim, 86
Johnson, William, xv
Jolly, Jon, 483
Joplin, Janis, 88
Joplin, Scott, 380
Jordan, Tom, 483, 485
junkyard effect, 141
Justh & Hunter, 482

Kastan, Brad, 483
Kellam, Richard B.:
    death of, 504
    on Galaxy case, 362, 367, 369, 371–72,
        374, 390–91
    on Gallaxy II claim, 482
    gold discovery and, 472
    in litigation over possession of gold,
        499–501, 503–5
Kelly, Bill:
    Central America project and, 176–77,
        218–19
    in filing admiralty claims, 309–11
    Galaxy case and, 365–66, 372
Kemble, John, xv
Kennedy, John F., 506
Key West, Fla., 99–113, 115, 121–22
    Atocha project in, 102–13, 169
    marine engineering opportunities in, 99
    treasure galleons in Caribbean off of,
        111, 141
Kirk, Gil, 413, 415–16
Kissner's restaurant, 80, 85
Kittredge, Almira:
    on *Marine*, 134–35
    rescue of, 64, 74
Koopman, Bernard, 215–16
Kreisle, Wally:
    Central America project of, 381, 387,
        390–92, 472
    contempt charges against, 391

Krupman, Victor, 483
Kuntz Drugstore, 85
Kutzleb, Robert, 183–89, 381, 387

La Concepción, 366–67
    salvaging of, 114–15, 117
Lake Erie, 97–98
Lamb, James, 474, 480–82
Lancaster, B. B., 498
Lape, Herb, 168–69
Laura, transfer of *Central America* sur-
    vivors to, 238
Leahy, Ellen, 89–90
Leonardo da Vinci, 93
Lettow, John:
    on *Pine River,* 243, 247, 249, 254, 256–
        57, 262, 267, 270, 273, 275–77
    Sidewheel and, 270, 273, 275–77
Lewis, Jack, 133
Liberty Star:
    in *Central America* project, 336–56,
        359–64, 367, 369–72, 381
    comparisons between *Nicor Navigator*
        and, 353–54
    contempt charge sought against, 364
    injunction against, 361–64, 369–70,
        372
    tracking movements of, 338–41, 347–
        50
Library of Congress, 205
Life, 299, 301
Liverpool Post, 203
Lloyd's of London, 499
Longitude Act, 208
Look, Prince, 496
Look, Samuel, 496
Los Angeles Times, 379
Lost Dutchman Mine, 299
Loveland, Curt:
    Central America project and, 176–77,
        218–19, 302, 309–11, 365, 369, 372

in filing admiralty claims, 309–11
Galaxy case and, 365, 369, 372

McGowan, John, 195–96, 198–99
McNeill, Annie:
    rescue of, 56, 60
    in storm, 40, 42, 56, 60
Maddock, John, 384
magnetometers, 139
Maguire's Opera House, 13
manganese, imaging systems in location
    of, 151
Manhattan Project, 137, 471
Manlove, Oliver Perry, 21–22, 496–97
    descendants of, 496
    in floating in sea, 130–31
    rescue of, 193
    in sinking of *Central America,* 124–25
    on *Sonora,* 14–16
    in storm, 26
man-of-war hawks, 190–91, 207
"Maple Leaf Rag" (Joplin), 380
Marine:
    Central America in sighting of, 51–53
    in Chesapeake Bay, 197–99, 482
    damages of, 51, 53, 72, 75, 197
    lifeboat of, 72–74
    in locating wreckage of *Central Amer-
        ica,* 207, 228–29, 246
    mothers and children reunited on, 74
    in rescuing passengers of *Central
        America,* 53–69, 71–75, 123–24,
        126, 128, 132–36, 159, 196–200, 202,
        482
    in sailing for Norfolk, 52, 136, 197–
        98
    in storm, 51–52, 63–66, 72–75, 134–35,
        197
    witnesses in sighting of, 207
marine engineering and technology:
    at Battelle, 138–39, 142–43

employment opportunities in, 99
    Thompson's interest in, 91–92, 96–99,
        106
    see also deep-ocean mining technology
Mariner's Museum, 205
Marine Technology Society, 500
Marne River, 93
Marshall, James, 3–6, 436
Martindale Hubbell, 310
Marvin, Mrs., 134
Mary, 238–39
Maryland, University of, 494
Marysville, Calif., 15
Maumee River, 79
Maury, Matthew Fontaine, 159, 203, 213
Max, Peter, 88
Mayport Naval Base, 257
Mediterranean Sea, 22
    crash of U.S. bomber in, 144–45
    divers in, 92
    dives of *Trieste* in, 294
Merchant Marine, U.S., 99, 321, 438
Metropolitan Hotel, 201
Mexico:
    Upper California ceded by, 4
    U.S. treaty with, 4–6
    U.S. war with, 4–7, 22
Miami, University of, 114, 239, 246
Miami Herald, 101
Miami University, 99
Mickey (cook), 486
Miller Brewing, 162–63
Mills, Darius Ogden, 13
Milosh, Mike, 475–77
Money Pit, 299
Monson, Alonzo Castle, 14, 160
    Herndon's friendship with, 27, 37–38,
        48, 62, 211
    in locating wreckage of *Central Amer-
        ica,* 210–13
    in rescue effort, 53
    rescue of, 62–63, 67